Praise
You Don't Know Us Negroes and Other Essays

"This volume enables readers both steeped in and new to Hurston to discover her acerbic wit, her crisp prose, and the breadth of her artistic ability and interests . . . an invaluable nonfiction companion to the collection of Hurston's short stories."

—*Booklist*

"I liked this book. . . . Reading Hurston, you always wonder what shape her dignity will take next. Her style and spark were her own."

—*New York Times*

"Hurston is bold, honest, and provocative, as always, whether she's pontificating on the ideological mirage of white feminism or insisting that school integration did less than we thought to improve Black students' educations. The lyrical and uncompromising prose in this collection offers a window into the world of one of our greatest literary minds."

—*Vulture*

"Dazzling . . . provocative, funny, bawdy, informative, and outrageous. Gates and West have put together a comprehensive collection that lets Hurston shine as a writer, a storyteller, and an American iconoclast."

—*The Washington Post*

"*You Don't Know Us Negroes and Other Essays* by Zora Neale Hurston creates a powerful and nuanced mosaic of Black culture."

—*The Christian Science Monitor*

"This is a carry-it-everywhere-with-you kind of book, perfect for times when you need some introspection as diversion. *You Don't Know Us Negroes* is like that, and that's just the way it is."

—*The Philadelphia Tribune*

"Vigorous writings from a controversial and important cultural critic."

—*Kirkus Reviews*

"*You Don't Know Us Negroes* adds immeasurably to our understanding of Hurston, who was a tireless crusader in all her writing, and ahead of her time. Though she was often misunderstood, sometimes maligned and occasionally dismissed, her words make it impossible for readers to consider her anything but one of the intellectual giants of the 20th century. Despite facing sexism, racism, and general ignorance, Hurston managed to produce a written legacy that, thanks to enduring collections like this one, will engage readers for generations to come."

—*New York Times Book Review*

"The depth and power of Hurston's prose continues to dazzle."

—*The Guardian*

"With much of her work having been released and rereleased posthumously, this collection recogni[z]es one of the finest writers of the 20th century."

—*Sunday Express* (UK)

"*You Don't Know Us Negroes and Other Essays* showcases the author's breadth in a thrilling, if also uncomfortable, journey."

—*The Atlantic*

YOU DON'T KNOW US NEGROES

AND OTHER ESSAYS

ALSO BY ZORA NEALE HURSTON

Jonah's Gourd Vine

Mules and Men

Their Eyes Were Watching God

Tell My Horse

Moses, Man of the Mountain

Dust Tracks on a Road

Seraph on the Suwanee

Mule Bone

The Complete Stories

Every Tongue Got to Confess

Barracoon

Hitting a Straight Lick with a Crooked Stick

YOU DON'T KNOW US NEGROES

AND OTHER ESSAYS

ZORA NEALE HURSTON

EDITED AND WITH AN INTRODUCTION BY
GENEVIEVE WEST AND HENRY LOUIS GATES JR.

AMISTAD

An Imprint of HarperCollinsPublishers

Some material has been previously published in another format.

Foreword, annotations, and footnotes. Copyright © 2022 HarperCollins Publishers.

YOU DON'T KNOW US NEGROES AND OTHER ESSAYS. Copyright © 2022 by Zora Neale Hurston Trust. All rights reserved. Printed in the United States of America. No part of this book may be used or reproduced in any manner whatsoever without written permission except in the case of brief quotations embodied in critical articles and reviews. For information, address HarperCollins Publishers, 195 Broadway, New York, NY 10007.

HarperCollins books may be purchased for educational, business, or sales promotional use. For information, please email the Special Markets Department at SPsales@harpercollins.com.

FIRST HARPERCOLLINS PAPERBACK PUBLISHED IN 2023

Designed by SBI Book Arts, LLC

Library of Congress Cataloging-in-Publication Data is available upon request.

ISBN 978-0-06-304386-2

23 24 25 26 27 LBC 5 4 3 2 1

Contents

Editors' Note   ix

Introduction   1

Part One
On the Folk

Bits of Our Harlem   25

High John de Conquer   28

The Last Slave Ship   38

Characteristics of Negro Expression   47

Conversions and Visions   66

Shouting   72

Spirituals and Neo-Spirituals   76

Ritualistic Expression from the Lips of the Communicants   81
   of the Seventh Day Church of God

Part Two
On Art and Such

You Don't Know Us Negroes   107

Fannie Hurst   117

Art and Such   123

Stories of Conflict   129

The Chick with One Hen   131

Jazz Regarded as Social Achievement   134

Review of *Voodoo in New Orleans* by Robert Tallant   138

What White Publishers Won't Print   143

Part Three
On Race and Gender

The Hue and Cry About Howard University   151

The Emperor Effaces Himself   173

The Ten Commandments of Charm   182

Noses   184

How It Feels to Be Colored Me   186

Race Cannot Become Great Until It Recognizes Its Talent   191

Now Take Noses   194

Lawrence of the River   196

My Most Humiliating Jim Crow Experience   204

The Lost Keys of Glory   206

The South Was Had   220

Take for Instance Spessard Holland   224

Part Four
On Politics

The "Pet Negro" System   233

Negroes Without Self-Pity   242

The Rise of the Begging Joints   245

Crazy for This Democracy   254

I Saw Negro Votes Peddled 259

Mourner's Bench 270

A Negro Voter Sizes Up Taft 284

Court Order Can't Make Races Mix 296

Which Way the NAACP? 300

Part Five

On the Trial of Ruby McCollum

Zora's Revealing Story of Ruby's 1st Day in Court! 315

Victim of Fate! 321

Ruby Sane! 324

Ruby McCollum Fights for Life 327

Bare Plot Against Ruby 329

Trial Highlights 332

Justice and Fair Play Aim of Judge Adams as Ruby Goes on Trial 334

McCollum-Adams Trial Highlights 336

Ruby Bares Her Love Life 337

Ruby's Story: Doctor's Threats, Tussle over Gun
Led to Slaying! 344

Ruby's Troubles Mount: Named in $100,000 Lawsuit! 351

The Life Story of Mrs. Ruby J. McCollum! 354

My Impressions of the Trial 387

Chronological List of Essays 403

Acknowledgments 405

Credits 409

Notes 412

Index 441

Editors' Note

Most of the essays collected here appeared in Hurston's lifetime, over the course of almost forty years. Some appeared in print in the years following Hurston's death. Still others appear here for the very first time. "Ritualistic Expression," "The Chick with One Hen," "Which Way the NAACP?," "You Don't Know Us Negroes," "The Lost Keys of Glory," "Take for Instance Spessard Holland," and "The South Was Had" have never before been published. "The Hue and Cry About Howard University," "Race Cannot Become Great Until It Recognizes Its Talent," "Noses," "Now Take Noses," and the Ruby McCollum series are all reprinted here for the first time since their initial publications. The essays are often grounded in the political and social context of Hurston's own time, so to make the volume more accessible to today's readers, the essays are annotated to supply information that her initial readers would have likely had, such as names and details about events that would have been in the media.

To establish the copy-texts for this volume (that is, the versions upon which the essays printed here are based), we took a two-pronged approach. First, if an essay appeared in print during Hurston's lifetime, we returned to the earliest known publication of that essay. In the case of the other essays, we used either the earliest extant typescript or manuscript, whichever exists.

We have made only minor editorial changes to Hurston's writings. Rather than imposing a universal standard on the volume, each essay has been edited as a discrete text. Thus, readers will find the variations of *folklore*, *folk-lore*, and *folk lore*. Apostrophes in contractions (*aint* versus *ain't*) also follow the pattern in the original text. Among the few "silent" emendations, we corrected misspelled proper nouns and obvious typographical errors. The punctuation is largely Hurston's own. The exceptions include adding an occasional comma to introduce quotations or in a few instances moving end punctuation to inside the quotation marks in keeping with today's standard practices; adjusting capitalization within quotations to reflect current practices; adding a comma when one comma (either opening *or* closing) around an appositive phrase is missing; and adding periods where end punctuation is entirely missing. In general, capitalization here follows the original, so readers will find *heaven* and *Heaven*, *hell* and *Hell*, as well as *southern* and *Southern*. Alternate spellings have been retained within essays rather than regularized across the volume. The rare significant change appears in brackets, or those brackets indicate the insertion of a letter or word where the manuscript or typescript is illegible. Hurston's last writings, "The South Was Had" and "Take for Instance Spessard Holland," exist only in Hurston's handwriting, and parts of the manuscripts were lost when staff at the nursing home in which she died began burning her belongings. The brackets there indicate text that has been lost.

The essays published in Hurston's lifetime, including her reportage on the Ruby McCollum case, were often accompanied by artwork, blurbs, and biographical descriptions of Hurston that publishers created to situate the author and her texts in their particular moment. However, including them here was beyond our scope. Likewise, we are unable to reproduce all of the textual features of the essays. For instance, some paragraphs in the Ruby McCollum series appear entirely in bold to attract readers to more salacious or surprising aspects of Ruby's life story. Those features of the text have been standardized. We have, however, maintained the asterisk marks between paragraphs in the series.

The essays are grouped loosely by a dominant theme or topic to support readers interested in exploring a single area, but often those distinctions are admittedly artificial, as any one essay might have appeared in two sections. For readers who want to approach the essays from a more historical perspective, there is a chronological list of essays at the end of the volume.

Introduction

I'm going to sit right here on this porch chair and prophesy that these are the last days of the know-nothing writers on Negro subjects. Both editors and readers are clamoring for something that makes their side meat taste like ham, for to tell the truth, Negro reality is a hundred times more imaginative and entertaining than anything that has ever been hatched up over a typewriter. From now on, the writers must back their rubbish with something more substantial than the lay-figure of the past decade. Go hard or go home. Instead of coloring up coconut grease in the kitchen, go buy a cow and treat the public to some butter.

Zora Neale Hurston, from "You Don't Know Us Negroes"

The witty rhyme with which Zora Neale Hurston ends the title essay of this collection—"Biddy, biddy, bend, my story is end,/ Turn loose the rooster and hold the hen"—can be taken as a sort of epitaph for her, certainly, but also as the naming of a key theme to which she returns again and again throughout the essays she wrote over almost four decades—monumental decades that saw the birth of the Harlem Renaissance and the launch of the classic period of the civil rights movement with the Montgomery Bus Boycott; the desegregation of the US military and the integration of Central High School in Little Rock, Arkansas. Through these essays, collected in one volume for the first time, Hurston takes her place as a major essayist of the twentieth century.

Hurston's words in the epigraph above would prove prophetic. The renowned Black psychiatrist from Martinique, Frantz Fanon, brilliantly observed that the West could never "understand the being of the black man, since it ignores [Black people's] lived experience."[1] Hurston dedicated her writings, especially her novels, to addressing this very shortcoming, which braids its way through so many of her political and aesthetic essays. Essentially, Hurston argues that "the Negro in fiction," as she said, was too often an artificial, two-dimensional construct. Both white and Black authors were guilty of creating a fictional Negro, the former to demean or exoticize, the latter as one more propaganda weapon in the war against white supremacy. What she wanted instead was a revelation of the richness and complexity of Black life behind "the Veil," as W. E. B. Du Bois famously put it in 1903 in *The Souls of Black Folk*.[2] "And so," she argues, "the writings that made out they were holding a looking-glass to the Negro had everything in them except Negroness. Some of the authors meant well. The favor was in them. They had a willing mind, but too light behind." Slavery, Jim Crow, white supremacy, and anti-Black racism, she explains, "intensified our inner life instead of destroying it." And rather than using literature to deflect the white gaze, Hurston maintained that the purpose of the Black writer was both to lift the Veil and to allow the Black experience to speak in its own voice, in all of its sublime resonance—good and bad, positive and negative.

Reading Hurston's reflections on the inner logic of Black cultural forms, social institutions, and behavior is a bit like overhearing an internal monologue in the same way that soliloquies function in Shakespeare. This is one of the innovations she makes in the history of the African American essay form. And throughout these essays, she argues that the full richness of the African American experience could only be realized in print if writers allowed the tradition to speak for itself, thus revealing "a genuine bit of Negroness" in the same way blues and jazz artists had done in the secular tradition; Black preachers and the "unknown bards" had done in their compositions of sermons, spirituals, and gospel music; and as she herself had done, much to the annoyance of Black male contemporaries such as Richard Wright and Alain Locke, in *Their Eyes Were Watching God* (1937). Her lesser-known essays "Race Cannot Become Great Until It Recognizes Its Talent," "You Don't Know Us Negroes," and "The Chick with One Hen" capture these lifelong aesthetic commitments to lift the Veil.

One of the delightful aspects of Hurston's nonfiction is the subtle way in which it serves as commentary on her practice of fiction writing in a relationship of theory to practice. Her foundational essay "Characteristics of Negro Expression," for example, is an attempt to define systematically, like a linguist would, the unique ways in which African Americans *speak*, the ways in which "the American Negro has done wonders to the English language." This essay is one of the first attempts to arrive at a typology of Black English, helping us to understand the principles behind her representation of Black speech in her novels. Hurston identifies "the Negro's greatest contribution to the language" as these three original usages: metaphor and simile ("You sho is propaganda"); the double-descriptive ("chop-axe"); and verbal nouns ("I wouldn't scorn my name all up on you"). These she groups under the larger Black aesthetic principle of "the will to adorn." Then she traces examples of this originality in traditional Black artistic forms, such as folklore, prayers, and sermons. These religious forms, she says, "are tooled and polished until they are true works of art," forged "in the frenzy of creation," a theme to which she returns several times in these essays. "The beauty of the Old Testament does not exceed that of a Negro prayer," she asserts. As rendered in print, "dialect," which Hurston distinguishes from "idiom," was the often racist representation of Black spoken English, widely dismissed as a sign of Black people's lack of intelligence, in minstrelsy, dialect poetry, and vaudeville. But when flowing from Hurston's pen to page, her use of idiom underscores the beauty and range of what is actually a poetic diction, a language within a language. She highlights the manner in which African Americans have fashioned, and continue to refashion, the English language in their own resplendent voices, investing in English new power, poetry, neologisms (colorful coinings of words and expressions), and originality of expression, a thing at which to marvel and not to mock.

For Hurston, Black Vernacular English and folk cultural forms are two of the African American people's most original contributions to American culture. Most importantly, she argues, the cultural artifacts produced by the enslaved community and their heirs are proof that the "will to adorn," in spoken English and storytelling, in the composition of sacred forms such as sermons, prayers, and the spirituals, in the blues and jazz, was one of the most salient signs of cultural vitality and survival and adaptation in the face of the horrors of enslavement and Jim Crow. These forms are parts or manifestations of what we might think of as a larger, organic "*culture of themselves*," one that Black people formulated behind the Veil. And in these essays, Hurston is determined, detail by detail, to lift that Veil for the world to see, and just as importantly, to *hear the sounds of African American cultural formations.* She mounts a defense of what we might think of as traditional Black culture against those who would disparage it, be they white or members of the Black middle class.

Hurston can be quite bold in her taxonomies of what she terms "Negro Folklore." For instance, she characterizes Jack or John and Brer Rabbit, Black culture's ultimate heroes, both with the wit and power to defeat the Master and, in John's case, even the Devil, as he is "often smarter than God." She also, in several asides, characterizes the Black Church as something of its own, a *sui generis* belief system: in the essay's "Culture Heroes" section, she daringly writes, "The Negro is not a Christian really," because of the vestiges of African religions still very much alive and patently manifest in traditional forms of worship, especially in the South. "We are not Christians really, but pagans," she repeats in "Full of Mud, Sweat and Blood," her review of David Cohn's novel *God Shakes Creation.* "It is true that we employ all of the outward symbols of Christianity, but it is a beating of drums before new altars and calling old gods by new names."[3]

The church, to Hurston's mind, is also the ultimate source of the most sublime Black poetry:

The finest poetry that has come out of the Negro race so far has come out of the church, out of the mouths of preachers. If a man announces that he is called to preach and cannot get up in the pulpit and call God by all His praise-giving names; cannot gild the sunrise; heighten the glory of the rainbow, he will soon find himself back at his plowing and digging. Like others we have that consciousness of the inexpressible and a hunger for beauty, and the preacher must fill that want.

Precisely when her contemporaries either wanted to render Black vernacular forms in standard English infused with the African American idiom (for example, James Weldon Johnson) or see them as reflections of economic exploitation and desperate cries for salvation (for example, Richard Wright), Hurston not only defends their sublimity, but subtly makes the case for an aesthetics based on these traditional forms themselves, a true "Black Aesthetic." This was a most radical act, a spirited declaration of the need to recover the essence of Black creativity in the sublime artifacts of the Southern, unreconstructed slave past.

Hurston returns to this idea repeatedly, particularly in "Mother Catherine" and "Ritualistic Expression," her perceptive analysis of the sacred cultural forms that define the Black Church, far too many to enumerate here.[4] But her keen observation is that Black religious practice really was what we might think of as a "cultural laboratory," because, as Hurston puts it, "the religious service is a conscious art expression," reflecting both strikingly original musical forms and neologisms. In "Spirituals and Neo-Spirituals," she explains that "[i]n the mouth of the Negro the English language loses its stiffness, yet conveys its meaning accurately." She offers marvelous examples: "'The booming bounderries of this whirling wind,' conveys just as accurate a picture as mere 'boundaries,' and a little music is gained besides. 'The rim bones of nothing' is just as truthful as 'limitless space.'" Here she summarizes the relation between art and the religious service in action:

> [A]ll religious expression among Negroes is regarded as art, an ability recognised as definitely as in any other art. The beautiful prayer receives the accolade as well as the beautiful song. It is merely a form of expression which people generally are not accustomed to think of as art. Nothing outside of the Old Testament is as rich in figure as a Negro prayer. Some instances are unsurpassed anywhere in literature.

This practice by Black people of reshaping Christian forms of worship in their own image is just one example of a cultural characteristic shared throughout African American culture. Hurston argues:

> So if we look at it squarely, the Negro is a very original being. While he lives and moves in the midst of a white civilisation, everything that he touches is re-interpreted for his own use. He has modified the language, mode of food preparation, practice of medicine, and most certainly the religion of his new country, just as he adapted to suit himself the Sheik hair-cut made famous by Rudolph Valentino.

Creativity and originality, Hurston argues, infuse every aspect of Black life.

Hurston valorizes traditional Black culture as a defense against the social insecurities of a rising Black middle class, whom she chastises for their "self-despisement," refusing "to do or be anything Negro. 'That's just like a Nigger' is the most terrible rebuke one can lay upon this kind." Her list of what psychologists would call cultural "self-loathing" among the Black middle class includes their mocking of traditional Black preaching, the blues, the spirituals, and essentially any of the other cultural forms created by enslaved people. Hurston's critique of this social class predated by decades E. Franklin Frazier's classic work, *Black Bourgeoisie* (1957), in which he would severely chastise the emergent middle class for these same cultural practices.[5] A special target for critique in "Spirituals and Neo-

Spirituals" is anyone who tries to rearrange the traditional forms of Black sacred music, "some daughter or son [who] has been off to college and returns with one of the old songs with its face lifted, so to speak." "But I say again," she continues, "that not one concert singer in the world is singing the songs as the Negro song-makers sing them. If anyone wishes to prove the truth of this let him step into some unfashionable Negro church and hear for himself."

In short, we might think of Hurston as a Black cultural nationalist, in contemporary Black political parlance, or as a Black cultural "conservative" or a "traditionalist," in the sense of valuing traditional forms of cultural expression in the forms in which we received them, as it were. She defends these traditional Black cultural forms against those who think they are too much echoes of the slave past to be "presentable" in an era defined by the primitive modernism of the Harlem Renaissance or its predecessor, turn-of-the-century "politics of respectability," to use the historian Evelyn Brooks Higginbotham's resonant term.

Let us be blunt: Hurston is engaged in a *war of representation*, defending "the race" against detractors both white and Black, on the one hand—against those who had long parodied and mocked Black speech, song, and sermons and other traditional cultural forms—and on the other hand, against the modernists who thought these forms needed to be "tidied up," given a "face lift," as Hurston put it, to be fit to be seated at the proverbial welcome table of American and, indeed, world civilization. Nor did "dialect" need to be abandoned and the traditional Black cultural forms transformed into standard English, as James Weldon Johnson had argued in his preface to *The Book of American Negro Poetry* (1922), and as he did in his monumental standard-English renditions of canonical Black sermons, *God's Trombones* (1927).[6] The spirituals were sublime as they were created; there was no need for them to be "concertized," a scornful word Hurston seems to have coined. The stakes in this battle over the face and voice of Black cultural representation were enormous. Hurston was one of the most articulate defenders of the tradition in its unadulterated purity, and the essays in the first two sections of this book prove that. There was, she argues again and again, absolutely nothing produced by the ancestors of which to be ashamed. To the contrary, these secular and sacred forms contained the heart and soul of the race. As she says of the spirituals, their "truth dies under training like flowers under hot water."

Hurston—the critic, the linguist, the cultural anthropologist—is at all points very much Hurston the novelist, implicitly outlining her theory of the novel as she practiced it, as well as perhaps the very first comprehensive theory of African American culture itself. She articulates an aesthetic theory "based on Negro idioms," as she puts it in her review essay "Jazz Regarded as Social Achievement." In her pioneering observations about the nature of Black culture, Hurston always insists that African Americans are a people, concerned by the full range of human emotions from love to death, just like every other people on the planet. As she writes in "Art and Such," written just a year after her masterpiece, *Their Eyes Were Watching God*, had been roundly criticized by a Richard Wright in the grip of social realist aesthetics:

> So the same old theme, the same old phrases get done again [in representations of African American characters in literature] to the detriment of art. To [writers such as these] no Negro exists as an individual—he exists only as another tragic unit of the Race. This in spite of the obvious fact that Negroes love and hate and fight and play and strive and travel and have a thousand and one interests in life like other humans. When his baby cuts a new tooth he brags as shamelessly as anyone else without once weeping over the prospect of some Klansman knocking it out when and if the child ever gets grown. The Negro artist knows all this but he conceives that a Negro can do nothing but weave something in his particular art form about the Race problem. . . . Anyway, the effect of the whole period has been to fix activities in a mold that precluded originality and denied creation in the arts.

At the end of this stunningly insightful essay, in which she writes about herself in third person, Hurston gives scholars and critics of her fiction a gift—her own theory of the novel—the proverbial "figure in the carpet" that shaped *Jonah's Gourd Vine* (1934) and reached its summit three years later in *Their Eyes Were Watching God*. The latter would be rediscovered by Alice Walker, and its aesthetic principles, if not its forms, refashioned and echoed by a generation of Black authors, especially Black female authors, for the last fifty years. Her explicit lining out of her own sense of the import and originality of her very own practice of the art of fiction writing, remarkably, stands the test of time.

The Black novel, she argues, will only rise to the sublime levels of Black vernacular cultural forms if writers create "a Negro story without bias," stories in which the "characters live and move," stories "about Negroes," certainly, "but [characters] that could be anybody." In other words, stories about the human condition, cast in the world that Black people live and breathe *behind the Veil*; stories, as it were, that Black authors allow readers, white and Black, to "overhear" in the same way that traditional Black storytellers, in the church, on the porch, in the juke joint, around the fireplace, told their stories as works of art without worrying about the political implications of their stories under the white gaze. We see this same thing in the same way that blues and jazz compositions emerged in a Black-on-Black world.

Writers, she argued, needed to mimic that mode of self-revelation, of voice expression, without self-censorship and without concern about someone else's politics of race and representation, without apology or shame, "without special pleading," and—quite boldly—with characters "seen in relation to themselves and not in relation to the whites as has been the rule." To watch characters in novels such as these unfold in their fictional worlds, as we do in *Their Eyes Were Watching God*, for instance, "one would conclude that there were no white people in the world." And just as important is how the story is told, in what language the story is rendered. "[T]he telling of the story [must be] in the idiom—not the dialect—of the Negro. The Negro's poetical flow of language, his thinking in images and figures," she argues in "Art and Such," are the hallmarks of her own novelistic practice of which she was most proud. She brilliantly labels her approach as "stewing the subject in its own juice." Only then, she prophesies, out of this intraracial "stewing," will the African American novelist find her or his own voice.

History has borne out Hurston's prediction in the works of so many of her literary heirs, male as well as female, who, taken together, since Hurston's literary recovery, have produced perhaps the richest field of fiction in the history of the African American literary tradition, all indebted, in one way or another, to the poetics and practices of Zora Neale Hurston. And in large part because of the boldness of her aesthetic theory and the novelty of her political critique of novels that centered white racism, readers white and Black have been "overhearing" the resplendent voices of the Black experience, within the Veil, in a rich variety of ways of which Hurston no doubt would have approved.

This implicit and explicit political approach to Hurston's art makes discussions of race and gender central to understanding her larger body of work. Her willingness to argue for Black vernacular artistic culture and her concomitant creation of strong female characters often made her a lightning rod for those who would have preferred to see depictions of unambiguously centered, barefaced white racism, or of predictably noble and praiseworthy Black characters striving for the middle class. Long before second-wave feminism proclaimed that the personal was political, Hurston created resilient female characters who dared speak their pieces, often in the faces of their male antagonists and partners. Courtship and marriage lie at the heart of most of her fiction, and in her nonfiction, we are similarly offered a glimpse at Hurston's views on whether romance and self-regard can coexist for women in their relationships with men. She wrote the essays "The Ten Commandments of Charm" and "The Lost Keys of Glory" nearly twenty years apart, the former about a year before her first marriage, the latter after the dissolution of her third. Her portraits of men are unflattering: they are childish, shallow, easily manipulated. The mock biblical verse of the first is humorous to be sure, but also biting—and shockingly resonant, still today. The first commandment—"Be cheerful. Let not thy smile come off"—will make any woman who has ever been "encouraged" to smile by a stranger grimace. The second essay contains elements that have come to be associated with Hurston—a folklore adaptation followed by an incisive, insightful analysis—but its tone and intent are difficult to nail down, particularly since its claims contradict today's feminist constructions of Hurston's identity as a strong, independent woman.

Hurston was herself married three times (to younger men), but she never stayed married for long. From a distance—in her letters and her other writings—she seems to have been happiest pounding away at her typewriter, puttering in her garden, or collecting folklore, but—"The Lost Keys of Glory" suggests—she wanted *both*, a long-lasting romantic relationship *and* an active intellectual life as a writer and anthropologist. She certainly resisted giving up her career as a writer, as she explains in her autobiography, *Dust Tracks on a Road* (1942), for traditional married life. Given the gendered norms of her own time, modern readers often see Hurston as a feminist who resisted pressures to marry and keep a home for her husband, but this essay certainly complicates those arguments. Contradictions between Hurston's own life and the essay's musings abound.

The central metaphor of "The Lost Keys of Glory" comes from a Black folktale in which Woman—with advice from the Devil—requests from God the keys to the kitchen, the bedroom, and Man's generations. With those in hand, she is able to rely on Man to work for her, but his physical strength, enhanced by God following Man's own request, can quite literally beat her into submission. Man and Woman have something the other needs. The examples Hurston offers seem to suggest that the happiest women are those who maintain a career but keep a husband at the center of their worlds, providing "steak and potatoes and apple pie on demand" and offering slippers "wherever he might choose to flop down and wallow in the house." Given Hurston's own biography, it is surprising to see her advising women that the path to happiness requires catering to men who "wallow," piglike. She even describes feminism as a "mirage," "the light that failed" to deliver what it promised. But let us be clear: none of this means that the essay endorses traditional gender roles as equitable or fair. Instead, it highlights the dishonesty of cultural norms that ostensibly permits women to pursue professional careers and then penalizes them for doing so. Women who "tear out after the free life of the males" find themselves alone or neurotic, damaged by the way others—especially men—see them: as "too mannish," echoing the well-known and phenomenally popular sermon of the Reverend J. M. Gates, "Manish Women," which he released on vinyl in 1930. She corrected the spelling, but the sentiment was the same.

Throughout Hurston's life, her aesthetic philosophy staunchly resisted the white gaze and distortions or "modernizations" of traditional Black culture and art, but this essay reveals a psychological fatigue in dealing with the male gaze. Was it disappointment that led Hurston to suggest that women accept the status quo? "No female careerist can avoid looking at the picture from time to time," she argues. "And the inevitable question arises inside her, how much is a career worth to a woman anyway? Are not the unknown women, bossing the man of her choice really happier than the career-woman, however famous outside her natural sphere?" As a career woman herself, Hurston certainly invites speculation that she, too, had "from time to time" regrets about having pursued a career, despite the demands of men in her life who assumed she would relinquish her professional ambitions when they wed. We might prefer to think of Hurston—our literary and cultural icon—as fiercely independent, happily waging her way in the world, committed to her art and research, but the essay here suggests she paid a steep price for that choice. It is an important perspective on Hurston, as a person, and on midcentury challenges facing career-minded women after World War II, when Rosie the Riveter was expected to return to her home. Remembering the context in which Hurston lived and wrote is critical to appreciating the complexity and subtlety of her argument here. It is not that she believes things *should* be the way that she describes. Rather, her approach accepts that they just *are*.

Hurston's writings on race and politics are no less complex, and despite the risks of offending powerful people or organizations, she never shrank from tackling both intraracial as well as interracial politics. She took on the intraracial dissention at Howard in "The Hue and Cry About Howard University," just as she does in "The Rise of the Begging Joints." In "The 'Pet Negro' System" she explores "the web of feelings and mutual dependencies" between Blacks and whites in the South, that "a lot of black folk . . . find . . . mighty cosy." The system, she suggests, benefits individuals and complicates blanket claims about race relations in the region. She notably even occasionally approaches such fraught topics with humor as in "Noses" and "Now Take Noses" where differences in Black and white noses reflect the races' different characters:

> The Roman nose, like all Gaul, is divided into three parts— the start, the bend, and the drop. It starts forthwith to rule the world, bends sharply to seek its means, and proceeds sharply after that to achieve its ends. This leads to conquest and law.
>
> The nose of Africa sits in the shade of its cheek bones and dreams. It points not upward, not downward, not anywhere. It sits and dreams and dreams.

Her satire on Marcus Garvey, "The Emperor Effaces Himself," takes a tongue-in-cheek approach to the man—to his uniforms and parades and contradictions—but never to the man's politics, which is an important distinction. "Crazy for This Democracy" centers the failings of American democracy by connecting racism at home to colonialism abroad. Even when, as she irreverently writes, she has "promised God and some other responsible characters, including a bench of bishops," to remain silent on *Brown v. Board of Education*, nevertheless, she voiced her deep reservations about abandoning all-Black institutions in the unreflecting frenzy of embracing— after systematically being excluded from them for so long—integrated ones.

Hurston's race pride permeates everything she writes. She exults in a "timeless" but feminine self even as she sharpens her oyster knife in "How It Feels to Be Colored Me." She takes a more serious look at racism in "My

Most Humiliating Jim Crow Experience," but she refuses to give up even a scintilla of pride or sense of self to a racist physician who has the gall to request payment after seeing her in his laundry closet, as if *she* were so much dirty linen to keep out of view. Absolutely certain of her worth as Lucy Hurston's daughter, she recalls,

> I got up, set my hat at a reckless angle and walked out, telling him that I would send him a check, which I never did. I went away feeling the pathos of Anglo-Saxon civilization.
> And I still mean pathos, for I know that anything with such a false foundation cannot last. Whom the gods would destroy, they first make mad.

Her pride in Florida's Black cattle rancher whose hard work, expertise, and unimpeachable character made him a community pillar permeates "Lawrence of the River." It also surfaces in "Race Cannot Become Great Until It Recognizes Its Talent," a neglected but important work, reprinted here for the first time. In Hurston's earliest explicit statement of her aesthetics, she draws parallels between her own commitments to folk culture and Black idiomatic expression and the works of Shakespeare and Chaucer. Hurston recognized that she was pushing boundaries in depicting folk or working-class characters, and, in doing so, sometimes made middle-class, New Negroes uncomfortable, but her pride in her culture assured her that what others found shameful or reduced to stereotype held tremendous promise for the arts—and for Black culture as a whole.

What, then, explains Hurston's well-known opposition to *Brown v. Board of Education*? Her letter to the editor bears careful reading—both for what it says and for what it does not. The inclusion of "Which Way the NAACP?," which appears here in print for the first time, also sheds light on her thinking. Hurston recoiled at the unintended message she saw lurking beneath the court case—that integrated schools with integrated teachers were without question somehow better than all-Black schools with all-Black teachers; that white teachers and students were "inherently" better than Black ones. Her "white mule" critique of the decision reflects her immense pride in Black educators and her knowledge that Black teachers in Black schools teach more than merely academic subjects. They serve as role models and shepherd students through a racist culture. At the same time, Hurston points out inequities in the way the schools are administered:

> The Supreme Court would have pleased me more if they had concerned themselves about enforcing the compulsory education provisions for Negroes in the South as is done for white children. The next 10 years would be better spent in appointing truant officers and looking after conditions in the homes from which the children come.

So why did she oppose the landmark Supreme Court decision?

Hurston saw the efforts to integrate schools—primary, secondary, and post-secondary—as a declaration that African Americans were not independent and needed the approval of, and social mingling with, whites. She wanted instead to see "[g]rowth from within." Hurston argues that it was conceding too much to declare that all-Black institutions, inherently, were unequal. Underfunded schools were inherently unequal, not Black schools. What she believed in was willing separation versus enforced segregation. She points out, quite rightly, that "[i]t is a contradiction in terms to scream race pride and equality while at the same time spurning Negro teachers and self-association." And, she could sometimes be naive, a trait we also see in the Ruby McCollum case below.

Along with her belief in the benefits of Black schooling for Black children by Black educators—by extension for and by the Black community—Hurston stakes out a controversial position that separate could be, practically speaking, equal. She argues,

> If there are not adequate Negro schools in Florida, and there is some residual, some inherent and unchangeable quality in white schools, impossible to duplicate anywhere else, then I am

the first to insist that Negro children of Florida be allowed to share this boon. But if there are adequate Negro schools and prepared instructors and instructions, then there is nothing different except the presence of white people.

Her claim is conditional, based on an *if*. One of the foundational premises of "Which Way the NAACP?" is that "in every state in the South the identical text-books are issued to White and Negro schools." History has not borne out this assumption by Hurston—which is at the foundation of her reasoning. Hurston's assertion that white is not necessarily better is true, but so is the reality that Black students and white students almost never used the same textbooks at the same time. The Black schools would get the used, outdated textbooks discarded from white schools. And new lab facilities were often reserved for white students only. Despite this flaw in her thinking, in many ways, what Hurston imagined came to pass. Even as schools desegregated, structural racism persisted and created new problems. Black teachers were fired so white teachers could remain, and truancy laws that existed prior to 1954 continued to be enforced selectively. Integrating schools did not necessarily improve the education of Black students.

"Which Way the NAACP?" also conveys anxieties about communism and government overreach, as well as her growing conservatism. Hurston abhorred communism and similar ideologies, because they cast African Americans in the role of objects, as a people doomed to react to forces beyond their control, as a race without agency. Hurston repeatedly returns to the theme of Black agency throughout her essays. To see Black people as victims, she deeply believes, is to succumb to the ultimate form of anti-Black racism. She was increasingly distrustful of outsiders—whether Communists influenced by Russia or white and NAACP leaders from the North. While white outsiders undermine the agency of Black people, so, too, she argues, do Black people aligned with the NAACP. She still sees an elitism in the organization's leadership, one she traces back to its founding, and she wonders why the organization has not "found a home" in the "hearts" of the larger Black population. She fails to see the "advancement" gained when Black students are assaulted by white students in integrated bathrooms. And she correctly points out that integrated schools in the North had not led to equality or social mingling.

Underlying the move to desegregate, she fears, is self-loathing, a failure among rising middle-class African Americans to love themselves, their traditions, their culture regardless of what whites think. The fight for desegregation, she believes, flies in the face of all she has argued for in her fiction and nonfiction alike. She explains:

> In close to a century of education and progress by the American Negro, self-consciousness of race and an inferiority complex stemming out of the past, we should have come to the place where notice by the Whites and the bolstering of proximity as a sign of tolerance would be utterly unnecessary. Not what counts with the majority in the nation, but what counts within ourselves should have arrived by now.

Pride in self and in traditional Black culture permeates her argument, much as it does in her all-Black novels, short stories, plays, and ethnographies. What white people think does not and should not matter. In this sense, remarkably, Hurston was a proto-Black cultural nationalist, a forerunner of an artistic and political philosophy that would become central tenets of the Black Arts Movement, born circa 1966. And rarely is Hurston credited with voicing these attitudes well before that movement commenced. In her fictional worlds, Hurston pointed people in this direction by demonstrating that Black people lived full, complex lives without white people. It was a depiction that countered the oversimplified two-dimensional Black-white opposition found in so much writing by both Black and white writers from the period. Here, in her opposition to court-ordered desegregation, she argues explicitly for the cultural pride and the cultural politics imbued in her fiction.

In essence, we recognize in Hurston's position a consistency in the whole, self-sufficient Black world she portrays in her fiction and the whole, self-sufficient Black world for which she advocates in her argument opposing *Brown v. Board*. It stemmed from a complicated blend of mourning for something being lost or undermined—Black agency—and umbrage at the insult the court decision implied. Black people wanted to end segregation, but they didn't want to admit, or cede to, the notion that anything all Black was inherently inferior, substandard, or downright bad. Hurston supported integration for Black people as equals, not as second-class citizens. In this, once again, she was making an argument that Black Nationalist political and aesthetic proponents would draw upon and elaborate upon a decade or so later. Integration had its place, and was a noble social goal, but meaningful integration could only take place between cultural equals.

Hurston was an intrepid observer and tireless reporter. Her fiction, nonfiction, and ethnographic work bears witness to this. In 1952, she essentially became a beat reporter for the *Pittsburgh Courier*, turning her formidable powers of observation to Ruby McCollum's murder trial and the community of Live Oak, Florida. On Sunday morning, August 3, 1952, Ruby McCollum, a wealthy, married African American mother of four, walked into Dr. C. LeRoy Adams's office and shot him dead. Adams was white, a medical doctor, and a recently elected state senator. McCollum admitted to the crime, but the question of motive and the disputed nature of their relationship made the trial a powder keg that attracted national and international headlines. While many questions remain unanswered, what seems clear is that McCollum and the physician had engaged in a yearslong sexual relationship. He fathered one of her children, and at the time of the murder she had become pregnant with a second. Hurston describes the notion that white men were entitled to use Black women for sexual gratification as "Paramour Rights," which dated back to enslavement. Whether their relationship was entirely consensual remains uncertain. Ruby's husband was an important local figure, but his illegal, lucrative gambling business depended upon bribes of powerful white men, like Adams, and Ruby's mental health was in decline.

At trial, the prosecution contended—and the community publicly concurred—that the murder resulted from a dispute over a $100 medical bill, but Hurston's writings make it clear that that was unlikely. She argues the community's response "amounted to a mass delusion of mass illusion. A point of approach to the motive for the slaying of the popular medico and politician had been agreed upon, and however bizarre and unlikely it might appear to the outside public, it was going to be maintained and fought for." It was a white narrative to which African Americans acquiesced, understandably, rather than put their own lives at risk. McCollum's attorney hoped for a second-degree murder conviction rather than a determination of premeditated murder, which required the death penalty. With strenuous objections from the prosecution to Ruby's testimony, only bits of her story were admitted into evidence. McCollum's trial by an all-white, all-male jury of her "peers" and subsequent sentence to the electric chair was a national travesty rooted in America's racist (and sexist) past that made it impossible for the accused to face a fair trial. McCollum's conviction was later overturned, but after being declared incompetent to stand trial a second time, she was committed to the Florida State Hospital, where she remained until 1974, at which time she was released on the grounds that she posed no threat. She died in 1992, claiming that she had retained no memory of the shooting.

Hurston describes the trial as unfolding beneath "a smothering blanket of silence." That silence may have been the consequence of Live Oak's Black community holding its collective breath. Less than a decade earlier, in 1944, Willie James Howard, a Black fifteen-year-old, had been lynched in Live Oak for the outrage of sending all of his coworkers, including a white girl, Christmas cards. When Ruby's husband, Sam McCollum, learned what she had done, he reportedly said, "Ruby's as good as dead . . . and so am I. They'll have to kill us now."[7] He immediately took his children and fled in hopes of escaping that violence.

Those in Live Oak who had been closest to the McCollum family were the most vocal in their condemnations of Ruby. It was a defensive pose, Hurston surmised, that would help safeguard those close associates from racial violence. Hurston, like other members of the press, was unable to speak directly with the accused. Who had the courage to talk to Hurston about Ruby and Sam? We don't know. Much of what Hurston relates about McCollum's early life had to have come from family and friends who clearly spoke on the condition of anonymity. At times Hurston even quotes people without attribution, concealing her sources, a protective decision for a reporter. In a more literary sense, these anonymous voices take on a communal choral quality, echoing the role of the chorus in a Greek tragedy similar to the store porch we see in her Eatonville fiction.

Hurston's reporting on McCollum's trial is restrained by the politics of race and gender in the South, so what boils under the surface, submerged out of view, is essential. Hurston was, after all, a Black woman reporting from a community she describes as "hostile."[8] She knew about Howard's lynching, too. Even she, as a reporter, was limited in what she could safely articulate, so she creates gaps—silences—to illustrate the racialized power dynamics at play. As with the questions about Ruby's motive, it is never clear why the prosecuting attorney and the judge (both of whom knew the victim well) wanted her silenced. Was it the implication that Adams had been manipulating (and even drugging) McCollum for years to preserve his sexualized power over her? Or did the drive to silence stem from Sam McCollum's illegal dealings? Ruby McCollum's husband had died of a heart attack the day after Ruby's arrest, so limiting her testimony protected everyone involved with Sam's illegal dealings—except Ruby.

As the prosecution tried to convince the all-white male jury that the dispute between McCollum and Adams was over a bill, Hurston tells us the prosecuting state's attorney said, "'You all know that a good way to get them (Negroes) mad is to try to collect. You've seen 'em mull' (sullen). Black [the prosecuting attorney] cautioned the jury not to be deceived by Ruby's behavior on the stand. He warned that she was highly intelligent, wealthy, but full of sly cunning." A representative of the state legal system, the prosecuting attorney, blatantly invoked racist stereotypes of African Americans without rebuke. No wonder Hurston, as she says in "Crazy for

This Democracy," wanted to sample true American democracy for herself. Like women's gender equality, it was still just a mirage.

Hurston took the judge at his word when he impaneled the jury—that McCollum deserved a fair trial. And yet Hurston watched, horrified, as events unfolded. She explains her changed thinking about Judge Hal W. Adams (of no relation to the doctor) this way: "He will never consent for any human being to be sent to their death without permitting the jury to hear their side of the story. He won't! Judge Adams won't! It was only when he exhibited anger and threatened Cannon [McCollum's attorney] with contempt of Court if he persisted, that I wilted back, first in my soul and then in my seat. My disillusionment was terrible. My faith had been so strong." Her naiveté—despite her lived experience—allowed her to believe that Ruby's story would eventually come out. Tragically, it did not.

Hurston reached conclusions about the shooting that did not make it into newspapers. She came to believe that the doctor was trying to end his affair with Ruby at the time of the shooting, but she was, in Hurston's words, "refus[ing] to fade out of the picture."[9] With LeRoy Adams's election to the State Senate and his eye on the governor's mansion, the physician may have needed to end things. There were other theories as well. Adams had ordered a pistol that fired bird shot rather than a conventional bullet that could be forensically tied back to the weapon.[10] The implication is that Adams may have been planning to use the gun to kill McCollum. At the same time, Ruby said she was caught between *two* guns. Whose guns? Is it possible she shot Adams in self-defense? Or was she a woman scorned? It is impossible to know. While Hurston's reporting made much of Black America an eyewitness to the trial, her writings tell us more about the cultural moment, the politics of race and gender, in northern Florida than they do about Ruby McCollum's motives. An all-white, all-male jury sat on the case of a Black woman who admitted to murdering one of the community's leading white citizens. Never was she able to tell her story in court. Her confession, it seems, was story enough. In today's era of the "#MeToo" and "Black Lives Matter" movements, Hurston's dispatches on the Ruby McCollum case serve as a blunt reminder of the long history of the intersection of race, gender, and violence in America and the steep price that Black Americans have paid—and continue to pay—in an unjust criminal justice system.

Zora Neale Hurston's politics have been criticized over the years. Her contemporaries Richard Wright, Sterling Brown, and Ralph Ellison accused her of pandering to racist stereotypes in her writings, but the volume you hold in your hands demonstrates that was simply not the case. In fact, it brings to the fore Hurston's lifelong attempt to reclaim traditional Black folk culture from racist and classist degradations, to share with her readers the "race pride" she felt, to build the race from within. She was often constrained—by her patron, by her publishers, by the limits of what white publishers *would* print. To acknowledge that material reality doesn't make her one of those "handkerchief-head[s]," as she put it in 1955. Rather, it grounds her writings in a particular cultural and historical moment, one full of closed doors for a Black woman who wanted to do more than clean homes. It makes her remarkable contributions to American arts and letters, her place in American culture as an icon of African American, modern, and women's literature, all the more inspiring. She wrote in her visually inspired vernacular that "every tub must sit on its own bottom." With a long-overdue collection of her best essays finally in the hands of readers, Hurston can now do just that.

HENRY LOUIS GATES JR. AND GENEVIEVE WEST

Part One

ON THE FOLK

Bits of Our Harlem

We looked up from our desk and he was standing before us, tall, gaunt and middle-aged. In his hand was one of those tin receptacles for charity-begging. Like all other long-suffering Harlemites we shuddered. Beggars with tin cups are so numerous. He smiled and stood there. We tried to look austere—some money-seekers may be easily intimidated—but not so our hero.

"Well, what can I do for you?" we asked, looking the visitor in the face for the first time.

"A few pennies for homeless children," he answered.

We felt that it was useless to struggle so we donated a dime. No sooner had the coin rattled to the bottom of the cup than we received a hearty "Thank yuh. God will shorely bless yuh."

We looked closely at his face this time and saw fanatic fires burning in the small eyes set in a thin freckled face. But our eyes rested longest on the mouth and environs.

The short, thin upper-lip showed his Caucasian admixture, but a full drooping under-lip spoke for the Negro blood in him. A fringe of scrubby rusty-red hair completely encircled the whole. When he spoke, four teeth showed forlornly in the bottom jaw. We are still wondering if there were any others scattered about in his aging gums.

"You don't know me, do you?" he asked.

"I am afraid I haven't had the pleasure," we answered.

"Well, they calls me th' black Longfellow."[1]

We brightened. "These be gray days, and a sweet singer in Israel is to be highly honored. Would you favor us with a selection or two?"

"Shorely, shorely; but drop in a few mo' pennies, please."

What are a few pennies against the songs of an immortal bard? We dropped in six cents.

The poet cleared his throat and sang:

"God Shall Without a Doubt Heal Every Nation"

"There shall be no sickness, no sorrow after while,
There shall be no sickness, no sorrow, after while,
There wil' be no more horror,
Watch for joy and not for sorrow,
God shall heal up every nation tomorrow, after while.
God will bring good things to view after while,
God shall make all things new,
Every child of God will without a doubt be called a Jew,
God will make us all one nation, after while."

"Ain't that beautiful, now?" the poet asked. I'll recite yuh another one."

Before we could protest he was in the midst of

"The Automobiles"

"Once horses and camels was the style,
Now they fly 'round in automobile,
They don't look at a policeman's sign,
Sometimes they run over chillun,
Sometimes over a divine,
When they are drunk with devils' wine,
They scoots—"

But we had fled into the inner office with our fingers in our ears.

. . .

The hurly-burly of Lenox Avenue fretted our soul.[2] The dirty corpses of yesterday's newspapers, flapping upon the pavement or lying supine in the gutter, together with the host of the unwashed and washed but glaringly painted, was too unlovely and we fled up 181st Street.[3] We were not really hungry but we longed for rest.

A little sign caught our eyes. "Odds and Ends," it read. A yellowish teapot was depicted in the midst of the inscription. A little hunger, a great weariness of spirit, and a sufficient amount of curiosity drew us into the basement dining room.

A raucous bell rang when we opened the door and a soft-footed attendant instantly appeared to take charge of our wraps.

Back of a green screen was a snug room full of odds and ends. Chairs from Colonial New England, bits of pottery from France and Spain, candlesticks from China, bric-a-brac from Nippon, samplers from England—the ends of the world brought together in a basement. The effect was pleasing, very pleasing.

And the guests. At one table was a woman writer of some ability in company with a wealthy realtor. In a corner, dining alone, a lawyer of national fame slowly sipped his coffee behind a red candle and nodded to a world-famous baritone and composer. A widely discussed editor was dining with a young woman who hopes to be an editor some day.

But the atmosphere is the most attractive thing about "Odds and Ends." We do not know whether it is the subtle lighting, the ingenious arrangement of the furnishings, or the spirit of the great number of celebrities that frequent the place. There IS a peace, a calm that falls like a benediction upon the guest who enters there. The food was delicious, but mere food does not create atmosphere. Perhaps it is the kindly spirit of the proprietor—we do not know him yet—that bids the weary rest.

High John de Conquer

Maybe, now, we used-to-be black African folks can be of some help to our brothers and sisters who have always been white. You will take another look at us and say that we are still black and, ethnologically speaking, you will be right. But nationally and culturally, we are as white as the next one. We have put our labor and our blood into the common causes for a long time. We have given the rest of the nation song and laughter. Maybe now, in this terrible struggle, we can give something else—the source and soul of our laughter and song.[1] We offer you our hope-bringer, High John de Conquer.

High John de Conquer came to be a man, and a mighty man at that. But he was not a natural man in the beginning. First off, he was a whisper, a will to hope, a wish to find something worthy of laughter and song. Then the whisper put on flesh. His footsteps sounded across the world in a low but musical rhythm as if the world he walked on was a singing-drum. The black folks had an irresistible impulse to laugh. High John de Conquer was a man in full, and had come to live and work on the plantations, and all the slave folks knew him in the flesh.

The sign of this man was a laugh, and his singing-symbol was a drum-beat. No parading drum-shout like soldiers out for show. It did not call to the feet of those who were fixed to hear it. It was an inside thing to live by. It was sure to be heard when and where the work was the hardest, and the lot the most cruel. It helped the slaves endure. They knew that something better was coming. So they laughed in the face of things and sang, "I'm so glad! Trouble don't last always." And the white people who heard them were struck dumb that they could laugh. In an outside way, this was Old Massa's fun, so what was Old Cuffy laughing for?[2]

Old Massa couldn't know of course, but High John de Conquer was there walking his plantation like a natural man. He was treading the sweat-flavored clods of the plantation, crushing out his drum tunes, and giving out secret laughter. He walked on the winds and moved fast. Maybe he was in Texas when the lash fell on a slave in Alabama, but before the blood was dry on the back he was there. A faint pulsing of a drum like a goat-skin stretched over a heart, that came nearer and closer, then somebody in the saddened quarters would feel like laughing and say, "Now, High John de Conquer, Old Massa couldn't get the best of *him*. That old John was a case!" Then everybody sat up and began to smile. Yes, yes, that was right. Old John, High John could beat the unbeatable. He was top-superior to the whole mess of sorrow. He could beat it all, and what made it so cool, finish it off with a laugh. So they pulled the covers up over their souls and kept them from all hurt, harm and danger and made them a laugh and a song. Night time was a joke, because daybreak was on the way. Distance and the impossible had no power over High John de Conquer.

He had come from Africa. He came walking on the waves of sound. Then he took on flesh after he got here. The sea captains of ships knew that they brought slaves in their ships. They knew about those black bodies huddled down there in the middle passage, being hauled across the waters to helplessness.[3] John de Conquer was walking the very winds that filled the sails of the ships. He followed over them like the albatross.[4]

It is no accident that High John de Conquer has evaded the ears of white people. They were not supposed to know. You can't know what folks won't tell you. If they, the white people, heard some scraps, they could not understand because they had nothing to hear things like that with. They were not looking for any hope in those days, and it was not much of a strain for them to find something to laugh over. Old John would have been out of place for them.

Old Massa met our hope-bringer all right, but when Old Massa met him, he was not going by his right name. He was traveling, and touristing around the plantations as the laugh-provoking Brer Rabbit. So Old Massa and Old Miss and their young ones laughed with and at Brer Rabbit and wished him well. And all the time, there was High John de Conquer playing his tricks of making a way out of no-way. Hitting a straight lick with a crooked stick. Winning the jack pot with no other stake but a laugh. Fighting a mighty battle without outside-showing force, and winning his war from within. Really winning in a permanent way, for he was winning with the soul of the black man whole and free. So he could use it afterwards. For what shall it profit a man if he gain the whole world, and lose his own soul?[5] You would have nothing but a cruel, vengeful, grasping monster come to power. John de Conquer was a bottom-fish. He was deep. He had the wisdom tooth of the East in his head. Way over there, where the sun rises a day ahead of time, they say that Heaven arms with love and laughter those it does not wish to see destroyed. He who carries his heart in his sword must perish. So says the ultimate law. High John de Conquer knew a lot of things like that. He who wins from within is in the "Be" class. *Be* here when the ruthless man comes, and *be* here when he is gone.

Moreover, John knew that it is written where it cannot be erased, that nothing shall live on human flesh and prosper.[6] Old Maker said that before he made any more sayings. Even a man-eating tiger and lion can teach a person that much. His flabby muscles and mangy hide can teach an emperor right from wrong. If the emperor would only listen.

II

There is no established picture of what sort of looking-man this John de Conquer was. To some, he was a big, physical-looking man like John Henry.[7] To others, he was a little, hammered-down, low-built man like the Devil's doll-baby. Some said that they never heard what he looked like. Nobody told them, but he lived on the plantation where their old folks were slaves. He is not so well known to the present generation of colored people in the same way that he was in slavery time. Like King Arthur of England, he has served his people, and gone back into mystery again. And, like King Arthur, he is not dead. He waits to return when his people shall call again. Symbolic of English power, Arthur came out of the water, and with Excalibur, went back into the water again. High John de Conquer went back to Africa, but he left his power here, and placed his American dwelling in the root of a certain plant.[8] Only possess that root, and he can be summoned at any time.

"Of course, High John de Conquer got plenty power!" Aunt Shady Anne Sutton bristled at me when I asked her about him. She took her pipe out of her mouth and stared at me out of her deeply wrinkled face. "I hope you ain't one of these here smart colored folks that done got so they don't believe nothing, and come here questionizing me so you can have something to poke fun at. Done got shamed of the things that brought us through. Make out 'tain't no such thing no more."

When I assured her that that was not the case, she went on.

"Sho John de Conquer means power. That's bound to be so. He come to teach and tell us. God don't leave nobody ignorant, you child. Don't care where He drops you down, He puts you on a notice. He don't want folks taken advantage of because they don't know. Now, back there in slavery time, us didn't have no power of protection, and God knowed it, and put us under watch-care. Rattlesnakes never bit no colored folks until four years after freedom was declared. That was to give us time to learn and to know. 'Course, I don't know nothing about slavery personal like. I wasn't born till two years after the Big Surrender. Then I wasn't nothing but a infant baby when I was born, so I couldn't know nothing but what they told me. My mama told me, and I know that she wouldn't mislead me, how High John de Conquer helped us out. He had done teached the black folks so they knowed a hundred years ahead of time that freedom was coming. Long before the white folks knowed anything about it at all.

"These young Negroes reads they books and talk about the war freeing the Negroes, but Aye, Lord! A heap sees, but a few knows. 'Course, the war was a lot of help, but how come the war took place? They think they knows, but they don't. John de Conquer had done put it into the white folks to give us our freedom, that's what. Old Massa fought against it, but us could have told him that it wasn't no use. Freedom just *had* to come. The time set aside for it was there. That war was just a sign and a symbol of the thing. That's the truth! If I tell the truth about everything good as I do about that, I can go straight to Heaven without a prayer."

Aunt Shady Anne was giving the inside feeling and meaning to the outside laughs around John de Conquer. He romps, he clowns, and looks ridiculous, but if you will, you can read something deeper behind it all. He is loping on off from the Tar Baby with a laugh.[9]

Take, for instance, those words he had with Old Massa about stealing pigs.

Old John was working in Old Massa's house at that time, serving around the eating table. Old Massa loved roasted young pigs, and had them often for dinner. Old John loved them too, but Massa never allowed the slaves to eat any at all. Even put aside the left-over and ate it next time. John de Conquer got tired of that. He took to stopping by the pig pen when he had a strong taste for pig-meat, and getting himself one, and taking it on down to his cabin and cooking it.

Massa began to miss his pigs, and made up his mind to squat for who was taking them and give whoever it was a good hiding. So John kept on taking pigs, and one night Massa walked him down. He stood out there in the dark and saw John kill the pig and went on back to the "big house" and waited till he figured John had it dressed and cooking. Then he went on down to the quarters and knocked on John's door.

"Who dat?" John called out big and bold, because he never dreamed that it was Massa rapping.

"It's me, John," Massa told him. "I want to come in."

"What you want, Massa? I'm coming right out."

"You needn't to do that, John. I want to come in."

"Naw, naw, Massa. You don't want to come into no old slave cabin. Youse too fine a man for that. It would hurt my feelings to see you in a place like this here one."

"I tell you I want to come in, John!"

So John had to open the door and let Massa in. John had seasoned that pig *down*, and it was stinking pretty! John knowed Old Massa couldn't help but smell it. Massa talked on about the crops and hound dogs and one thing and another, and the pot with the pig in it was hanging over the fire in the chimney and kicking up. The smell got better and better.

Way after while, when that pig had done simbled down to a low gravy, Massa said, "John, what's that you cooking in that pot?"

"Nothing but a little old weasly possum, Massa. Sickliest little old possum I ever did see. But I thought I'd cook him anyhow."

"Get a plate and give me some of it, John. I'm hungry."

"Aw, naw, Massa, you ain't hongry."

"Now, John, I don't mean to argue with you another minute. You give me some of that in the pot, or I mean to have the hide off your back tomorrow morning. Give it to me!"

So John got up and went and got a plate and a fork and went to the pot. He lifted the lid and looked at Massa and told him, "Well, Massa, I put this thing in here a possum, but if it comes out a pig, it ain't no fault of mine."

Old Massa didn't want to laugh, but he did before he caught himself. He took the plate of brownded-down pig and ate it up. He never said nothing, but he gave John and all the other house servants roast pig at the big house after that.[10]

III

John had numerous scrapes and tight squeezes, but he usually came out like Brer Rabbit. Pretty occasionally, though, Old Massa won the hand. The curious thing about this is, that there are no bitter tragic tales at all. When Old Massa won, the thing ended up in a laugh just the same. Laughter at the expense of the slave, but laughter right on. A sort of recognition that life is not one-sided. A sense of humor that said, "We are just as ridiculous as anybody else. We can be wrong, too."

There are many tales, and variants of each, of how the Negro got his freedom through High John de Conquer. The best one deals with a plantation where the work was hard, and Old Massa mean. Even Old Miss used to pull her maids' ears with hot firetongs when they got her riled. So, naturally, Old John de Conquer was around that plantation a lot.

"What we need is a song," he told the people after he had figured the whole thing out. "It ain't here, and it ain't no place I knows of as yet. Us better go hunt around. This has got to be a particular piece of singing."

But the slaves were scared to leave. They knew what Old Massa did for any slave caught running off.

"Oh, Old Massa don't need to know you gone from here. How? Just leave your old work-tired bodies around for him to look at, and he'll never realize youse way off somewhere, going about your business."

At first they wouldn't hear to John, that is, some of them. But, finally, the weak gave in to the strong, and John told them to get ready to go while he went off to get something for them to ride on. They were all gathered up under a big hickory nut tree. It was noon time and they were knocked off from chopping cotton to eat their dinner. And then that tree was right where Old Massa and Old Miss could see from the cool veranda of the big house. And both of them were sitting out there to watch.

"Wait a minute, John. Where we going to get something to wear off like that. We can't go nowhere like you talking about dressed like we is."

"Oh, you got plenty things to wear. Just reach inside yourselves and get out all those fine raiments you been toting around with you for the last longest. They is in there, all right. I know. Get 'em out, and put 'em on."

So the people began to dress. And then John hollered back for them to get out their musical instruments so they could play music on the way. They were right inside where they got their fine raiments from. So they began to get them out. Nobody remembered that Massa and Miss were setting up there on the veranda looking things over. So John went off for a minute. After that they all heard a big sing of wings. It was John come back, riding on a great black crow. The crow was so big that one wing rested on the morning, while the other dusted off the evening star.

John lighted down and helped them, so they all mounted on, and the bird took out straight across the deep blue sea. But it was a pearly blue, like ten squillion big pearl jewels dissolved in running gold. The shore around it was all grainy gold itself.

Like Jason in search of the golden fleece, John and his party went to many places, and had numerous adventures.[11] They stopped off in Hell where John, under the name of Jack, married the Devil's youngest daughter and became a popular character. So much so, that when he and the Devil had some words because John turned the dampers down in old Original Hell and put some of the Devil's hogs to barbecue over the coals, John ran for High Chief Devil and won the election. The rest of his party was overjoyed at the possession of power and wanted to stay there. But John said no. He reminded them that they had come in search of a song. A song that would whip Old Massa's earlaps down. The song was not in Hell. They must go on.

The party escaped out of Hell behind the Devil's two fast horses. One of them was named Hallowed-Be-Thy-Name, and the other, Thy-Kingdom-Come. They made it to the mountain. Somebody told them that the Golden Stairs went up from there. John decided that since they were in the vicinity, they might as well visit Heaven.

They got there a little weary and timid. But the gates swung wide for them, and they went in. They were bathed, robed, and given new and shining instruments to play on. Guitars of gold, and drums, and cymbals and wind-singing instruments. They walked up Amen Avenue and down Hallelujah Street, and found with delight that Amen Avenue was tuned to sing bass and alto. The west end was deep bass, and the east end alto. Hallelujah Street was tuned for tenor and soprano, and the two promenades met right in front of the throne and made harmony by themselves. You could make any tune you wanted to by the way you walked. John and his party had a very good time at that and other things. Finally, by the way they acted and did, Old Maker called them up before His great workbench, and made them a tune and put it in their mouths. It had no words. It was a tune that you could bend and shape in most any way you wanted to fit the words and feelings that you had. They learned it and began to sing.

Just about that time a loud rough voice hollered, "You Tunk! You July! You Aunt Diskie!" Then Heaven went black before their eyes and they couldn't see a thing until they saw the hickory nut tree over their heads again. There was everything just like they had left it, with Old Massa and Old Miss sitting on the veranda, and Massa was doing the hollering.

"You all are taking a mighty long time for dinner," Massa said. "Get up from there and get on back to the field. I mean for you to finish chopping that cotton today if it takes all night long. I got something else, harder than that, for you to do tomorrow. Get a move on you!"

They heard what Massa said, and they felt bad right off. But John de Conquer took and told them, saying, "Don't pay what he say no mind. You know where you got something finer than this plantation and anything it's got on it, put away. Ain't that funny? Us got all that, and he don't know nothing at all about it. Don't tell him nothing. Nobody don't have to know where us gets our pleasure from. Come on. Pick up your hoes and let's go."

They all began to laugh and grabbed up their hoes and started out.

"Ain't that funny?" Aunt Diskie laughed and hugged herself with secret laughter. "Us got all the advantage, and Old Massa think he got us tied!"

The crowd broke out singing as they went off to work. The day didn't seem hot like it had before. Their gift song came back into their memories in pieces, and they sang about glittering new robes and harps, and the work flew.

IV

So after a while, freedom came. Therefore High John de Conquer has not walked the winds of America for seventy-five years now. His people had their freedom, their laugh and their song. They have traded it to the other Americans for things they could use like education and property, and acceptance. High John knew that that was the way it would be, so he could retire with his secret smile into the soil of the South and wait.

The thousands upon thousands of humble people who still believe in him, that is, in the power of love and laughter to win by their subtle power, do John reverence by getting the root of the plant in which he has taken up his secret dwelling and "dressing" it with perfume, and keeping it on their person, or in their houses in a secret place. It is there to help them overcome things they feel that they could not beat otherwise, and to bring them the laugh of the day. John will never forsake the weak and the helpless, nor fail to bring hope to the hopeless. That is what they believe, and so they do not worry. They go on and laugh and sing. Things are bound to come out right tomorrow. That is the secret of Negro song and laughter.

So the brother in black offers to these United States the source of courage that endures, and laughter. High John de Conquer. If the news from overseas reads bad, and the nation inside seems like it is stuck in the Tar Baby, listen hard, and you will hear John de Conquer treading on his singing-drum. You will know then, that no matter how bad things look now, it will be worse for those who seek to oppress us. Even if your hair comes yellow, and your eyes are blue, John de Conquer will be working for you just the same. From his secret place, he is working for all America now. We are all his kinfolks. Just be sure our cause is right, and then you can lean back and say, "John de Conquer would know what to do in a case like this, and then he would finish it off with a laugh."

White America, take a laugh out of our black mouths, and win! We give you High John de Conquer.

The Last Slave Ship

At low tide the hull of the *Clotilda* can be seen a little even now, in the marsh of Bayou Corne in Alabama, where she was scuttled and sunk.[1] She was the last ship to bring a cargo of "black ivory" to the United States—stealing into Mobile Bay on a sultry night in August, 1859, only two years before Abraham Lincoln was elected and only five years before Emancipation. The progeny of those last-minute slaves today still live in Alabama, mostly in the untidy clapboard village of Plateau, long also known as African Town.

They have shed their accent, and even the oldest among them, in an eagerness to be like other American Negroes, conceal differences and emphasize sameness. To strangers, especially, they will not let on that their memories of Africa are fresher than those of their neighbors. Yet the last survivor of the last boat-load of black men and women who had been set down stark naked on American soil—one who called himself Cudgo Lewis[2] and whose tribal name was Kossula-O-Lo-Loo-Ay—did not die until 1935, at the age of ninety-five.[3] I myself talked to him often about his early capture and the fateful voyage on the *Clotilda*.

Only one physical symbol distinguished Plateau from any other American Negro village. That was the eight gates around Cudgo Lewis' half-acre plot—built in memory of the Takkoi home from which he had been torn as a youth of eighteen. Through those original eight gates had poured the King of Dahomey's warriors more than seventy years before, to massacre most of Cudgo's tribesmen and to start the remainder on their forced journey overseas.[4] Before then, Cudgo—or Kossula-O-Lo-Loo-Ay—had known a happy life as a member of his peaceful Nigerian tribe.

The people of Takkoi formed an industrious, agricultural community. They had remained on good terms with their neighbors for many years before the King of Dahomey sent the first message which was to spell their eventual doom. That message demanded tribute in the form of fruits and vegetables to be delivered regularly to the King of Dahomey.

The King of Takkoi was a fearless man. He sent back word that he had no garden produce other than what was needed for his own household. The rest, he said, belonged to his people, and he had no right to rob them for the sake of feeding Dahomey. Kossula told me that the people of Takkoi were proud of their king's courageous reply. The King of Dahomey was widely known as a powerful chieftain whose main occupation had long been the capture and sale of slaves to cross the water. He commanded huge legions who were equipped with guns and swords from Europe. He had ships, and his palaces were richly filled with foreign luxuries. His capitol at Abomey was protected by cannon.[5]

It would seem, from the records, that the King of Dahomey was a merciful and long-suffering man. He tried to overlook the fact that other tribes refused to send him farm products, or made malicious remarks about his person and his family. He would bear these iniquities until urged by his generals to avenge the insults heaped upon him. Then, and only then, would he order his armies to march upon the culprits and punish them for their evil deeds.

This he did with admirable thoroughness. All who offered resistance were instantly beheaded—along with those who were too old, too young or otherwise unfit for the slave markets of America. He made sure, in this way, that their no-good breed would no longer be able to insult or abuse a merciful and patient man like the King of Dahomey.

The king's legions included not only every able-bodied man in Dahomey but also a large female army, known to the Occident as Amazons. By all accounts, these female warriors were strong, courageous and even more bloodthirsty than the men. Their generals held equal rank with the leaders of the male army. When the king sat under his state umbrella on important occasions, the male generals sat on his right and the Amazon leaders at his left.

To foster thoroughness in battle, the king paid an extra bonus for every head brought in by his soldiers. Travelers reported that the palace at Abomey was made of bleached skulls. The gateposts of the royal compound was a mass of skulls, and the drive leading to the palace was marked out by skulls stuck on posts. Apparently the nations and tribes of West Africa were much addicted to slander of the King of Dahomey, and to mulishness in the matter of farm produce.

II

When the King of Takkoi's reply arrived, the Dahomey chieftain was naturally outraged. Such impudence, from so petty a kingdom, had to be punished—and the King of Dahomey waited only for the rainy season to end before dealing with this latest impertinence. But fate intervened. The King of Dahomey died.

All the other chiefs and kings sent flowery messages of condolence at news of his death. But the King of Takkoi did not believe in concealing the hatred that they all felt. He sent word that the death of the King of Dahomey was a good thing. Now, he said, the ocean was dried up, and one could see the bottom of the sea. This slur on the King of Dahomey's source of income fell harshly upon the surviving son's ears. As soon as the long period of mourning was over, he struck. It was the season for wars, anyway.

Kossula said that the blow fell just before dawn. A few people of Takkoi had already come out of doors and were on their way to the nearby fields. Suddenly there was a fiendish yelling, and then a horde of the dreaded Amazons burst through the main gate of the town. They rushed down the avenue, brandishing their broad, short blades, and seizing all who ran out to meet them. Some of the Amazons dashed into the houses and dragged out those who still remained inside. The heads of all who offered the slightest resistance fell at one stroke from the flashing blades.

Kossula said that he had just gotten up from bed when he heard the terrible cry of the Amazons. He rushed out and fled to one of the farthest gates. "I runnee to de gate, but it somebody there. I runnee to 'nodder one, but somebody there too. I runnee and I runnee, but somebody at every gate.

"One big woman, she grab hold of me and tie me. No, I can't do nothing with her. She too strong. She tie me. I cry for my mamma and papa and my brothers and sisters, and try to see where they at. I no see 'em. I never see them no more. O Lor'! People dead everywhere with no head. Plenty like me tied. De King, he fight till they ketchee him and tie him too. They take him outside the big gate and there de King of Dahomey sit in a great big chair. He say 'bring our king before him.' Our king, he standee there and de King of Dahomey lookee at him and lookee. But our king, he no lookee scared. He standee there straight up.

"De King of Dahomey say: 'I take you to Dahomey, so people can see what I do with the man who insult my father.' But our king, he say: 'I am king in Takkoi. What for I go to Dahomey? No, I no go to Dahomey.'

"Many soldiers of Dahomey stand aroundee there to watchee and do what the King of Dahomey say. He make a sign, and one soldier grabbee the right hand of our king. Another grabbee the left hand, and one she-soldier, she raisee her big knife and chop off our king head. One cut—and it off. They pickee up the head and give it to the King of Dahomey. He take it in a basket. They pick up his chair, and he go."

The captives of war were bound, and a forked stick placed around their necks. In a long line, tied together, they began their march to Dahomey. Kossula said that Dahomey was three sleeps from Takkoi. During their march they were forced to behold the severed heads of their friends and relatives dangling by the waistbands of the Dahomans. By the second day, the heads began to give off an offensive odor. The Dahomans stopped and smoked the heads, holding them on long sticks over a smoldering fire.

Near sundown of that second day they passed near a Dahoman village at a bend in the road. They caught sight of a great number of fresh heads raised on poles above the huts. Some huts had many heads, as a sign of the inhabitant's valor. Others had only one. Kossula remembers that they passed through the towns of Ekko, Bardrigay and Adache, and finally arrived at Whydah, only seaport of the Dahomey kingdom.

At Whydah, there was a big white house near the waterfront. Behind it were the barracouns, or stockades, which the King of Dahomey had built for holding supplies of "black ivory" until sold.[6] Captured tribes were each held in a separate barracoun, to prevent fighting. "Do for do bring black man here from Guinea," says the Jamaican proverb—which means that fighting among themselves caused black folks to be in slavery.

Kossula and his fellow Takkoi tribesmen were placed in one barracoun by themselves. For three weeks they waited here, as fate spun the threads of their future wanderings.

III

News of their round-up had already found its way across the oceans and into the consciousness of four men in Mobile, Alabama. The news came by way of a dispatch in the Mobile *Register* for November 9, 1858.[7] A brief item described conditions on the West Coast of Africa and concluded with the information: "The King of Dahomey is driving a brisk trade in slaves at from fifty to sixty dollars apiece at Whydah. Immense numbers of Negroes are collected along the coast for export."

The four men who read this news with particular attention were the Meaher brothers—Tom, Tim and Burns—and their business associate, Captain Bill Foster.[8] They were not the kind of men to stand aloof when an exciting opportunity presented itself. The Meaher brothers had a lumber mill and shipyard three miles north of Mobile, at the mouth of the Chickasabogue (now known as Three Mile Creek). There they built fine, fast sailing vessels for blockade-running, river trade or any other filibustering expeditions that might arise. They had just completed the *Clotilda* for

Captain Foster. She was very fast, handled beautifully, and merely awaited a proper occasion to show what she could do.

None of the four men was a Southerner. The Meaher brothers were originally from Maine, and Captain Foster had been born in Nova Scotia, of English parentage. There was a Federal law against importing additional slaves from Africa, but this was no more of an obstacle than prohibition laws in later years would be to a bootlegger. The danger might be great, but the profits were huge for men daring enough to run the blockade by both the British and American navies.

Captain Foster and the Meaher brothers were daring and able. They immediately fitted out the *Clotilda*, hired a crew of all Northern sailors and slipped out of Mobile harbor and headed straight for Whydah and the kingdom of Dahomey. Not many days later Foster was anchored safely in the Gulf of Guinea before Whydah. Dressed in his best, and bearing presents for the native rulers, he was met on shore by six robust black soldiers, who carried him in a hammock to the presence of one of the Dahoman princes.

The prince was tall and heavy, weighing around three hundred pounds. He was very hospitable to Foster, showing him the snake collections, Juju houses and other sacred objects of Whydah. His musicians played for Foster after dinner, and one drummer performed a solo upon a special drum which the prince called to the captain's particular attention. The skin of the drum had once belonged to a trader who had made the mistake of trying to capture his own slaves instead of buying them from the prince. Foster agreed with the prince that the drum was worth listening to.[9]

After two or three days of ceremonial politeness, the prince was ready for business. He took Foster to the barracouns, where the captives were gathered in circles of ten men or ten women each. Foster picked out 130 of the best-looking ones and closed his deal. He was carried in a hammock back to the beach. His newly-acquired purchases, chained one behind the other, trudged after him and began piling into the boats that were to take them out to the *Clotilda*. As they stepped into the boats, the waiting Dahomans rushed out and snatched their garments off their backs. The captives cried out in shame and outrage, but the Dahomans shouted consolingly: "You will get clothes where you are going." Men and women alike were left entirely nude.

One hundred and sixteen of his purchases had been brought aboard, when Captain Foster suddenly noticed that the harbor seemed to come alive with activity. He climbed the riggings and was startled to see that all the Dahomey ships were flying the black flag of piracy. The Dahomans were obviously preparing to bear down on the *Clotilda*, recapture the slaves and take Foster and his crew prisoners. The captain scrambled down hurriedly, gave orders to weigh anchor at once, and to abandon all the slaves not already on board. The Dahomey ships were almost upon her when the *Clotilda* got under way. But her fleetness, coupled with expert handling, soon left the pursuers behind.

IV

Captain Foster had a pleasant and lucky voyage home. He treated his captives kindly, although the food was scarce. He gave them water with vinegar in it, to ward off scurvy. On the thirteenth day at sea, they were taken on deck to stretch their limbs. The crew supported them and walked them about the decks until their cramped muscles were rested. "We lookee, and lookee!" said Kossula. "We see nothing but water. Whar we come (from) we not know. Whar we go, we no know neither. One day we see islands." Except for one occasion a week later—when Foster had to elude a British man-of-war on the lookout for slavers—the captives were allowed to stay on deck as much as possible.

The *Clotilda* slipped in behind some islands in the Mississippi Sound, at the lower end of Mobile Bay, on a Thursday night in August 1859. The return trip from the Slave Coast had taken just seventy days. At Mobile the Meaher brothers were eagerly waiting for Foster.

Disposal of the cargo, naturally, had to be carried out with great secrecy and caution. The *Clotilda* was taken directly to Twelve Mile Island, a very lonely spot. There another ship waited to take on the cargo. Some say it was the *June*, belonging to the Meahers. Others insist it was the *R. B. Taney*, named for Chief Justice Taney of Dred Scott Decision fame.[10] In the darkness, the captives were quickly and quietly transferred to the steamboat, taken up the Alabama River, and landed the following day at John Dabney's plantation just below Mount Vernon.[11]

The Africans, newly clothed, were kept hidden at Dabney's plantation for eleven days. Then they were put aboard the *S. S. Commodore* and carried to Burns Meaher's plantation at the bend in the Alabama River where the Tombigbee joins it. Meaher sent word to various slave owners, and prospective buyers were led to the secret hideaway by the Negro, Dennison.[12] Some of the captives were bought and carried to Selma, Alabama. The remaining sixty were divided between the Meahers and Captain Foster. Sixteen men and sixteen women went to Tim Meaher. Burns Meaher took five couples, and another five couples went to Foster. The remaining eight slaves were apportioned to Tim Meaher.[13]

Thus began five years of slavery in America for the 116 black people who had journeyed from the beach at Whydah, Dahomey. Kossula-O-Lo-Loo-Ay went to Captain Tim Meaher, and was immediately put to work. "We astonish to see de mule behind de plow to pull," he said to me. His naked state upon arriving in America was still a matter of humiliation to Kossula. Apparently the American-born slaves thought that the new arrivals always went naked in Africa. Kossula's feeling about this episode was so profound that tears came to his eyes and his voice broke when he told me of it, after more than seventy years.

Upon gaining their freedom with Emancipation, Kossula and the rest resolved to return to Africa. They worked in mills and shipyards, made baskets and grew vegetables, and saved every possible penny. But their plan proved impractical, and after a year or two they gave up their dream of going home. They bought a tract of land from Tim Meaher, instead, and settled down to life in America.

African Town was and is an orderly community. Kossula, who with the rest became converted to Christianity, tried to assure me that his Takkoi tribe were practically Christians before their capture. He would never talk to me about his native religion—which had apparently been the butt of much joking when he first came to America—but he did say by way of explanation: "We know in Africa dat it a God, but we no know He got a Son."

Kossula was erect in his carriage even at ninety-five, and remained cheerful to the end. He felt his uprooting deeply. If I ever went to Africa, he said, I must tell his people where he was. Kossula never knew that not one survivor of Takkoi was known to be alive. Like numerous other tribes and nations of West Africa, Takkoi had disappeared from the earth, its name to be recalled only at the yearly state ceremony to celebrate the might and majesty of Dahomey—when the names of all the tribes and nations destroyed by Dahomey were called as testimony to Dahoman valor.

Two commercial travelers to the Court of Dahomey saw the skull of the King of Takkoi there on such an occasion. It was mounted in a beautiful ship model, highly ornamented and occupying a place of honor. Thus was tribute paid to the memory of a brave man. When asked about the relic's elaborate decoration, the King of Dahomey said: "It is what I would have wished if I had fallen before him. It is the due of a king."

Characteristics of
Negro Expression

DRAMA

The Negro's universal mimicry is not so much a thing in itself as an evidence of something that permeates his entire self. And that thing is drama.

His very words are action words. His interpretation of the English language is in terms of pictures. One act described in terms of another. Hence the rich metaphor and simile.

The metaphor is of course very primitive. It is easier to illustrate than it is to explain because action came before speech. Let us make a parallel. Language is like money. In primitive communities actual goods, however bulky, are bartered for what one wants. This finally evolves into coin, the coin being not real wealth but a symbol of wealth. Still later even coin is abandoned for legal tender, and still later for cheques in certain usages.

Every phase of Negro life is highly dramatised. No matter how joyful or how sad the case there is sufficient poise for drama. Everything is acted out. Unconsciously for the most part of course. There is an impromptu ceremony always ready for every hour of life. No little moment passes unadorned.

Now the people with highly developed languages have words for detached ideas. That is legal tender. "That-which-we-squat-on" has become "chair." "Groan-causer" has evolved into "spear," and so on. Some individuals even conceive of the equivalent of cheque words, like "ideation" and "pleonastic." Perhaps we might say that *Paradise Lost* and *Sartor Resartu*s are written in cheque words.[1]

The primitive man exchanges descriptive words. His terms are all close fitting. Frequently the Negro, even with detached words in his vocabulary—not evolved in him but transplanted on his tongue by contact—must add action to make it do. So we have "chop-axe," "sitting-chair," "cook-pot" and the like because the speaker has in his mind the picture of the object in use. Action. Everything illustrated. So we can say the white man thinks in a written language and the Negro thinks in hieroglyphics.

A bit of Negro drama familiar to all is the frequent meeting of two opponents who threaten to do atrocious murder one upon the other.

Who has not observed a robust young Negro chap posing upon a street corner, possessed of nothing but his clothing, his strength and his youth? Does he bear himself like a pauper? No, Louis XIV could be no more insolent in his assurance.[2] His eyes say plainly, "Female, halt!" His posture exults, "Ah, female, I am the eternal male, the giver of life. Behold in my hot flesh all the delights of this world. Salute me, I am strength." All this with a languid posture, there is no mistaking his meaning.

A Negro girl strolls past the corner lounger. Her whole body panging and posing.[3] A slight shoulder movement that calls attention to her bust, that is all of a dare. A hippy undulation below the waist that is a sheaf of promises tied with conscious power. She is acting out "I'm a darned sweet woman and you know it."

These little plays by strolling players are acted out daily in a dozen streets in a thousand cities, and no one ever mistakes the meaning.

WILL TO ADORN

The will to adorn is the second most notable characteristic in Negro expression. Perhaps his idea of ornament does not attempt to meet conventional standards, but it satisfies the soul of its creator.

In this respect the American Negro has done wonders to the English language. It has often been stated by etymologists that the Negro has introduced no African words to the language. This is true, but it is equally true that he has made over a great part of the tongue to his liking and has had his revision accepted by the ruling class. No one listening to a Southern white man talk could deny this. Not only has he softened and toned down strongly consonanted words like "aren't" to "aint" and the like, he has made new force words out of old feeble elements. Examples of this are "ham-shanked," "battle-hammed," "double-teen," "bodaciously," "muffle-jawed."

But the Negro's greatest contribution to the language is: (1) the use of metaphor and simile; (2) the use of the double descriptive; (3) the use of verbal nouns.

1. METAPHOR AND SIMILE

One at a time, like lawyers going
    to heaven.
You sho is propaganda.
Sobbing hearted.
I'll beat you till: (*a*) rope like okra,
    (*b*) slack like lime, (*c*) smell
    like onions.
Fatal for naked.
Kyting along.
That's a lynch.

That's a rope.
Cloakers—deceivers.
Regular as pig-tracks.
Mule blood—black molasses.
Syndicating—gossiping.
Flambeaux—cheap cafe (lighted
    by flambeaux).
To put yo'self on de ladder.

2. THE DOUBLE DESCRIPTIVE

High-tall.
Little-tee-ninchy (tiny).
Low-down.
Top-superior.
Sham-polish.
Lady-people.
Kill-dead.

Hot-boiling.
Chop-axe.
Sitting-chairs.
De watch wall.
Speedy-hurry.
More great and more better.

3. VERBAL NOUNS

She features somebody I know.

Funeralize.

Sense me into it.

Puts the shamery on him.

'Taint everybody you kin
confidence.

I wouldn't friend with her.

Jooking—playing piano or guitar
as it is done in Jook-houses
(houses of ill-fame).

Uglying away.

I wouldn't scorn my name all up
on you.

Bookooing (beaucoup) around—
showing off.

NOUNS FROM VERBS

Won't stand a broke.

She won't take a listen.

He won't stand straightening.

That is such a compelment.

That's a lynch.

The stark, trimmed phrases of the Occident seem too bare for the voluptuous child of the sun, hence the adornment.[4] It arises out of the same impulse as the wearing of jewelry and the making of sculpture—the urge to adorn.

On the walls of the homes of the average Negro one always finds a glut of gaudy calendars, wall pockets and advertising lithographs. The sophisticated white man or Negro would tolerate none of these, even if they bore a likeness to the Mona Lisa. No commercial art for decoration. Nor the calendar nor the advertisement spoils the picture for this lowly man. He sees the beauty in spite of the declaration of the Portland Cement Works or the butcher's announcement. I saw in Mobile a room in which there was an over-stuffed mohair living-room suite, an imitation mahogany bed and chifferobe, a console victrola.[5] The walls were gaily papered with Sunday supplements of the *Mobile Register*. There were seven calendars and three wall pockets. One of them was decorated with a lace doily. The mantel-shelf was covered with a scarf of deep home-made lace, looped up with a huge bow of pink crêpe paper. Over the door was a huge lithograph showing the Treaty of Versailles being signed with a Waterman fountain pen.[6]

It was grotesque, yes. But it indicated the desire for beauty. And decorating a decoration, as in the case of the doily on the gaudy wall pocket, did not seem out of place to the hostess. The feeling back of such an act is that there can never be enough of beauty, let alone too much. Perhaps she is right. We each have our standards of art, and thus are we all interested parties and so unfit to pass judgment upon the art concepts of others.

Whatever the Negro does of his own volition he embellishes. His religious service is for the greater part excellent prose poetry. Both prayers and sermons are tooled and polished until they are true works of art. The supplication is forgotten in the frenzy of creation. The prayer of the white man is considered humorous in its bleakness. The beauty of the Old Testament does not exceed that of a Negro prayer.

ANGULARITY

After adornment the next most striking manifestation of the Negro is Angularity. Everything that he touches becomes angular. In all African sculpture and doctrine of any sort we find the same thing.

Anyone watching Negro dancers will be struck by the same phenomenon. Every posture is another angle. Pleasing, yes. But an effect achieved by the very means which an European strives to avoid.

The pictures on the walls are hung at deep angles. Furniture is always set at an angle. I have instances of a piece of furniture in the *middle* of a wall being set with one end nearer the wall than the other to avoid the simple straight line.

ASYMMETRY

Asymmetry is a definite feature of Negro art. I have no samples of true Negro painting unless we count the African shields, but the sculpture and carvings are full of this beauty and lack of symmetry.

It is present in the literature, both prose and verse. I offer an example of this quality in verse from Langston Hughes:[7]

> I aint gonna mistreat ma good gal any more,
> I'm just gonna kill her next time she makes me sore.
>
> . . . .
>
> I treats her kind but she don't do me right,
> She fights and quarrels most ever' night.
>
> . . . .
>
> I can't have no woman's got such low-down ways
> Cause de blue gum woman aint de style now'days.
>
> . . . .
>
> I brought her from the South and she's goin on back,
> Else I'll use her head for a carpet track.

It is the lack of symmetry which makes Negro dancing so difficult for white dancers to learn. The abrupt and unexpected changes. The frequent change of key and time are evidences of this quality in music. (Note the St. Louis Blues.)[8]

The dancing of the justly famous Bo-Jangles and Snake Hips are excellent examples.[9]

The presence of rhythm and lack of symmetry are paradoxical, but there they are. Both are present to a marked degree. There is always rhythm, but it is the rhythm of segments. Each unit has a rhythm of its own, but when the whole is assembled it is lacking in symmetry. But easily workable to a Negro who is accustomed to the break in going from one part to another, so that he adjusts himself to the new tempo.

DANCING

Negro dancing is dynamic suggestion. No matter how violent it may appear to the beholder, every posture gives the impression that the dancer will do much more. For example, the performer flexes one knee sharply, assumes a ferocious face mask, thrusts the upper part of the body forward with clenched fists, elbows taut as in hard running or grasping a thrusting blade. That is all. But the spectator himself adds the picture of ferocious assault, hears the drums and finds himself keeping time with the music and tensing himself for the struggle. It is compelling insinuation. That is the very reason the spectator is held so rapt. He is participating in the performance himself—carrying out the suggestions of the performer.

The difference in the two arts is: the white dancer attempts to express fully; the Negro is restrained, but succeeds in gripping the beholder by forcing him to finish the action the performer suggests. Since no art ever can express all the variations conceivable, the Negro must be considered the greater artist, his dancing is realistic suggestion, and that is about all a great artist can do.

NEGRO FOLKLORE

Negro folklore is not a thing of the past. It is still in the making. Its great variety shows the adaptability of the black man: nothing is too old or too new, domestic or foreign, high or low, for his use. God and the Devil are paired, and are treated no more reverently than Rockefeller and Ford.[10]

Both of these men are prominent in folklore, Ford being particularly strong, and they talk and act like good-natured stevedores or mill-hands. Ole Massa is sometimes a smart man and often a fool. The automobile is ranged alongside of the oxcart. The angels and the apostles walk and talk like section hands. And through it all walks Jack, the greatest culture hero of the South; Jack beats them all—even the Devil, who is often smarter than God.

CULTURE HEROES

The Devil is next after Jack as a culture hero. He can out-smart everyone but Jack. God is absolutely no match for him. He is good-natured and full of humour. The sort of person one may count on to help out in any difficulty.

Peter the Apostle is the third in importance. One need not look far for the explanation. The Negro is not a Christian really. The primitive gods are not deities of too subtle inner reflection; they are hard-working bodies who serve their devotees just as laboriously as the suppliant serves them. Gods of physical violence, stopping at nothing to serve their followers. Now of all the apostles Peter is the most active. When the other ten fell back trembling in the garden, Peter wielded the blade on the posse.[11] Peter first and foremost in all action. The gods of no peoples have been philosophic until the people themselves have approached that state.

The rabbit, the bear, the lion, the buzzard, the fox are culture heroes from the animal world. The rabbit is far in the lead of all the others and is blood brother to Jack. In short, the trickster-hero of West Africa has been transplanted to America.

John Henry is a culture hero in song, but no more so than Stacker Lee, Smokey Joe or Bad Lazarus.[12] There are many, many Negroes who have never heard of any of the song heroes, but none who do not know John (Jack) and the rabbit.

Examples of Folklore and the Modern Culture Hero

WHY DE PORPOISE'S TAIL IS ON CROSSWISE

Now, I want to tell you 'bout de porpoise. God had done made de world and everything. He set de moon and de stars in de sky. He got de fishes of de sea, and de fowls of de air completed.

He made de sun and hung it up. Then He made a nice gold track for it to run on. Then He said, "Now, Sun, I got everything made but Time. That's up to you. I want you to start out and go round de world on dis track just as fast as you kin make it. And de time it takes you to go and come, I'm going to call day and night."

De Sun went zoonin' on cross de elements. Now, de porpoise was hanging round there and heard God what he tole de Sun, so he decided he'd take dat trip round de world hisself. He looked up and saw de Sun kytin' along so he lit out too, him and dat Sun!

So de porpoise beat de Sun round de world by one hour and three minutes. So God said, "Aw naw, this aint gointer do! I didn't mean for nothin' to be faster than de Sun!" So God run dat porpoise for three days before he run him down and caught him, and took his tail off and put it on crossways to slow him up. Still he's de fastest thing in de water.

And dat's why de porpoise got his tail on crossways.

ROCKEFELLER AND FORD

Once John D. Rockefeller and Henry Ford was woofing at each other. Rockefeller told Henry Ford he could build a solid gold road round the world. Henry Ford told him if he would he would look at it and see if he liked it, and if he did he would buy it and put one of his tin lizzies on it.

ORIGINALITY

It has been said so often that the Negro is lacking in originality that it has almost become a gospel. Outward signs seem to bear this out. But if one looks closely its falsity is immediately evident.

It is obvious that to get back to original sources is much too difficult for any group to claim very much as a certainty. What we really mean by originality is the modification of ideas. The most ardent admirer of the great Shakespeare cannot claim first source even for him. It is his treatment of the borrowed material.

So if we look at it squarely, the Negro is a very original being. While he lives and moves in the midst of a white civilisation, everything that he touches is re-interpreted for his own use. He has modified the language, mode of food preparation, practice of medicine, and most certainly the religion of his new country, just as he adapted to suit himself the Sheik hair-cut made famous by Rudolph Valentino.[13]

Everyone is familiar with the Negro's modification of the whites' musical instruments, so that his interpretation has been adopted by the white man himself and then re-interpreted. In so many words, Paul Whiteman is giving an imitation of a Negro orchestra making use of white-invented musical instruments in a Negro way.[14] Thus has arisen a new art in the civilised world, and thus has our so-called civilisation come. The exchange and re-exchange of ideas between groups.

IMITATION

The Negro, the world over, is famous as a mimic. But this in no way damages his standing as an original. Mimicry is an art in itself. If it is not, then all art must fall by the same blow that strikes it down. When sculpture, painting, acting, dancing, literature neither reflect nor suggest anything in nature or human experience we turn away with a dull wonder in our hearts at why the thing was done. Moreover, the contention that the Negro imitates from a feeling of inferiority is incorrect. He mimics for the love of it. The group of Negroes who slavishly imitate is small. The average Negro glories in his ways. The highly educated Negro is the same. The self-despisement lies in a middle class who scorns to do or be anything Negro. "That's just like a Nigger" is the most terrible rebuke one can lay upon this kind. He wears drab clothing, sits through a boresome church service, pretends to have no interest in the community, holds beauty contests, and otherwise apes all the mediocrities of the white brother. The truly cultured Negro scorns him, and the Negro "farthest down" is too busy "spreading his junk" in his own way to see or care. He likes his own things best. Even the group who are not Negroes but belong to the "sixth race," buy such records as "Shake Dat Thing" and "Tight Lak Dat."[15] They really enjoy hearing a good bible-beater preach, but wild horses could drag no such admission from them. Their ready-made expression is: "We done got away from all that now." Some refuse to countenance Negro music on the grounds that it is niggerism, and for that reason should be done away with. Roland Hayes was thoroughly denounced for singing spirituals until he was accepted by white audiences.[16] Langston Hughes is not considered a poet by this group because he writes of the man in the ditch, who is more numerous and real among us than any other.

But, this group aside, let us say that the art of mimicry is better developed in the Negro than in other racial groups. He does it as the mockingbird does it, for the love of it, and not because he wishes to be like the one imitated. I saw a group of small Negro boys imitating a cat defecating and the subsequent toilet of the cat. It was very realistic, and they enjoyed it as much as if they had been imitating a coronation ceremony. The dances are full of imitations of various animals. The buzzard lope, walking the dog, the pig's hind legs, holding the mule, elephant squat, pigeon's wing, falling off the log, seabord (imitation of an engine starting), and the like.

ABSENCE OF THE CONCEPT OF PRIVACY

It is said that Negroes keep nothing secret, that they have no reserve. This ought not to seem strange when one considers that we are an outdoor people accustomed to communal life. Add this to all-permeating drama and you have the explanation.

There is no privacy in an African village. Loves, fights, possessions are, to misquote Woodrow Wilson, "Open disagreements openly arrived at."[17] The community is given the benefit of a good fight as well as a good wedding. An audience is a necessary part of any drama. We merely go with nature rather than against it.

Discord is more natural than accord. If we accept the doctrine of the survival of the fittest there are more fighting honors than there are honors for other achievements. Humanity places premiums on all things necessary to its well-being, and a valiant and good fighter is valuable in any community. So why hide the light under a bushel?[18] Moreover, intimidation is a recognised part of warfare the world over, and threats certainly must be listed under that head. So that a great threatener must certainly be considered an aid to the fighting machine. So then if a man or woman is a facile hurler of threats, why should he or she not show their wares to the community? Hence the holding of all quarrels and fights in the open. One relieves one's pent-up anger and at the same time earns laurels in intimidation. Besides, one does the community a service. There is nothing so exhilarating as watching well-matched opponents go into action. The entire world likes action, for that matter. Hence prize-fighters become millionaires.

Likewise love-making is a biological necessity the world over and an art among Negroes. So that a man or woman who is proficient sees no reason why the fact should not be moot. He swaggers. She struts hippily about. Songs are built on the power to charm beneath the bed-clothes. Here again we have individuals striving to excel in what the community considers an art. Then if all of his world is seeking a great lover, why should he not speak right out loud?

It is all in a view-point. Love-making and fighting in all their branches are high arts, other things are arts among other groups where they brag about their proficiency just as brazenly as we do about these things that others consider matters for conversation behind closed doors. At any rate, the white man is despised by Negroes as a very poor fighter individually and a very poor lover. One Negro, speaking of white men, said, "White folks is alright when dey gits in de bank and on de law bench, but dey sho kin lie about wimmen folks."

I pressed him to explain. "Well you see, white mens makes out they marries wimmen to look at they eyes, and they know they gits em for just what us gits em for. 'Nother thing, white mens say they goes clear round de world and wins all de wimmen folks away from they men folks. Dat's a lie too. They don't win nothin, they buys em. Now de way I figgers it, if a woman don't want me enough to be wid me, 'thout I got to pay her, she kin rock right on, but these here white men don't know what to do wid a woman when they gits her—dat's how come they gives they wimmen so much. They got to. Us wimmen works just as hard as us does an come home an sleep wid us every night. They own wouldn't do it and its de mens fault. Dese white men done fooled theyself bout dese wimmen.

"Now me, I keeps me some wimmens all de time. Dat's whut dey wuz put here for—us mens to use. Dat's right now, Miss. Y'all wuz put here so us mens could have some pleasure. Course I don't run round like heap uh men folks. But if my ole lady go way from me and stay more'n two weeks, I got to git me somebody, aint I?"

THE JOOK

Jook is the word for a Negro pleasure house. It may mean a bawdy house. It may mean the house set apart on public works where the men and women dance, drink and gamble. Often it is a combination of all these.

In past generations the music was furnished by "boxes," another word for guitars. One guitar was enough for a dance; to have two was considered excellent. Where two were playing one man played the lead and the other seconded him. The first player was "picking" and the second was "framming," that is, playing chords while the lead carried the melody by dextrous finger work. Sometimes a third player was added, and he played a tom-tom effect on the low strings. Believe it or not, this is excellent dance music.

Pianos soon came to take the place of the boxes, and now player-pianos and victrolas are in all of the Jooks.

Musically speaking, the Jook is the most important place in America. For in its smelly, shoddy confines has been born the secular music known as blues, and on blues has been founded jazz. The singing and playing in the true Negro style is called "jooking."

The songs grow by incremental repetition as they travel from mouth to mouth and from Jook to Jook for years before they reach outside ears. Hence the great variety of subject-matter in each song.

The Negro dances circulated over the world were also conceived inside the Jooks. They too make the round of Jooks and public works before going into the outside world.

In this respect it is interesting to mention the Black Bottom. I have read several false accounts of its origin and name. One writer claimed that it got its name from the black sticky mud on the bottom of the Mississippi river. Other equally absurd statements gummed the press. Now the dance really originated in the Jook section of Nashville, Tennessee, around Fourth Avenue. This is a tough neighborhood known as Black Bottom—hence the name.

The Charleston is perhaps forty years old, and was danced up and down the Atlantic seaboard from North Carolina to Key West, Florida.

The Negro social dance is slow and sensuous. The idea in the Jook is to gain sensation, and not so much exercise. So that just enough foot movement is added to keep the dancers on the floor. A tremendous sex stimulation is gained from this. But who is trying to avoid it? The man, the woman, the time and the place have met. Rather, little intimate names are indulged in to heap fire on fire.

These too have spread to all the world.

The Negro theatre, as built up by the Negro, is based on Jook situations, with women, gambling, fighting, drinking. Shows like "Dixie to Broadway" are only Negro in cast, and could just as well have come from pre-Soviet Russia.[19]

Another interesting thing—Negro shows before being tampered with did not specialise in octoroon chorus girls.[20] The girl who could hoist a Jook song from her belly and lam it against the front door of the theatre was the lead, even if she were as black as the hinges of hell. The question was "Can she jook?" She must also have a good belly wobble, and her hips must, to quote a popular work song, "Shake like jelly all over and be so broad, Lawd, Lawd, and be so broad." So that the bleached chorus is the result of a white demand and not the Negro's.

The woman in the Jook may be nappy headed and black, but if she is a good lover she gets there just the same. A favorite Jook song of the past has this to say:

SINGER: It aint good looks dat takes you through dis world.

AUDIENCE: What is it, good mama?

SINGER: Elgin movements in your hips

Twenty years guarantee.[21]

And it always brought down the house too.

Oh de white gal rides in a Cadillac,
De yaller gal rides de same,
Black gal rides in a rusty Ford
But she gits dere just de same.

The sort of woman her men idealise is the type that can put forth in the theatre. The art-creating Negro prefers a not too thin woman who can shake like jelly all over as she dances and sings, and that is the type he put forth on the stage. She has been banished by the white producer and the Negro who takes his cue from the white.

Of course a black woman is never the wife of the upper class Negro in the North.[22] This state of affairs does not obtain in the South, however. I have noted numerous cases where the wife was considerably darker than the husband. People of some substance, too.

This scornful attitude towards black women receives mouth sanction by the mud-sills.[23]

Even on the works and in the Jooks the black man sings disparagingly of black women. They say that she is evil. That she sleeps with her fists doubled up and ready for action. All over they are making a little drama of waking up a yaller wife and a black one.[24]

A man is lying beside his yaller wife and wakes her up. She says to him, "Darling, do you know what I was dreaming when you woke me up?" He says, "No honey, what was you dreaming?" She says, "I dreamt I had done cooked you a big, fine dinner and we was setting down to eat out de same plate and I was setting on yo' lap jus huggin you and kissin you and you was so sweet."

Wake up a black woman, and before you kin git any sense into her she be done up and lammed you over the head four or five times. When you git her quiet she'll say, "Nigger, know whut I was dreamin when you woke me up?"

You say, "No honey, what was you dreamin?" She says, "I dreamt you shook yo' rusty fist under my nose and I split yo' head open wid a axe."

But in spite of disparaging fictitious drama, in real life the black girl is drawing on his account at the commissary.[25] Down in the Cypress Swamp as he swings his axe he chants:

Dat ole black gal, she keep on grumblin,
New pair shoes, new pair shoes,
I'm goint to buy her shoes and stockings
Slippers too, slippers too.

Then adds aside: "Blacker de berry, sweeter de juice."

To be sure, the black gal is still in power, men are still cutting and shooting their way to her pillow. To the queen of the Jook!

Speaking of the influence of the Jook, I noted that Mae West in "Sex" had much more flavor of the turpentine quarters than she did of the white bawd.[26] I know that the piece she played on the piano is a very old Jook composition. "Honey let yo' drawers hang low" had been played and sung in every Jook in the South for at least thirty-five years.[27] It has always puzzled me why she thought it likely to be played in a Canadian bawdy house.

Speaking of the use of Negro material by white performers, it is astonishing that so many are trying it, and I have never seen one yet entirely realistic. They often have all the elements of the song, dance, or expression, but they are misplaced or distorted by the accent falling on the wrong element. Every one seems to think that the Negro is easily imitated when nothing is further from the truth. Without exception I wonder why the black-face comedians *are* black-face; it is a puzzle—good comedians, but darn poor niggers. Gershwin and the other "Negro" rhapsodists come under this same axe.[28] Just about as Negro as caviar or Ann Pennington's athletic Black Bottom.[29] When the Negroes who knew the Black Bottom in its cradle saw the Broadway version they asked each other, "Is you learnt dat *new* Black Bottom yet?" Proof that it was not *their* dance.

And God only knows what the world has suffered from the white damsels who try to sing blues.

The Negroes themselves have sinned also in this respect. In spite of the goings up and down on the earth, from the original Fisk Jubilee Singers down to the present, there has been no genuine presentation of Negro songs to white audiences.[30] The spirituals that have been sung around the world are Negroid to be sure, but so full of musicians' tricks that Negro congregations are highly entertained when they hear their old songs so changed. They never use the new style songs, and these are never heard unless perchance some daughter or son has been off to college and returns with one of the old songs with its face lifted, so to speak.

I am of the opinion that this trick style of delivery was originated by the Fisk Singers; Tuskegee and Hampton followed suit and have helped spread this misconception of Negro spirituals. This Glee Club style has gone on so long and become so fixed among concert singers that it is considered quite authentic. But I say again, that not one concert singer in the world is singing the songs as the Negro song-makers sing them.

If anyone wishes to prove the truth of this let him step into some unfashionable Negro church and hear for himself.

To those who want to institute the Negro theatre, let me say it is already established. It is lacking in wealth, so it is not seen in the high places. A creature with a white head and Negro feet struts the Metropolitan boards. The real Negro theatre is in the Jooks and the cabarets. Self-conscious individuals may turn away the eye and say, "Let us search elsewhere for our dramatic art." Let em search. They certainly won't find it. Butter Beans and Susie, Bo-Jangles and Snake Hips are the only performers of the real Negro school it has ever been my pleasure to behold in New York.[31]

DIALECT

If we are to believe the majority of writers of Negro dialect and the burnt-cork artists, Negro speech is a weird thing, full of "ams" and "Ises."[32] Fortunately we don't have to believe them. We may go directly to the Negro and let him speak for himself.

I know that I run the risk of being damned as an infidel for declaring that nowhere can be found the Negro who asks "am it?" nor yet his brother who announces "Ise uh gwinter." He exists only for a certain type of writers and performers.

Very few Negroes, educated or not, use a clear clipped "I." It verges more or less upon "Ah." I think the lip form is responsible for this to a great extent. By experiment the reader will find that a sharp "I" is very much easier with a thin taut lip than with a full soft lip. Like tightening violin strings.

If one listens closely one will note too that a word is slurred in one position in the sentence but clearly pronounced in another. This is particularly true of the pronouns. A pronoun as a subject is likely to be clearly enunciated, but slurred as an object. For example: "You better not let me ketch yuh."

There is a tendency in some localities to add the "h" to "it" and pronounce it "hit." Probably a vestige of old English. In some localities "if" is "ef."

In story telling "so" is universally the connective. It is used even as an introductory word, at the very beginning of a story. In religious expression "and" is used. The trend in stories is to state conclusions; in religion, to enumerate.

I am mentioning only the most general rules in dialect because there are so many quirks that belong only to certain localities that nothing less than a volume would be adequate.

Conversions and Visions

The vision is a very definite part of Negro religion. It almost always accompanies conversion. It always accompanies the call to preach.

In the conversion the vision is sought. The individual goes forth into waste places and by fasting and prayer induces the vision. The place of retirement chosen is one most likely to have some emotional effect upon the seeker. The cemetery, to a people who fear the dead, is a most suggestive place to gain visions. The dense swamps with the possibility of bodily mishaps is another favorite.

Three days is the traditional period for seeking the vision. Usually the seeker is successful, but now and then he fails. Most seekers "come through religion" during revival meetings, but a number come after the meeting has closed.

Certain conversion visions have become traditional, but all sorts of variations are interpolated in the general framework of the convention, from the exceedingly frivolous to the most solemn. One may go to a dismal swamp, the other to the privy house. The imagination of one may carry him to the last judgement and the rimbones of nothing, the vision of another may hobble him at washing collard greens.[1] But in each case there is an unwillingness to believe—to accept the great good fortune too quickly. So God is asked for proof. One man told me that he refused to believe that he had truly been saved and said: "Now, Lord, if you have really saved my soul, I ask you to move a certain star from left to right." And the star shot across the heavens from the left hand to the right. But still he wouldn't believe. So he asked for the sun to shout and the sun shouted. He still didn't believe. So he asked for one more sign. But God had grown impatient with his doubtings and told him sharply that if he didn't believe without further proof that He'd send his soul to hell. So he ran forth from his hiding and proclaimed a new-found savior.

In another case, a woman asked that a tree be moved and it stepped over ten feet, and then she asked for the star and God told her He had given her one sign and if she couldn't believe and trust Him for the balance He'd send her soul to torment.

Another woman asked for a windstorm and it came. She asked for the star to move and it did. She asked for the sun to shout and God grew angry and rebuked her like the others.

Still another woman fell under conviction in a cow lot and asked for a sign. "Now, Lord, if you done converted my soul, let dat cow low three times and I'll believe. A cow said, 'Mooo—oo, moo—oo—oo, moo—ooo—ooo'—and I knowed I had been converted and my soul set free."

Three is the holy number and the call to preach always comes three times. It is never answered until the third time. The man flees from the call, but is finally brought to accept it. God punishes him by every kind of misfortune until he finally acknowledges himself beaten and makes known the call. Some preachers say the spirit whipped them from their heads to their heels. They have been too sore to get out of bed because they refused the call. This never ceased until the surrender. Sometimes God sends others to tell them they are chosen. But in every case the ministers refuse to believe the words of even these.

We see that in conversion the sinner is first made conscious of his guilt. This is followed by a period called "lyin' under conviction" which lasts for three days. After which Jesus converts the supplicant, and the supplicant refuses to believe without proof, and only gives in under threat of eternal damnation. He flees from this to open acknowledgement of God and salvation. First from the outside comes the accusation of sin. Then from within the man comes the consciousness of guilt, and the sufferer seeks relief from Heaven. When it is granted, it is at first doubted, but later accepted. We have a mixture of external and internal struggles.

The call to preach is altogether external. The vision seeks the man. Punishment follows if he does not heed the call, or until he answers.

In conversion, then, we have the cultural pattern of the person seeking the vision and inducing it by isolation and fasting. In the call to preach we have the involuntary vision—the call seeking the man.

COMING THROUGH RELIGION

I went out to pray in my back yard. I had done prayed and prayed but didn't know how to pray. I had done seen vision on top of vision, but still I wouldn't believe. Then I said: "Lord, let my head be a footstool for you." He says: "I plant my feet in the sea, follow after me. Your sins are forgiven and your soul set free. Go and tell the world what a kind savior you have found." I broke out the privy and went running and the voice kept following: "I set your feet on the rock of eternal ages; and the wind may blow and the storm may rise, but nothing shall frighten you from the shore."

I carried them messages.

"Jesus." "I am Jesus." "Father!" "I am the Father, and the Father is in me." It just continued and He sent me to the unconverted. I had some more visions. In one of them I laid down and a white man come to me all dressed in white and he had me stretched out on de table and clipped my breath three times and the third time I rose and went to a church door and there was a weeping willow. There was a four-cornered garden and three more knelt with me in the four corners. And I had to pray, to send up a prayer.

The next vision I had a white woman says: "I am going home with you." I didn't want her to go. My house was not in order. Somebody stopped her. When I got home, a tall white man was standing at my door with a palmetto hat. I noticed he was pale-like. He was looking down on my steps.

They was washed with redding.[2] He says: "I have cleaned your house. How do you like it?" I looked down on it and after he was gone I said: "I don't like that. It looks too much like blood." After I was converted it come to me about the blood and I knew it was Jesus, and my heart was struck with sorrow that to think I had been walking upon His precious blood all this time and I didn't believe.

(MRS. SUSANNA SPRINGER.)

I was a lad of a boy when I found Jesus sweet to my ever-dying soul. They was runnin p'tracted meetin and all my friends was gettin religion and joinin de church; but I never paid it no mind. I was hard. But I don't keer how hard you is, God kin reach you when He gits ready for you. One day, bout noon, it was de 9th day of June, 1886, when I was walkin in my sins, wallerin in my sins, dat He tetched me wid de tip of His finger and I fell right where I was and laid there for three long days and nights. I layed there racked in pain under sentence of death for my sins. And I walked over hell on a narrer foot log so I had to put one foot right in front de other, one foot right in front de other wid hell gapped wide open beneath my sin-loaded and slippery feet. And de hell hounds was barkin on my tracks and jus before dey rushed me into hell and judgement I cried: "Lawd, have mercy," and I crossed over safe. But still I wouldn't believe. Then I saw myself hangin over hell by one strand of hair and de flames of fire leapin up a thousand miles to swaller my soul and I cried: "Jesus, save my soul and I'll believe, I'll believe." Then I found myself on solid ground and a tall white man beckoned for me to come to him and I went, wrapped in my guilt, and he 'nointed me wid de oil of salvation and healed all my wounds. Then I found myself layin on de ground under a scrub oak and I cried: "I believe, I believe." Then Christ spoke peace to my soul and de dungeon shook and my chains fell off, and I went shoutin in His name and praising Him. I put on de whole armor of faith and I speck to stay in de fiel till I die.

(DEACON ERNEST HUFFMAN.)

First thing started me—it come to me dat I had to die. And worried me so I got talkin wid an old Christ man—about seventy years old. I wasn't but twenty-one. And I started out from his instruction and I heered people say in my time dat de speerit would command you to de graveyard (to pray). And I ast de Lawd not to send me dere cause I wuz skeered uh de graveyard. But every answer I got commanded me to the graveyard.

One cold night, March de twentieth, 1867, at night, de speerit command me to de graveyard and I didn't go. And de Lawd sent Death after me and when I knowed anything I was on my way to de graveyard. And when I got dere I fell. I fell right between two graves and I saw Him when He laid me upon a table in my vision. I was naked and He split me open. And there was two men there—one on each side of de table. I could hear de knives clicking in me, inside. And after dey got through wid me, they smoothed they hand over de wound and I wuz healed. And when I found myself I wuz standin naked beside de table and there was three lights burnin on de table. De one in de middle wuz de brightest. I wuzn't between de two graves no more. When I got up from between de two graves, I tracked my guide by de drops of blood. I could hear de blood dripping from Him before me. It said as it dropped: "Follow me." And I looked at de three lights and dey tole me to reach forth wid my right hand and grasp de brightest one and I did. It wuz shining like de Venus star. And they tole me it wuz to be my guidin star. I found myself before I left de table wid five white balls in each hand. "Them is the ten tablets I give you." And I put my hand to my breast and I put the balls inside me. Then He slapped something on my breast and said: "Now, you are breastplated and shielded." He pointed: "Go to yonder white house. You will find there one who will welcome you." And when I got to de steps I thowed my foot on de first step and de house rang and a lady come out and welcomed me in. And when I got inside, as far as mortal eye could behold, the robes was hanging level and touched my head as I passed under. Then I found myself robed in the color of gold. Then I commenced shouting. And when I commenced shouting I found myself leaving the graveyard. And He told me that was my robe for me bye and bye. In dat swamp where dat graveyard was there was catamounts and panters [sic] and wild beasts but not a one of 'em touched me and I laid there all night.[3]

Now He tole me, He said: "You got the three witnesses. One is water, one is spirit, and one is blood. And these three correspond with the three in heben—Father, Son and Holy Ghost."

Now I ast Him about this lyin in sin and He give me a handful of seeds and He tole me to sow 'em in a bed and He tole me: "I want you to watch them seeds." The seeds come up about in places and He said: "Those seeds that come up, they died in the heart of the earth and quickened and come up and brought forth fruit. But those seeds that didn't come up, they died in the heart of the earth and rottened.[4]

"And a soul that dies and quickens through my spirit they will live forever, but those that don't never pray, they are lost forever."

(REV. JESSIE JEFFERSON.)

Shouting

There can be little doubt that shouting is a survival of the African "possession" by the gods. In Africa it is sacred to the priesthood or acolytes, in America it has become generalised. The implication is the same, however. It is a sign of special favor from the spirit that it chooses to drive out the individual consciousness temporarily and use the body for its expression.

In every case the person claims ignorance of his actions during the possession.

Broadly speaking, shouting is an emotional explosion, responsive to rhythm. It is called forth by (1) sung rhythm; (2) spoken rhythm; (3) humming rhythm; (4) the foot-patting or hand-clapping that imitates very closely the tom-tom.

The more familiar the expression, the more likely to evoke response. For instance, "I am a soldier of the cross, a follower of the meek and lowly lamb. I want you all to know I am fighting under the blood-stained banner of King Jesus" is more likely to be amen-ed than any flourish a speaker might get off. Perhaps the reason for this is that the hearers can follow the flow of syllables without stirring the brain to grasp the sense. Perhaps it is the same urge that makes a child beg for the same story even though he knows it so well that he can correct his parents if a word is left out.

Shouting is a community thing. It thrives in concert. It is the first shout that is difficult for the preacher to arouse. After that one they are likely to sweep like fire over the church. This is easily understood, for the rhythm is increasing with each shouter who communicates his fervor to someone else.

It is absolutely individualistic. While there are general types of shouting, the shouter may mix the different styles to his liking, or he may express himself in some fashion never seen before.

Women shout more frequently than men. This is not surprising since it is generally conceded that women are more emotional than men.

The shouter always receives attention from the church. Members rush to the shouter and force him into a seat or support him as the case might be. Sometimes it is necessary to restrain him to prevent injury to either the shouter or the persons sitting nearest, or both. Sometimes the arms are swung with such violence that others are knocked down. Sometimes in the ecstasy the shouter climbs upon the pew and kicks violently away at all; sometimes in catalepsis he falls heavily upon the floor and might injure himself if not supported, or fall upon others and wound. Often the person injured takes offense, believing that the shouter was paying off a grudge. Unfortunately this is the case at times, but it is not usual.

There are two main types of shouters: (1) Silent; (2) Vocal. There is a sort of intermediary type where one stage is silent and the other vocal.

The silent type take with violent retching and twitching motions. Sometimes they remain seated, sometimes they jump up and down and fling the body about with great violence. Lips tightly pursed, eyes closed. The seizure ends by collapse.

The vocal type is the more frequent. There are all gradations from quiet weeping while seated to the unrestrained screaming while leaping pews and running up and down the aisle. Some, unless restrained, run up into the pulpit and embrace the preacher. Some are taken with hysterical laughing spells.

The cases will illustrate the variations.

1. During sermon. Cried "well, well," six times. Violent action for forty seconds. Collapsed and restored to her seat by members.

2. During chant. Cried "Holy, holy! Great God A'mighty!" Arose and fell in cataleptic fit backwards over pew. Flinging of arms with clenched fists, gradually subsiding to quiet collapse. Total time: two minutes.

3. During pre-prayer humming chant. Short screams. Violent throwing of arms. Incoherent speech. Total time: one minute thirty seconds.

4. During sermon. One violent shout as she stood erect: two seconds. Voiceless gestures for twenty-nine seconds. She suddenly resumed her seat and her attention to the words of the preacher.

5. During sermon. One single loud scream: one and one-half seconds.

6. During singing. Violent jumping up and down without voice. Pocket book cast away. Time: one minute forty seconds.

7. During prayer. Screaming: one second. Violent shoulder-shaking, hat discarded: nineteen seconds.

8. During sermon. Cataleptic. Stiffly back over the pew. Violent but voiceless for twenty seconds. Then arms stiff and outstretched, palms open stark and up. Collapse. Time: three minutes.

9. During sermon. Young girl. Running up and down the aisle: thirty seconds. Then silence and rush to the pulpit: fourteen seconds; prevented at the altar rail by deacon. Collapse in the deacon's arms and returned to seat. Total time: one minute fifteen seconds.

10. During chant after prayer. Violent screams: twelve seconds. Scrambles upon pew and steps upon the back of pew still screaming: five seconds. Voiceless struggle with set teeth as three men attempt to restore her to seat. She is lifted horizontal but continues to struggle: one minute forty-eight seconds.

Decreasing violence, making ferocious faces: two minutes. Calm with heavy breathing: twenty-one seconds.

11. During sermon. Man quietly weeping: nineteen seconds. Cried "Lawd! My soul is burning with hallow-ed fire!" Rises and turns round and round six times. Carried outside by the deacons.

12. During sermon. Man jumping wildly up and down flat-footed crying "Hallelujah!": twenty-two seconds. Pulled back into his seat. Muscular twitching: one minute thirty-five seconds. Quiet weeping: one minute. Perfect calm.

Spirituals and Neo-Spirituals

The real spirituals are not really just songs. They are unceasing variations around a theme.

Contrary to popular belief their creation is not confined to the slavery period. Like the folk-tales, the spirituals are being made and forgotten every day. There is this difference: the makers of the songs of the present go about from town to town and church to church singing their songs. Some are printed and called ballads, and offered for sale after the services at ten and fifteen cents each. Others just go about singing them in competition with other religious minstrels. The lifting of the collection is the time for the song battles. Quite a bit of rivalry develops.

These songs, even the printed ones, do not remain long in their original form. Every congregation that takes it up alters it considerably. For instance, *The Dying Bed Maker*, which is easily the most popular of the recent compositions, has been changed to *He's a Mind Regulator* by a Baptist church in New Orleans.[1]

The idea that the whole body of spirituals are "sorrow songs" is ridiculous. They cover a wide range of subjects from a peeve at gossipers to Death and Judgment.

The nearest thing to a description one can reach is that they are Negro religious songs, sung by a group, and a group bent on expression of feelings and not on sound effects.

There never has been a presentation of genuine Negro spirituals to any audience anywhere. What is being sung by the concert artists and glee clubs are the works of Negro composers or adaptors *based* on the spirituals. Under this head come the works of Harry T. Burleigh, Rosamond Johnson, Lawrence Brown, Nathaniel Dett, Hall Johnson and Work.[2] All good work and beautiful, but *not* the spirituals. These neo-spirituals are the outgrowth of the glee clubs. Fisk University boasts perhaps the oldest and certainly the most famous of these. They have spread their interpretation over America and Europe. Hampton and Tuskegee have not been unheard. But with all the glee clubs and soloists, there has not been one genuine spiritual presented.

To begin with, Negro spirituals are not solo or quartette material. The jagged harmony is what makes it, and it ceases to be what it was when this is absent. Neither can any group be trained to reproduce it. Its truth dies under training like flowers under hot water. The harmony of the true spiritual is not regular. The dissonances are important and not to be ironed out by the trained musician. The various parts break in at any old time. Falsetto often takes the place of regular voices for short periods. Keys change. Moreover, each singing of the piece is a new creation. The congregation is bound by no rules. No two times singing is alike, so that we must consider the rendition of a song not as a final thing, but as a mood. It won't be the same thing next Sunday.

Negro songs to be heard truly must be sung by a group, and a group bent on expression of feelings and not on sound effects.

Glee clubs and concert singers put on their tuxedoes, bow prettily to the audience, get the pitch and burst into magnificent song—but not *Negro* song.[3] The real Negro singer cares nothing about pitch. The first notes just burst out and the rest of the church join in—fired by the same inner urge. Every man trying to express himself through song. Every man for himself. Hence the harmony and disharmony, the shifting keys and broken time that make up the spiritual.

I have noticed that whenever an untampered-with congregation attempts the renovated spirituals, the people grow self-conscious. They sing sheepishly in unison. None of the glorious individualistic fights that make up their own songs. Perhaps they feel on strange ground. Like the unlettered parent before his child just home from college. At any rate they are not very popular.

This is no condemnation of the neo-spirituals. They are a valuable contribution to the music and literature of the world. But let no one imagine that they are the songs of the people, as sung by them.

The lack of dialect in the religious expression—particularly in the prayers—will seem irregular.

The truth is, that the religious service is a conscious art expression. The artist is consciously creating—carefully choosing every syllable and every breath. The dialect breaks through only when the speaker has reached the emotional pitch where he loses self-consciousness.

In the mouth of the Negro the English language loses its stiffness, yet conveys its meaning accurately.

"The booming bounderries of this whirling world" conveys just as accurate a picture as mere "boundaries," and a little music is gained besides. "The rim bones of nothing" is just as truthful as "limitless space."

Negro singing and formal speech are breathy. The audible breathing is part of the performance and various devices are resorted to to adorn the breath taking. Even the lack of breath is embellished with syllables. This is, of course, the very antithesis of white vocal art. European singing is considered good when each syllable floats out on a column of air, seeming not to have any mechanics at all. Breathing must be hidden. Negro song ornaments both the song and the mechanics. It is said of a popular preacher, "He's got a good straining voice." I will make a parable to illustrate the difference between Negro and European.

A white man built a house. So he got it built and he told the man: "Plaster it good so that nobody can see the beams and uprights." So he did. Then he had it papered with beautiful paper, and painted the outside. And a Negro built him a house. So when he got the beams and all in, he carved beautiful grotesques over all the sills and stanchions, and beams and rafters. So both went to live in their houses and were happy.

The well-known "ha!" of the Negro preacher is a breathing device. It is the tail end of the expulsion just before inhalation. Instead of permitting the breath to drain out, when the wind gets too low for words, the remnant is expelled violently. Example: (inhalation) "And oh!"; (full breath) "my Father and my wonder-working God"; (explosive exhalation) "ha!"

Chants and hums are not used indiscriminately as it would appear to a casual listener. They have a definite place and time. They are used to "bear up" the speaker. As Mama Jane of Second Zion Baptist Church, New Orleans, explained to me: "What point they come out on, you bear 'em up."

For instance, if the preacher should say: "Jesus will lead us," the congregation would bear him with: "I'm got my ha-hands in my Jesus' hands." If in prayer or sermon, the mention is made of nailing Christ to the cross: "Didn't Calvary tremble when they nailed Him down."

There is no definite post-prayer chant. One may follow, however, because of intense emotion. A song immediately follows prayer. There is a pre-prayer hum which depends for its material upon the song just sung. It is usually a pianissimo continuation of the song without words. If some of the people use the words it is done so indistinctly that they would be hard to catch by a person unfamiliar with the song.

As indefinite as hums sound, they also are formal and can be found unchanged all over the South. The Negroised white hymns are not exactly sung. They are converted into a barbaric chant that is not a chant. It is a sort of liquefying of words. These songs are always used at funerals and on any solemn occasion. The Negro has created no songs for death and burials, in spite of the sombre subject matter contained in some of the spirituals. Negro songs are one and all based on a dance-possible rhythm. The heavy interpretations have been added by the more cultured singers. So for funerals fitting white hymns are used.

Beneath the seeming informality of religious worship there is a set formality. Sermons, prayers, moans and testimonies have their definite forms. The individual may hang as many new ornaments upon the traditional form as he likes, but the audience would be disagreeably surprised if the form were abandoned. Any new and original elaboration is welcomed, however, and this brings out the fact that all religious expression among Negroes is regarded as art, and ability recognised as definitely as in any other art. The beautiful prayer receives the accolade as well as the beautiful song. It is merely a form of expression which people generally are not accustomed to think of as art. Nothing outside of the Old Testament is as rich in figure as a Negro prayer. Some instances are unsurpassed anywhere in literature.[4]

There is a lively rivalry in the technical artistry of all of these fields. It is a special honor to be called upon to pray over the covered communion table, for the greatest prayer-artist present is chosen by the pastor for this, a lively something spreads over the church as he kneels, and the "bearing up" hum precedes him. It continues sometimes through the introduction, but ceases as he makes the complimentary salutation to the deity. This consists in giving to God all the titles that form allows.

The introduction to the prayer usually consists of one or two verses of some well-known hymn. "O, that I knew a secret place" seems to be the favorite.[5] There is a definite pause after this, then follows an elaboration of all or parts of the Lord's Prayer. Follows after that what I call the setting, that is, the artist calling attention to the physical situation of himself and the church. After the dramatic setting, the action begins.

There are certain rhythmic breaks throughout the prayer, and the church "bears him up" at every one of these. There is in the body of the prayer an accelerando passage where the audience takes no part. It would be like applauding in the middle of a solo at the Metropolitan. It is here that the artists come forth. He adorns the prayer with every sparkle of earth, water and sky, and nobody wants to miss a syllable. He comes down from this height to a slower tempo and is borne up again. The last few sentences are unaccompanied, for here again one listens to the individual's closing peroration. Several may join in the final amen. The best figure that I can think of is that the prayer is an obligato over and above the harmony of the assembly.

Ritualistic Expression from the Lips of the Communicants of the Seventh Day Church of God*

COMMANDMENT KEEPER CHURCH

Rev. George Washington, Pastor

. . .

Communicants Interviewed:
Bishop R. A. R. Johnson
Rev. George Washington
Mrs. Julia Jones
Mrs. Izora Robinson
Mr. Hugh Washington
Mrs. Hattie Mae Washington
Miss Carrie Washington
Mr. Henry Moore

---

*This text on deposit at the Library of Congress is followed by a transcription of a religious service, but the punctuation and pronouns in that text suggest it was to some degree coauthored with Jane Belo, so it has been excluded here.

Note:

This Seventh Day Church of God, holiness, is a further or newer step in the sanctified church in the United States.[1] To the Interviewer this church seems to be a protest against the stereotype form of Methodist and Baptist churches among Negroes. It is a revolt against the white man's view of religion which has been so generally accepted by the literate Negro, and is therefore a version to the more African form of expression.

Its keynote is rhythm. In this church they have two guitars, three symbols, two tambourines, one pair of rattle goers, and two washboards. Every song is rhythmic as are their prayers and their sermons. The unanimous prayer is one in which every member of the church prays at the same time but prays his own prayer aloud, which consists of exotic sentences, liquefied by intermittent chanting so that the words are partly submerged in the flowing rising and falling chant. The form of prayer is like the limbs of a tree, glimpsed now and then through the smothered leaves. It is a thing of wondrous beauty, drenched in harmony and rhythm.

Explanation of a TARRY SERVICE

This service is one of consecration beginning at 4:30 and going until sundown. The procedure is:

a. Read the Bible

b. Sing great number of songs

c. All the members, starting with the pastor, kneel at the bench that they use for an altar, before which they have spread a piece of ragged carpet

d. Sister Izora anoints the pastor

They pray; when the oil touches their head, they go into a tremendous ecstasy which lasts several minutes, ending in a trance. When the ecstasy and trance is [sic] over, they get up and take their seats and start singing again. Just before the sun goes down the preacher preaches a short sermon. They say the doxology and go out just as the sun sets.[2]

When we entered the sun was at the top of a long unshaded window of the 24 rectangular panes. The kneeling and anointing of heads from a bottle of oil began. After the prayer and ecstasy had ended the sun hung in the last low pane. Then there was whispered harmony, intense but soothing. The sun dropped out of the window and the service was at an end.

Communicant Senior Bishop R. A. R. Johnson (about 82 yrs. old)[3]

INTERVIEWER: "Bishop Johnson, will you tell me the religious experience that you had in getting religion and becoming sanctified?"[4]

BISHOP JOHNSON: "Yes. I don't mind telling you for I have had plenty of visions. First I will tell you of the beginning of the church.

"I could not be satisfied with none of the different dogmas as was established among the different sects. So I would go to all the different churches to hear the great men preach their doctrinal sermon. None satisfied my spiritual longing. Some were very flowery, some consistent and rhetrical (rhetorical). Therefore I would remain by myself in secret seclusion, inquiring of God all wisdom and understanding with His intelligence. He expected (summoned) me at the spring one day with the natural voice: 'I've ordained you to teach the people My laws, statutes and judgements. They stand fast and endure forever. Nothing can be added or taken away. I will make you wise, a great teacher of the old and the new, and you will find My commandments like a golden thread running from the beginning to the end.' He said there was NO church is His name. My name must have the preference. 'You will have to turn two or three pages of the book and literature of the different so-called churches before you can find My name.'

"I decided then to establish a church to be known as THE EVAN-GELISTICAL METHODIST CHURCH OF GOD.[5] I wrote a wonderful manuscript and turned it over to a committee for criticism and correction. They approved it as being timely and grammatical. I felt good at their recommendation. I left then joyful on my way to the printers.

"My feet got heavy, a weakness befell me, and I could not move. I cried unto the Lord to know what is this and why. I felt bound in weakness. He seemed to scold me with a loud voice saying, 'That is not the church I want you to organize. I want a church that is fully in My name, no surnames.' I turned around to go back home. My feet got light, the weakness (spell) left me, and new life and strength penetrated my body. I begin to step lively. I left the book with Elder B. A. Dugger. I don't know what became of it.

"We fasted seven days for God to give us the name of His church. At the end of the fast we saw in a vision on a large window glass: HOLY CHURCH OF THE LIVING GOD, THE PILLAR AND THE GROUND OF THE TRUTH, AND A HOUSE OF PRAYER FOR ALL PEOPLE. We shouted with joy and received with one voice of adopting the name. (Seven fasted.) And we looked upon a large window on the right side of the temple and there was an image of a swift flying eagle. I asked of the Lord what did that image mean and represent. The answer came immediately: 'Power and swiftness is the growth and prosperity of My Kingdom.'

"We then organized the Holy Church with the name that the Lord had given. To our great surprise, in our second general convocation in Philadelphia, a white gentleman from Bermuda listening to my explanatory notes in my American address, arose and asked might he speak. I granted his request. He said: 'I know that you and I are one because the very church that you have outlined, I have set up the same in Bermuda. Can we be as one body?' We answered with a loud shout and glorious applauges (applauses): 'Yes! Yes! Yes! We are one.' Well, to our surprise we found the church in Abyssinia with an organization nearly 4,000 years old.[6]

"We then began to study the origin of the true church of the Living God. And its dates goes back beyond the flood,—even to the Garden of Eden where God Himself shepherded the church. And in the Holy Church of the Living God, Adam and Eve and their first child were members.[7] So was Enoch who was translated, because he walked with God in perfection.[8]

We can prove that Noah, Abraham, Isaac and Jacob, all prophets, apostles and Jesus Christ were members of this Holy Church."[9] (End of the beginning of the church.)

"When I come to religion it was like this: The night lit up as bright as day. The earth looked new and glorious. Even my hands looked new. I shouted and praised God in the highest strains. And the Devil he came and said: 'You think you have something but you ain't got nothing.'

"I promised I wouldn't say anything about it, but when I entered home, Mama asked what was the matter with me, and when I knew anything I was telling all in the house. (The Devil told him not to tell it.) And great crowds came to hear me talk. Many wept. Mother asked: 'Do you want to join the church?' I said yes. She carried me to church on Sabbath (Saturday) and I joined. We had another big shout on the first day, called Sunday. They carried me to the creek and baptized me, and returned back to the church and fellowshipped me into the membership. But my mood and manner of teaching and of speaking were peculiar to that old dogma (what he said would not agree with the conventional Baptist ritual). Therefore they began to persecute me and to reject me. Therefore I had to look to Jesus for protection. I condemned wine for the use of sacrament. I condemned the doctrine of 'Once in Christ, never out.' I condemned the doctrine that parents were responsible for the sins of their children until 12. That doctrine inspired my brother Charles and I to read the Bible through. After that we found no support in the script, for such heresies.

"Well, they did not believe in children conversion. They thought Papa and Mama should beat those strange ideas out of me. But I was as hard to be understood by my parents as I was to the rest of the people.

"My birth was most peculiar. I came in the time of the fight. My mother and a white man was stroving together and in the struggle I came. They had two doctors as quick as possible. They gave her stimulants. They examined me and said: 'He cannot live.' I was not quite a six months child. They advised the nurse to give me fresh cow milk boiled with no sweetening.

"Out of that precarious condition God caused me to grow into maturity. I sit on the floor or wherever they would place me for five years. My brother

William was born two years later. He learned to walk and then learnt me. White and black said I was a curiosity and a miracle. How that God caused me to live and then to grow. My sister Hannah, when she came to her death, she told Mother why God had raised me up. So peculiar was it. It was because He had ordained me from the foundation of the world to be a great leader and emancipator of His people. She said that I would reign over nearly all the different nations spiritually. Her prognostications have come to pass. I now have churches in 35 states; in British West Indies, in Cuba, Puerto Rico, Bermuda, Bahamas, South India, Gold Coast, Liberia, and Abyssinia."

(VISION 2)

"There was a minister who had a spirit of pride. He loved to drive fine horses, with fine vehicles. At last he bought what was called a Texas pony. One Sunday morning he hitched his new horse to the buggy. He asked his wife, weighing 240 pounds, to go for a ride. She said yes and got into the buggy. The reverend got in and took the reins and bid the pony move. The pony being high spirited jumped off with full running speed to the heights of his strength. The preacher hollered whoa! The horse got faster. His wife saw danger heaving before her. She leaped out, and was unhurt. Her husband had the reins twisted around his hands when the horse broke loose from the buggy with a long, strong certain pull and leap. The preacher was thrown out on some rocks. His right leg was broken. The doctors came and set the leg and put it in stock. They told him to be quiet for nine days and not to try to walk. After a few days the doctor relieved the stock and told him to stay in bed until doctors got back.

"The preacher thought it was foolish to stay in bed as good as he was feeling. So he started to get up and walk around. His wife begged him not to. As soon as she went out, he gets up and begins to walk around. The weak knitted growth of bone broke loose and he fell. He hollered: 'Wife! Come here!' She had the doctor to come. The doctor was out done so he scolded the preacher. The preacher relapsed and seemed to be dying. I had a vision then.

"I saw many people going up the King's Highway.[10] It seemed the Milky Way or path of the sky. The good passed in white garments, while others were coming down dressed in black. I asked: 'Why are they dressed in black?' I heard a voice saying: 'These are they that was once near Heaven's gate. They become [sic] discouraged and turned back.' I looked down beneath me into a large dark valley, spread out in incomprehensible circumstance [sic] and in diameter. The ground shook and trembled like jelly. I saw thousands of souls prostrate on the jelly ground. As I was walking around looking at those lying there so miserable, in hunger and thirst, to my surprise I found the preacher in that dark habitation with the damned casted out from the presence of God.

"I awoke and went to the preacher and told him my vision. His reply was: 'I am lost. I will soon be where you saw me.' And soon after that he passed away."

(VISION 3)

"This vision was in the open day when the sun was shining bright. There was an unclouded sky.

"I were in the field alone by myself. I was meditating on the great Judgment Day and wondering about the great conflagerated flames what shoot and sweep everything wicked from off the face of the earth.[11] I was caused to stand still. I was told to look towards the south, west, north and east, I saw out of the four corners of the earth that a great volume of smoke came up and angry flames lashed here and there upon the firmament of Heaven until all seemed to have met in mid Heaven. I was told to look west. I saw great green orchard trees, shrubs and vegetation. I was told to look south. I looked and saw cities and weakened men's houses. They were in the houses, gambling drinking of whiskey, wine, and beer. They were in pool rooms and moving picture shows. The fire were [sic] raging in the Heavens and the wind were blowing in great si-rock-ledams [sic], so much so that I said: 'Lord how can *I* escape?' He said: 'The righteous will not be hurt and I will show you how I will save My people.' Bout that time the weakened people saw the flames. They run and scream. The fire swept them and their houses away, leaving the righteous standing on a Mesa (tableland).

"The valley had been lifted up. Mountains were leveled down and the earth was covered with beautiful flowers and trees and vines. God said: 'This is the way the Judgment will be executed.'"

(VISION 4)

"One evening Reverend Butts Justice and I studied this text: 'BLESSED ARE THE DEAD THAT DIETH IN THE LORD. THEY SHALL REST FROM THEIR LABOR AND THEIR WORK SHALL FOLLOW THEM.' We could not adjust what was meant by their works following them. The Lord took me away into a vision. He brought me out into a large field where folks was chopping cotton and plowing corn and so forth. There was one woman who was very mean to her children. She expected them to do as much work as a trained man. She died before the Lord give me this vision, but I saw her in the field on a rocky plain. Her and her children were working as hard as they could. Now and then she would scold her children and threaten to bust their heads open with a hoe.

"The day got so hot that it was dry. The crops looked green and promising. All at once everything dried up and become powder. I saw this being carried into stores. Folks were selling and buying. Goods turned to ashes, money to running water. I was then carried to flour and corn mills, but nothing came of the grinding. People who did good in their life time was still getting the good. The Lord said to me: 'Look, and see this is what I mean when I say their works shall follow them.'

"Then I come out of my trance and was sitting by my friend again."

(THE FEELING OF A TRANCE)

"I feel drowsy, as if something were giving away in me and I drift off from my ownself and (a) start to hear, (b) see things, (c) both at the same time. I awake into normality suddenly and reality comes slowly back. Sometimes it takes a few minutes, sometimes hours."

INTERVIEWER: "Do you hold yourself away from the embraces of women?"

BISHOP JOHNSON: "I can say that I lived holy and free from sin today, but I can't make no declaration for tomorrow.

"We all could be judged at our death if our works would cease with our death, but our work lives on in the souls of men from generation to generation." (Bishop classes himself with the following: 1. John Wesley 2. Roger Williams 3. Martin Luther 4. John Calvin 5. Bishop R. A. R.)

Communicant Rev. George Washington, 304 Prince Street

(Family)

1. Hugh Washington (24) Saint

2. Elise Washington No Saint

3. Eloise Washington No Saint

4. Carrie Belle Washington Saint

5. May Belle Washington Saint

6. George Washington Saint

Rev. Washington is pastor of Commandment Keeper church which has a membership of sixteen.[12]

Rev. Washington explained that there are two visions—open and closed, the latter coming in one's sleep.

"When I was connected I had a call in the spirit. I had been praying three weeks and four days. During this time I was godly sorry over my sins and the way I have gone in sin. I et no pleasant bread during this period of time. And the four days of this time I et *no* bread and no water [sic]. And I come [sic] then being happy down in my soul.

"I saw a great light shining in my room and that lit my whole room up like day, even though it was night and a voice spoke to me, saying: 'You must preach the gospel.' And the voice give me the text from the Fifth Galatians and 1st verse: 'Stand fast in thy liberty, where with Christ hath made me free. And be not entangled again with the yoke of bondage.'

"And I went on and [founded] a Baptist church by the name of Macedonia and I preached as a local preacher for six months; and they called me and give me a church named New Zion.[13] I pastored there for nine years.

"Then I had another call in the spirit to leave that place to go to another city, but some how, or rather I disobeyed the Lord, I didn't want to leave my home town and my people. The Lord smoted me with a terrible sickness and that sickness the doctor said was consumption.[14] I were illing along until the doctor stated one evening that he was going to take a picture of my body. He took the picture and said my right lung was completely gone and just half inch was holding the left lung. They said they couldn't do me no good and that I would have to die. They said money wouldn't do me no good and not necessary to go to another hospital. They all said the same thing. So I reached the conclusion that I'd talk to the Lord about this matter. That night lying on my bed I called the Lord in question about my ways. I felt like I had disobeyed the Lord when He told me to go and I refused and I felt like He was the one that smoted me, according to the 28th chapter of Deuteronomy, and I knowed He smoted me.[15] I told the Lord 'I'm a young man, only 26, and the doctor says I got to die, but Thou, O Lord, you made me and also give the doctor their wisdom.' That was my prayer. 'And you can heal me if you will. And you promised a man three scores and ten years and I'm only 26, having a wife and two children, and I don't want to go now because I have just started life. But if you will heal me just one time—try me—and wheresoever you say go, I will go; whatsoever you say be, I will be; whatsoever you say, I will say.'

"Then the Lord spoke again. I heard that same voice and knowed it. 'Don't take another drop of that medicine, and you get up in the morning at the rising sun.' He didn't say He was going to heal me, but I had done made the covenant with Him. Then I begin to wonder deep—was He going to heal me. I had been confined to the bed for three months and was helpless. I couldn't get up and down excepting somebody helped me.

"When the medicine hour rolled around I refused to take it from the nurse. She tried hard to open my mouth with a silver spoon handle. I closed my toothes and they couldn't get it in. She tried even the third time and failed so they give up from trying. They thought I was dying and said it was no use. 'He'll be dead soon.' I laid quiet on the bed all night long and didn't had to call nobody. The next morning about the rising of the sun, my strength come all at once and I was just as well as I am right now. The first attempt I made to sit up, the nurse tried to make me lay down. I told them I was well.

"The *Lord* did come. I wanted them to get my clothes so I could dress myself and walk out. I always keep my Bible near. So I dressed myself and went on the porch that was fronting the east, so that I could see the sun coming up out of the east one more time. Then I began thanking God for healing me through the course of the night.

"About the third hour the doctors came in—all three together—when they saw me sitting up with my Bible they called to me before they got out of [the] car: 'This is not the *dead* man sitting up here.' (Here Rev. Washington goes into ecstasy saying hallelujah, amen etc.) I said: 'Doctor, I don't need you anymore. The Lord has come and healed me last night.' (Ecstasy again, saying yes Lord, amen etc.)[16]

"Dr. Tooten, Dr. Warner, Dr. Fulge, they all three called me in. They took another picture of my body that morning. When they got it, my lungs was [sic] just as clear as their own—made whole again. They pronounced me well and sound and published it in the papers of Augusta, Georgia.

"So I came on back to Allendale where I was living and running a two horse farm (Glory!) and three weeks later, one evening just before the sun went down, I was walking over the farm by myself alone, and that same voice called to me again and said: 'You go to Beaufort, South Carolina, and I will show thee what I will have thee to do.' (Ecstasy—glory hallelujah, yes Lord.) So I stopped for a moment and thought over present conditions. I remember where I was three weeks before then—lying on a sick bed and couldn't move, I remembered when I were given up to die by three of the best doctors and that the Lord healed me in one night. I said: 'Lord, I will go to Beaufort.'

"I come on back to the house and told my wife that I had to go to Beaufort in the morning because there was no train to take me that night. Next morning I boarded the train for Beaufort. On my way I met a man coming from Charleston. We were riding together. I told him my errand and he rented me three rooms in his house for $10.00 unfurnished, at 610 Bladen Street. I paid him in advance and spent one night and returned to Allendale.

"Two weeks later I moved to Beaufort, turning my farm over to Mother and Brother. I didn't brought anything only my house furniture. After I arrived here in Beaufort, December 26, 1920, I preached from one church to another, but it seemed like I wasn't yet doing what God wanted. I felt a hungry thirst deep down in my soul. So I read [the] Bible for consolation. Yet still I wasn't happy.

"Three months after I got here I was on my way from [the] Baptist Church called Pilgrim.[17] I preached there that night. On [my] way home I heard singing and a man preaching and it all seemed joyful to me. I told Deacon Edward Allen: 'Let us stop in here.' So we went in and Dolley Jemison was preaching and he said things that I was looking for. He was preaching on the power of God and I knewed it. He said when he saw me: 'I see some strange brother.' I rise on my feet and said: 'I don't know what you call yourselves, but I see you have a blessing that I haven't got and I want it.'

"That was my call into holiness right there. Then it seemed like they all rejoiced to hear me. Then they told me what I would have to do. I would have to go to the altar and seek salvation from the earnest part of my heart. That night I fell on my knees to seek the Lord. I called Jesus (repeating it until my tongue could say it no more). They sang beautiful songs over me. I didn't get blessed that night. They held a protracted meeting for ten days. I went every night but failed.

"I had so many things to give up which was in the pride of life that it took me 34 days and 8 hours before I got the Holy Ghost. When I gave up my wife, children, friends, acquaintances, just the same as if I was going to die, the Lord come in and blessed my soul, speaking in other tongues as the spirit give utterance. (Acts 2:4.)[18] My wife went running out of the house calling neighbors to come and see what was the matter with me. It was strange to them. In the spirit, I viewed the beautiful city.

"In a moment of time He took me in an exceedingly high mountain; seem like it was in the night, a fair, beautiful night with the stars shining bright, like moons. There was a man standing on the right hand side of me who said: 'Look east!' When I looked east, I could look way down on the earth and see a city white as the dripping snow, shining so brightly that it sparkled like diamonds. Then the Lord spoke and said: 'Abraham saw that city. Daniel saw that city. John saw that city.'[19] Then I began to thank God that He had let me see that city and in a moment of time He took me back to my place again.

"From that day on and on I could tell the wonderful story of Jesus. Then the Lord told me: 'This is the work I want you to do. Cry aloud. Spare not. Lift up your voice.'

"That is the reason I am preaching this great gospel."

(ANOTHER VISION)

"On the fourth of July I went nine miles from Beaufort to pick huckleberries which were plentiful. I heard a great something under my feet. It was a great big rattlesnake. He hit me in the arm. I turned to run and the snake hit me in the hip. The snake's tooth hung in my overalls and I run one hundred and fifty feet with the cold snake whipping against my leg. As I was running the spirit spoke to me and said: 'Read Mark 16:17. Where is your faith?' I rode the bicycle the ten miles home. I got well without going to a doctor. I know the Lord healed me. This was the second healing of sickness, the first being tuberculosis."

Members of church:

1. Rev. George Washington
2. Izora Robinson (sister) one

3. Willie Robinson (her son)

4. Reuben Stephenson

5. Julian Jones (prophet) two

6. Jefferson Bates

7. Franke Bates

8. Liza Robinson (no relation)

9. Carrie Roberts

10. Melrose Range

11. Georgia Smith

12. Hattie Washington (daughter-in-law)

(CONVERSATION IN CHURCHYARD BETWEEN INTERVIEWER AND CHURCH SISTER, WHILE INTERVIEWER SITTING IN THE YARD)

SISTER MARY: "I didn't know that you could get yourself another man and just keep on having children. Sure I would have had more children if I had known that. I would have had forty or fifty—raise up a nation of your own."[20]

Communicant Mrs. Julia Jones

"I live the life and am free from sin. I was born in a sanctified family. My parents told me to be sanctified. When I got to be grown I was told not to marry an unsaved man." (Those not sanctified are unsaved.) But she married a man of bad havages (habits). She loved him and lived with him a long time.

"I asked the Lord about him. The Lord told me we were unequally yoked. The spirit told me I was going to be a widow. I finally told my husband that I couldn't live with him any longer.

"I sought the Lord and went higher and higher. In my vision the Lord come to me one night and touched me and woke me up. He told me that

He give me the gift of prophecy. He told me He give me a new name, Elizabeth, and He give me victory, redemption, higher power and a consecrated life. He gave me the gift of healing, the gift of teaching, the gift of dancing, the dance of the Children at the Red Sea, the gift of preaching and the gift of praying.[21]

"Look like to me I seen a high place. It seemed like it circled round and round the world and kept going higher. It was a big circle around the world (like a spherical ramp). People were there. And then He showed me three floors; in the last one it was a room look like. He showed me I was living clean. The symbol was a white pigeon that flew on me. He showed me I was in a room—a place—and the Bible was on that table. The power fell on till it looked like I couldn't hardly hold that Bible."

(LAST VISION)

"The Lord picked me up and carried me off and set me down in Beaufort. And I knowed [sic] I had to come to Beaufort to preach and prophesy. I was at Miles Mill, South Carolina, before that. He set me down at the corner of Izora Robinson (where she lives). After He put me down there and then He showed me how we was going to travel and carry the word." (Meaning by we, Izora Robinson, Rev. Washington and Julia.) (She went into Ecstasy for about three minutes.)

"In a trance things change. Your eyes are changed. You see and hear. The Lord have come to my bed and poured the power out on me. The Lord have showed me some of the most wonderful things I ever seen in my life.

"I have seen mules in a pen running round. Cows were in my vision also and some of my children would be after them, running them around. (I have five children.) I've seen a vision of water. I would be wading and going on across. The water never hurt me. I never feared that it would over-power me.

"I saw fire raining down and it was Judgment. I saw people running and screaming and hollering. The fire was coming down like drops of rain.

"Since I been in holiness, the Lord has showed me how I was going to die. I was just going to sleep away. No fear at all. He carried me away in a vision once. It must have been Hell. I was going down the pathway and there was people in all those places. And there was mules. They was in a close place just turning round. The Lord turned me loose—said I was a Sabbath keeper, so He turned me loose. I saw where I come on but I got back up on top of the earth. (The people were in the pit doing various things.)

"He give me Matthew 10, how to go and heal the sick.[22] Then He told me that if I couldn't forsake everybody, I couldn't follow Him. So I forsook Him. My children were left with my ex-husband. They's all about married now. Their names are Ada, Ida, Geneva, Wyman, and Willie Henry.

"Five years ago holiness come in this part and I heard about it and I humbled myself down and went to seeking holiness. I heard of it from my sister. The Lord stands over my bed for hours. He is a great comfort. (She is a prophet. The prophecy is very important in this ritual. Note the similarity of African witch smelling.)[23] The Lord said to me: 'The harvest have come.' I asked the Lord what He meant and He said: 'Harvest means what you sow, it will reap the same way.' Then He showed me in a vision of some sick folks and showed me they was sick from neglecting to serve the Lord.

"No pork in my mouth in five years. No saints eat pork.

"Before the war started my sister and I prophesied that war was coming and we was sent from city to city to warn people. The vision came to both of us at once. She is now in Jacksonville, Florida. Her name's Mary Jane Dunbar. We was in the same room when it come and we was seeking. The voice come at the same moment. It said: 'Go tell the people a world war is coming here too. We will hold out a long time, but it is coming.'"

During the ceremonies, Julia Jones goes first into ecstasy then into a trance, after which she sometimes utters prophecies aloud to the whole congregation. Other times she kneels by various members and whispers prophecies in their ears. Her eyes are half closed and her movements are like a sleep walker.

There is no appeal from her prophecy. It is the last word. It is to be accepted as the gospel truth. It is not always pleasant to members of the congregation. For instance, she uttered at one time: "I feel hate in here. Hate!! They say that old prophet, she tells too much. She talks too much. They hate the prophecy but the Lord sends the prophecy."

There is a suggestion of the African witch doctor smelling out evil doers in this phase.

Note:

The reason that preachers in church No. 2 cannot last is because they are not prepared right under the spirit, so they cannot stand; but the people in church No. 1 come on under deep teaching and are bound to prevail.

Communicant Mrs. Izora Robinson

"I have one son, Willie R. Robinson, 13 years old. When I got converted, I felt the change. Reverend Mazon preached a wonderful sermon, describing how a sinner stood before God. So I decided to pray. My sister prayed too. I was hindered because my father did not want me to join the A.M.E. Church, where I wanted to join. My father was a M.E. Church member.[24] I don't remember how old I was.

"So I got crossed and wouldn't pray till three years later. Then I had a dream. The Lord showed me in a dream my condition. He showed me that if I died, then I would have been lost. He showed me a vision.

"There were two men that I knew that was chained down in a dark house. I stood at the door and watched them and they grabbed me and dragged me on my stomach on those mow teeth (mowing machine blades) but did not tear my skin. It just hurt. But I managed to get loose and run and I got away from them. They ran after me but I got away. So that dream caused me to start to pray.

"I told my ma the dream. She told me I was old enough to pray for my ownself. I went to church and give my hand to Reverend W. M. Stony. He told me to pray. Next night I went out in the garden to pray and I felt the change. I went away in spirit and when I came back, I was light as a feather (ecstasy, glory Jesus). I felt the change but I didn't hear the voice as I had heard others say they heard. (Period of silent Ecstasy.)

"I prayed on. Then I told the pastor and he said I was done converted because he felt me. I could sing, felt good, felt like shouting, but I missed the voice that I expected.

"I was received in the church and was baptized. I had several travels in spirit after that. Every time I got slack in praying the Lord would warn me. I would see myself either playing around dangerous water or falling through bridges, etc. One time He showed me again. It seems I went in a big building,—the farthest one to the north—of pretty buildings. At the end of that street there was a deep well, one hundred feet or more. I went and peep in that well and there were wheels turning in that well; no water but black grease. And the wheel was turning and the people were hollering: 'This is Hell.' You could hear the big wheel turning like machinery. I went in [a] gray building and [a] white woman was there teaching a whole lot of colored children. I ran to another house, a pink shell building, and in that building a man was in an office. He had naked pretty men laying down on shelves on the wall. It was a pretty place. Something attractive! I ran to go out. That man who was in the office had two little horns on his head and they started to growing larger. He said: 'I am the Devil.' And he told the boys to 'Shoot after her! Catch her!' But I got away and ran towards the east.

"I ran to a new fence gate. (Dogs began to pursue her after she left shell building.) Just as I got there, the pursuing dogs got there. But I got through and closed the gate. They opened the gate and came on behind me. When I got to the third gate, the dogs couldn't come no further. They had been running me all night long.

"It was now a pretty sunshine morning and I was led on to a big old beautiful white building. With everything glistening I ran up long white steps from the east side and they led right to a window with curtains. Then a lady invited me in and dressed me. She had to dress me and she showed me different parts. I said: 'How lovely.' It was a little child called coopid (cupid) following me everywhere. The lady dressed me and put me on a long white pretty dress, and everywhere the rooms were cut off by curtains and the winds blew and the curtains just waved gently. How pretty! How pretty! And then I wake up. I don't remember leaving that place."

(MRS. ROBINSON'S EXPERIENCE WHEN SHE CAME TO HOLINESS)

"I was teaching school at Crocketville, South Carolina, in 1922. I came to Beaufort on a visit on the 3rd Saturday in January and went to my brother's house on King Street. He told me about sanctification. First, my sister told me about it, whose name is Isadora Ford. She told me it was right and I ought to get it. She said that she believed that I *would* get it. My brother got the Bible and wanted to read to me. I wouldn't let him. I said: 'No!' I said that if I was wrong God would show me. I went back to my work on Monday (Crocketville).

"A man came to me before day. I knowed it was Jesus, long white hair—. It seemed like an avenue opened up from the east, just like the morning before the sunrise. He come. He wasn't taking a step—He come gliding along. I was laying in the bed and He come touch me on my left arm. He said to me: 'You are wrong.' He pointed after me (gesture), and said: 'There is more for you to get. When you go back let your brother tell you.'

"I got up that morning and told the lady I was boarding with. She said: 'Well, honey, that's right. You let Him tell you.'

"When my school closed I came home to my mother, Pleasant Green. She and my brother Charles were fixing to go to Charleston to a convocation under Elder P.S. Jenkins. So when she come back home that next week, she come back praising God. She got out of the back (also brother George). She was giving God the praise. She said she had been in church for thirty-odd years, but didn't know that there was more for her to get until now. She said it was just as much difference in being sanctified as it was in a sinner life and a converted life. I stood still—couldn't move out of my track—just the same as a gun had shot me. I went over what the man told to me in the spirit. If Mother just find there was more, there was more for me.

"I started to church on Sunday. Brother told us to come by after we go to our church. I went instead to [a] sanctified church and found myself on my knees. After he read the Bible and explained, and after I read too, I saw my shortcomings. So I went down to seek for [the] Holy Ghost. That was in April.

"So I seek right on and on, working too with wicked people. I could not consecrate my mind to the service. Brother sent for some saints, Elder R. D. Jamison, his wife and sister Ellen Bailey. Then I didn't get sanctified. I seek right on. In July I was working at the tomato factor (factory). Still the girls would tease me, so I decided to give up work. I went to Walterboro.

"Reverend Jamison run a meeting for me special, and on the third Monday night in August 1922, I got sanctified. I was saying hallelujah and everything went black. When that power strikes you, you get cold all over and you forget everything you were thinking about. Your mouth gets open but you can't say nothing.

"I seeked on for [the] Holy Ghost. A man come to me in the spirit. It seem to me like it was an old Jew preacher and he brought a book look like a 5th grade geography and he told me to take that book and study it. I said: 'No. I done finished that book.' He told me: 'That's all right. You just take it and study it.' I took it out of his hand and I told him: 'You begin this book in the 4th grade and go to page 49. And you go from 49 and finish the book in the 5th grade. Then you take up the commercial complete geography in the 6th grade. Why I done finished this book a long time.' He told me again that it was all right and for me to take it. He pulled out a paper from the book written in blue large letters. In print was 'P R O G R A M' and it had a little set of scratches like a child learning to write but you couldn't understand it. He said: 'I must finish that book' (which there was no printing in it. There was nothing but relief maps). He said he would then have me to fill that blank out. 'Then you will be alright,' he said. I took the book out of his hand and put it under my arm and he vanished away.

"Then I told the sisters and the Elder that I was going back home. 'I'm not going to get the Holy Ghost now,' I said. (She went into ecstasy and said: 'Give me witness. They know I'm telling the truth.') Some of them said: 'Oh yes you will!' I said: 'No, I'm not going to get it now. There is more work for me to do.' And Sister Irene Lockwood said: 'You sure a man tell you that?' I said: 'Yes.' Then she said: 'No, you won't get it now.' Some said: 'Look here sister, you mustn't know so much.' I said: 'No, but I won't get it now.'

"All the desire to seek had done leff me. I wanted to go back home. I came back home and seek off and on, but Brother was not keeping the commandment yet. He was discouraged.

"But on the 4th Sunday in June, 1925 (I was keeping the commandment then) I received the baptism of the Holy Ghost. My brother George came from the meeting with some saints from Florida (Sally Johnson and Sister Clara Goodwin). They came in that day I was in the bed sick with a fever. They came to pray with me. Brother stopped praying and went to seeking—calling Jesus, and I begin to call Jesus too and that's how I get the Holy Ghost. I wasn't calling Jesus ten minutes before I was gone away in the spirit.

"I went to the east under beautiful shade trees and the green was like a carpet on the ground. There was a band of people. They all had different musical instruments. All was standing up and all was small children. I was a child myself. There was no grown folks at all. All had on khaki suits. I stood outside the ring. The director, he was a man marking time, giving the measures to the music. He just handed me a box (guitar) and didn't stop beating time. I played it. (Ecstasy—'Glory, sweet! Better felt than told! Glory!') I just went to play and went to laughing. Came through in the holiness, laughing and shouting in the bed. That was the glorious time of my life. That was the day!

"When I came through there was a huge round head with 'see-me' eyes on me and it was just laughing. I come through speaking in tongues. Before, when I opened my mouth nothing would come out, but now the spirit spoke through me as it gave utterance. (Ecstasy—'Glory, hallelujah! Unknown tongue and see-me eyes.')"

Communicant Mr. Hugh Washington (23 years of age)

"In August 1935 I entered holiness, following the meetings for several years.

"One day I went to a meeting at Port Royal. There was a great crowd. I always had it in my mind that I was going to seek but I couldn't date the time. They asked me when I would decide to come in the word. I would track behind him to see if the preacher was telling the truth or not. They sang a song: 'Most Too Late to Wait on You.' I loved it. That night, with several more, I got down with my whole mind on God and I seek for several hours. Many things came to my mind about the combats of life, but when they came I stopped and prayed to God to take them away. And when I became earnest in my seeking, the burden of sin rolled away. I got happy and felt new all over, but I didn't stop seeking. I seeked right on.

"When I was seeking it, it seemed as if the spirit was tapping me on my shoulders. Then I was taken away in the power for many hours. I saw the Heavenly hosts. Everything was white and glittery. It was a beautiful city and when I came through (No action, just people), I felt new after that. And I knew I had on my traveling shoes. And I had the *evidence of* speaking in unknown tongues as the spirit give me utterance. That is the real evidence. (He had been seeking from 9 a.m. to 2 p.m.)

"We have the custom of seeking for higher power every Sabbath. I ask to have things taken away and my mind cleared. Many things are heard and seen. Many times when you talk in unknown tongues you can understand what is said, but nobody else can. If there is an interpretation of the message it is through an interpreter, if there is one in the house. If the spirit is talking directly to me, I understand, but if the word is meant for someone else, I cannot understand it.

"I always pay strict attention to the prophecy." (At Port Royal the prophet told Hugh and his pal what they were talking about before they entered church. Also told him that he was going to be a widow etc.) (The prophecy comes at any service, when the power comes on.)

Hugh has never had an open vision. It has always been closed.

Communicant Mrs. Hattie Mae Washington (Age—19)

(OPEN VISION)

"I had wanted to be sanctified since a small girl. In seeking I saw a large church (in vision) with music. Went in and to [the] altar to be prayed for and I saw myself coming through. I went to shouting and singing. Many evil thoughts came to my mind and I went to crying. They went away and I was rejoicing.

"Songs that helped me to seek:

1. 'This World Is a Mean World'

2. 'Lead-uh Me On.'

"After that I begin feeling better and better every time. (When she went to church.) So I sang the song 'What a Friend We Have in Jesus.'[25] The prophet says that I was to be a teacher. I have not had no desire of turning back since I joined October 16, 1939."

Communicant Carrie Washington (15 years of age)

"While I was seeking I saw what looked like a dog in front of a cave. It looked like I saw a man with a shepherd staff. I ax him for the staff and he give it to me and I thanked him. While I was seeking many things come to my mind: dancing, hog meat, card playing, and so forth. I gave them up and I seek and I seek. I don't know how many hours, but I seek for a long time before I get the power. When I was seeking, I start talking in other tongues and then that same night I kept on talking in other tongues.

"Some others was seeking the same time I was but they didn't get it because they seek awhile and then they open their eyes. *You have* to keep your eyes closed to keep your mind clear and call the name of Jesus monotonously.

"And you keep on calling until your tongue gets heavy, and that is the time that the other tongues come. You don't know what you are saying and nobody else knows. The reason you know when the power strikes you is it moves you. You do not move yourself. You can tell when somebody moves themselves because they can't move as fast as when the spirit move them.

"You cannot seek in love. You have got to give up love."

Communicant Henry Moore (Box 103, Beaufort, S.C.)

"I was trying to make love to a young woman and I found it just tore me all up. I couldn't seek and make love. I seeked from 10 a.m. until sundown. There was not a dry thread on me. (Sweat.) After this the spirit of the Lord visited me and uh it appeared to me that someone was speaking behind a curtain, saying: 'Just steal away and go to another city and seek the Lord, and you'll receive the baptism of [the] Holy Ghost.'

"Whilst being under the power the Lord showed me a large field. There was quite a host; I couldn't mention the numbers. The spirit bid me to go in and tell them and show them what to do. Since that time I have been in holiness.

"A man came in, opened my mouth and fixed my tongue, and led me to a church and told me to open my services always with a spiritual."

Part Two

ON
ART AND SUCH

You Don't Know Us Negroes

The decade just past was the oleomargarine era in Negro writing.[1] Oleomargarine is the fictionized form of butter. Like the gopher in the Negro folk fables about God, the Devil, and the high land turtle, it will "go for (gopher)" butter.

Margarine is yellow, it is greasy, it has a taste that paraphrases butter. It even has the word "butter" printed on the label often. In short, it has everything butterish about it except butter.

And so the writings that made out they were holding a looking-glass to the Negro had everything in them except Negroness. Some of the authors meant well. The favor was in them. They had a willing mind, but too light behind.

Two hundred and forty-six years of outward submission during slavery time got folks to thinking of us as creatures of tasks alone. When in fact the conflict between what we wanted to do and what we were forced to do intensified our inner life instead of destroying it. We developed [a] turtle shell. So when folks come feeling around they find something smooth and round and simple on the outside. Like the six blind men who felt all over the elephant.[2] But if more had been known about us, this mistaken simplicity never would have got abroad.

So I'm going to take some folks out of the Bible and make a parable so folks can understand and know. Pictures were made before words, and signs before talking.

Hagar's Ishmael was yet a lad of a boy when old Sarah pulled her miracle by birthing Isaac.[3] It wasn't proper then no more than it is now for ninety year old ladies to be giving birth to babies. But Sarah wasn't in a position where manners were apt to seem important to her. The birth of Isaac was like that tree-climbing rabbit in the Uncle Remus tales. She was "Jus' 'bleeged to do it."

Here she was the wife of Abraham, one of the richest men of his time, playing second fiddle to her former handmaiden. And for why? Because the young and luscious Hagar was the mother of a son by Abraham. Sarah never realized how gripping the thing was going to be when she made that arrangement so that her husband might be a father. It took her nearly twelve years to understand what any modern married woman could have told her in a minute. But 'way along she got sensed into the fact that nothing but a baby and a boy baby at that was going to help her case. So Sarah just up and had Isaac in spite of all the odds against it. Nobody knows where old Sarah got Isaac from and I don't care. We didn't come here to talk about that old time tale nohow. We aim to talk about Negro writings, and Ishmael, Hagar and Sarah are just stuff to make a parable out of in the case.

Well, now, if Ishmael had been a man grown when Isaac was born, Isaac never would have become Abraham's heir. He would have just been another new born baby that nobody ever heard about. Ishmael, as a man, could have defended his rights. But Sarah timed her miracle well. Ishmael was too young and weak to contend for the place that was his by birth right. So Sarah had him declared a bastard and sent out upon the desert to die. There was to be no coming back to dispute her power and prestige.

II

And that is just what happened to the po' Ish of Negro literature. Shoved off into the desert to dry up and die. Before he even got dry behind the ears here come Brer Isaac pushing and shoving. So we have the Margarine Era in Negro writings.

You see, a lots of folks have never tasted real butter that a cow laid . . . been raised entirely on margarine. If somebody set sure 'nough butter before 'em they'd be just like the little boy from the city slums who was sent to the country for a week. He didn't like fresh eggs because he said they didn't have the twang he was used to at home.

Show some folks a genuine bit of Negroness and they rear and pitch like a mule in a tin stable. "But where is the misplaced preposition?" they wail. "Where is the Am it and I'se?"

Nobody didn't tell me, but I heard.

And then again you know they held a convention about this Negro in print and made out resolutions and by-laws. So no writer won't have to go thinking up things by themselves and be shocking editors with stories that are so far from the laws and statutes made and provided that they just can't be true.

The rules and regulations of this Margarine Negro calls for two dumb Negroes who chew up dictionaries and spit out grammar. They will and *must* go into some sort of business, and mess it all up. The big one always taking advantage of the little one. A fanatical religious scene, or a hoodoo dance (though there is no such thing in the popular Negro concept in America) is dragged in for no other reason than to prove that the author has gone deep into Negro life, and in the end the little fellow triumphs and the big fellow departs the scene urging his feet to more speed. If a villain is needed, go catch a mulatto and stick him into the plot. Yaller niggers being all and always wrong. This is the same plot that I have been reading under different titles and by various authors for ever so long.

Oh yes, and another thing. All Negro characters must have pop eyes. The only time when they are excused from popping is when they are rolling in fright. Authors must be orthodox on their eye-pops. Editors evidently keep special research workers to investigate the past lives of authors who turn in Negro stories. Special delivery rejection slips for those found unsound on this fright-and-eye business. Somewhere there must be some schismatic writers in this matter which seems so fraught with peril to the American nation, but they have been iron-fisted down. No editor is bound to burn the heretic on the spot but he must send him a rejection slip as cold as a milk-shake or take the consequences.

From all this I learned that most white people have seen our shows but not our lives. If they have not seen a Negro show they have seen a minstrel or at least a black-face comedian and that is considered enough. They know all about us. We say, "Am it?" And go into a dance. By way of catching breath we laugh and say "Is you is, or is you ain't" and grab our banjo and work ourselves into a sound sleep.[4] First thing on waking we laugh or skeer ourselves into another buck and wing, and so life goes.

All of which may be very good vaudeville, but I'm sorry to be such an image-breaker and say we just don't live like that. I can see that Miller and Lyles made a greater impression upon the nation than is generally admitted.[5] *Shuffle Along* and *Runnin' Wild* were both excellent entertainment and admirably adapted to the purpose for which they were written. But when their humorous distortions are set down in serious stories as actual Negro life, you have something else again. A cross between the duck-billed platypus and a dictionary gone crazy through the hips.

Whenever I pick up one of the popular magazines and read one of these mammy cut tales I often wonder whether the author actually believes that his tale is probable or whether he knows it is flapdoodle and is merely concerned about the check. Anyway, the deformity reminds me of that Negro folk rhyme about:

Li' boy, li' boy, who made yo' britches?
Mama did de cuttin' and pop did de stitches.

So I look with interest upon its warm reception and am reminded of a story of Hans Andersen about the shadow that finally took the place of the man who threw it.[6]

III

Needless to say, this attitude on the part of editors and producers is deforming the Negro writers themselves.

They naturally want to escape the solitude and suffering of the desert, so Hagar goes not forth to hunger and thirst with Ishmael. She sits outside the patriarch's tent door and fawns upon Isaac.

This near-Negro literature is the consequence of the hasty generalization that we Negroes are obvious and simple because, at a glance, we seem to be so. It is assumed by most outsiders that we were the very first commodity to be wrapped in cellophane. They think that if a person opens his mouth to laugh, the casual bystander may glance down the laughing throat and see all the deep-set emotions, including the liver. After that glance, of course, the peeper knows all, and tells more. But roughly to paraphrase Josh Billings, it's better not to know so much, than to know so much that ain't so.[7] Even Belasco was wed to the formula.[8] He commissioned Wallace Thurman to dramatize a Negro novel, but when Thurman turned in the finished work, Belasco rejected it with the statement: "It does not contain the simple Negro that *we all know*."

What *is* actually known about us? Very little. Certainly little that we do not wish to tell. Because we do not refuse to answer questions does not mean that we welcome probings. We are a polite people. So we say something, and usually what we say is what is expected of us, rather than the truth. The Indian resists curiosity with silence, but we offer the feather-bed resistance. That is, the probe enters, but never comes out. Gets smothered under irrelevant detail and laughter. The questioner leaves us feeling very pleased with him or herself.

You see, we have experienced the white man long enough to know that nothing pleases him more than to find out what he thought all along was the truth. So that's what makes the answers easy. And then again, we say a white man is always wanting to know into other folks's business. All right, mold yourself an image and set it outside the doorway. When he comes he will seize it and drag it away—thinking he has the real you. Then when he is gone, you can say your say and sing your song.

It is tremendously interesting to note that there are some white writers like what the rabbit said when he heard a dog barking way down the creek right after the dogs and rabbits had held a convention and agreed that dogs wasn't to run no more rabbits. When Brer Dog urged Brer Rabbit not to pay no attention to the barking because of the agreement, Brer Rabbit said, "Yeah, dat's so, but you know all de dogs ain't been to dat convention. Some of 'em ain't got no better sense than to run all over dat convention and tear up all dem laws."

There is a group of writers who evidently didn't attend the convention, so the rules don't worry them. DuBose Heyward, and that author of the *Congaree Sketches*, Julia Peterkin, T. S. Stribling and Paul Green.[9] Each of these writers has looked further than the too-obvious outside. DuBose Heyward has come nearest the true inside of Negro life. He is positively startling in his accuracy at times. It is obvious that Julia Peterkin has made a collection of Negro sayings and folk ways, but despite her Carolina plantation she does not assemble her material in a pattern to give a true picture. There is more to the life of every people than striking aspects. One could collect the highest peaks of the whole world, but put together they would not be a continent. Just a collection of bristling peaks. No nuances of valleys and plains. One may hold all of the elements of Negro life within the hand, but proportion and stress are important. Indeed it is these two elements that differentiate individuals as well as races.

Stribling is another of those souls who have sought beyond the laugh along with that author of the *Congaree Sketches*, he is going somewhere. All these whom I have mentioned are earnest seekers, halted only by the barrier that exists somewhere in every Negro mind for the white man.

Then there is the school of the wise-crack. The best known of these writers are Roark Bradford and Octavus Roy Cohen.[10] I always think of them in connection with minstrel shows and black-face comedians. For their work looks just as much like Negro life as Al Jolson looks like a Negro.[11] The same sort of thing, too.

We will pass over Cohen and his nothing else but slapstick, for he makes no pretense of reality. He is merely concerned with telling an amusing story.

But Roark Bradford is something else again. He runs some journalistic wise-cracks through his typewriter, then wraps himself in the robes of a savant and passes it out as deep inside stuff on the Negro.

Take his *Ol' Man Adam an' His Chillun*, for instance. It's got just about as much Negro folk-lore in it as Cal Coolidge had in him. I must give him credit, though. He hinted at one. Where he says something about Peter and the rock that was turned to bread. So he must have passed by some Negroes who were talking. It's just about as much as he would have heard in passing. It's a very good story, the way we tell it, but the way he's got it all mommucked up, it's not worth doodly squat.

The Green Pastures is a swell spectacle because Marc Connelly is a great dramatist and Richard B. Harrison is a great actor.[12] The best scenes owe nothing to the Negro at all, and certainly nothing to Roark Bradford unless Mr. Connelly wants to thank him for reminding him that it was in the Bible. It is good Old Testament and Marc Connelly. The reverence comes not from the wise-cracking lines, but from the dignity of Richard B. Harrison's acting. The Negro's idea of heaven is certainly not dusting out a plantation boss's office with aprons on their wings. Nothing like work and bossy white folks in our heavenly concept. If Roark Bradford had even spoken to one Negro preacher, even one sister in the Amen corner—he would have heard something about the gold-gilded Hallelujah Avenue, the diamond-studded Amen Street, the mile-high diamond vases, the ten-mile hanging lamps, the golden shoes that sing sol, me sol, do, at every step of the silver-shining wearer.

Oh no, if he puts what he knows about Negro religion on paper, he's a got-dat-wrong all down the line.

And then take this John Henry business that he is strowing all over. All wrong.

John Henry is no more a culture hero in Negro folk-lore than Casey Jones is in White. There is nothing to the legend except the ballad, and it is even more recent than Casey Jones. It is the celebration in song—not of a legendary character, but of an event. Judge of its age by the fact that he attempts to beat something as recent as a *steam-drill* driving spikes, and drops dead from heart-strain. John Henry is a railroad song and incident, and trying to transplant him to the Mississippi plantation and levees is just as ridiculous as making Casey Jones out a football hero. Not a thing back of it but Roark Bradford's typewriter.

And the stories of John Henry's miraculous birth according to Bradford—his being a forty pound baby, and so on, are not bits of Negro folk-lore at all. They are well known Paul Bunyan stories from the lumber camps of the Northwest. Consult Dr. Ruth Benedict, Department of Anthropology, Columbia University, and she will point out the proper numbers of the *Journal of American Folk-Lore* to those who might want to know really.

Now, there *is* a John, sometimes called Jack, who is the great human culture hero in our folk-lore. But he is at least two hundred years old in tradition. He is always beside Ole Massa, sometimes with him, sometimes against him, but always the smarter of the two. Ole Massa out-thinks Ike, Big Boy and Fat Sam, but never John. He even out-smarts the Devil, and in Negro folk-lore the Devil is smarter than God. John or Jack is evidently our slavery time wish fulfillment ideal. That's why we say Jack beat the Devil. But this Bradford version of John Henry is just a big old Georgia something-ain't-so.

IV

And now, there's another side to this fictionized butter, too. While the white writers have been putting it in the street that we laugh and laugh and hold no malice, the Negro writers have set out to prove that we can pout.

With slight exception the novels have been sociological. At the lowest, a prolonged wail on the tragedy of being a Negro, filled with incidents of escape, and in a varying key, a catalog of incidents intended to show starkly the pity of it all. A forlorn pacing of a cage barred by racial hatred.

This, too, is an insincere picture of the American scene. The number of mixed bloods who pass for white is highly over-stated. Because of many complications they are necessarily few. The passing is usually temporary and usually has to do with earning a living. At that, many who could pass with ease have never done so for even a day, and would feel most uncomfortable if called upon to do it at all. Only a few self-conscious Negroes feel tragic about their race, and make a cage for themselves. The great majority of us live our own lives and spread our jenk in our own way, unconcerned about other people. Sufficient unto ourselves. At any rate, very few grief-hackled bodies have been found. So any literature that proposes to point out to the world fourteen million frustrated Negroes is also insincere. There is certainly more outspoken racial prejudice in the South than elsewhere, but it is also the place of the strongest inter-racial attachment. The situation is so contradictory, paradoxical and what not, that only a Southerner could ever understand it. And Northern Negroes, unless they have spent years in residence and study, know no more about Negro life in the South than Northern white folks do. Thus a great deal of literary postures and distortion has come from Negro pens.

V

I don't need to go to the wilderness like Elijah and get bread brought to me by the ravens.[13] In the first place, I can cook better biscuits than any old black bird I ever expect to see. And then again I wouldn't like to depend on those ravens. I notice that birds are mighty forgetful creatures at times. I don't need to cross the sea in a whale's belly like Jonah; I don't need to lay on one side for forty days like Isaiah and I don't need to cry like Jeremiah.[14] I'm going to sit right here on this porch chair and prophesy that these are the last days of the know-nothing writers on Negro subjects.

Both editors and readers are clamoring for something that makes their side meat taste like ham, for to tell the truth, Negro reality is a hundred times more imaginative and entertaining than anything that has ever been hatched up over a typewriter. From now on, the writers must back their rubbish with something more substantial than the lay-figure of the past decade. Go hard or go home. Instead of coloring up coconut grease in the kitchen, go buy a cow and treat the public to some butter.

Biddy, biddy, bend, my story is end,
Turn loose the rooster and hold the hen.

Fannie Hurst

Somebody said one day in writing that Fannie Hurst was the only woman writer in America who looked like what she was.[1] Right away I knew what he meant. It is true that she is a stunning wench, but that was not all that he saw with his eyes. He meant that she had something more to show than a head, two legs, two arms, and what holds them together. He was saying that she looked like a Somebody to him. And that is right. She does.

It is not just her grooming either, though she knows very well what to do with those kind of eyes and her white skin and black hair. She knows what black and white and red can do for her looks and she does it. She and the Queen of Sheba both know what to look like when calling on the king.[2] She has got something else besides clothes that sprangle out from her when she moves. It is something like a rainbow wrapped and tied around her shoulder that glints and gleams.

And if you stay around Fannie Hurst you are not going to take a good look at her and go on off to sleep either. No, you will hop from one emotion to the other so fast that not one suspicion of sleep will dim your eyes. She is apt to wring you dry and be bored with you long before you are through with her. You will pay attention, for Fannie Hurst is a person of the most contradictory moods and statements of anyone in public life. But if you study her out, she is the very essence of consistency. She is the repository of talents and not the usual assembly of female parts and thoughts. What the gods within her direct today may be absolutely repudiated by what they dictate tomorrow, and that is as it should be. So it is ridiculous to go and get your mob-size measure and attempt to take her size.

She is a writer because she had to be one. She had money when she was born, so she did not need to try for that. She had looks and sex appeal so she did not need a career as a mating device. She had place in the social scheme. She writes because she must. But that must is inside of her, and has nothing to do with a publisher's order. If everything she writes is not a *Humoresque*, *Back Street*, a *Lummox*, or a *Vertical City*, it is not a play to the gallery.[3] It is because the gods inside have failed her for the moment.

Personally, Fannie Hurst is a little girl who is tall for her age. You can just see her playing doll-house with grown-up tools. One moment a serious worker controlled by her genii; the next instant playing make-believe with all her heart. Playing it so that it is impossible for you to doubt that for her it is true while it lasts. Then you can understand her. You can just see the child in St. Louis wandering around the big house with no other children to play with. She could not run loose in the streets because her people were never poor. There were too many good carpets and lace curtains in her house for it to be over-run with just anybody's children, either. She did not even have a cousin near her own age. She had her little life all to herself. She was the little girl with the long curls who looked out of the big window until she got tired. Then she would fiddle with the lace curtains until she was told to stop. She put up with adult company as long as she could stand it, and you can see her when she hit upon the glorious device of making up her playmates out of her own head.

She did not feel any great yearning for degrees, but she went to college anyway. She took her bachelor's degree at a college in St. Louis, but prefers to think of Columbia University as her Alma Mater because there she felt a spiritual tie that she never felt in St. Louis. By the time she entered college she was putting her fancies on paper. She sent things out to the publications, but no success to speak of in those days when she was learning to use her tools, so she decided to come East and fight it out with these editor-people at close range. That brought her to New York and to Jacques Danielson.[4] Not before she had tried briefly the career of waitress, elevator operator, and a few other odd jobs like that to gain experience. That period was short because she met Jacques Danielson, and that sensitive musician wanted to take her under a roof that he paid for and keep her. But at the same time he understood her dreams and agreed to the separate domiciles that the writer might not be destroyed by the musician and vice versa. So they made that agreement of three nights a week together which has been given so much publicity. However I have known him to sneak up the back stairs to spend some extra evenings with his wife and I have never seen her make the first move to drive him off.

She had his full sympathy and cooperation in trying her wings in literature. She tried and failed and tried again and he was always full of encouragement and faith. He told her she had genius over and over, and after "Humoresque" there was no doubt in the minds of the critics that she was among the unforgettables. The short story or the book length piece were all one to her. The length had nothing to do with her creation. She had learned to write prose. She was asked why she never attempted a play and she said that the drama was life in high relief and she preferred to work in the round.

Miss Hurst loves beauty. You can tell that the moment you enter her cathedral-like studio apartment. There is the air of the Medicis about.[5] There are beautiful bits of this and that from here and there in the world, but mostly from Italy and Italy of the great art period. She might seek here and there for a bargain in soda biscuits to eat with cheese, and then pay an enormous sum for a beautiful plate to eat it from.

But to go back to the little girl in Fannie Hurst. She is extremely intelligent and nobody's fool in business. I have heard some editor or publisher call her up and try a little strong arm stuff. Try to sort of push and shove her around a bit. Just so much of that and The Fannie is all over him like gravy over rice. Then I have seen her annoyed by some petty incident, drop everything and rush to the telephone to tell her husband all about it. And the way she says "Jack" is the most helpless sound in the whole world. "Jack, they did this and thus to me. What must I do now, Jack? When are you coming home, Jack?" You can just hear him on the other end of the wire figuratively patting her curls and wiping her eyes, and she is all mollified again. She returns to her desk with the air of, "There now! I told Jack on you and I bet he'll fix you good. Goody, goody, goody!" It is like the cave woman who probably took a rock and beat a sabre-toothed tiger to death for sticking his head in her cave. Then when her big hairy husband came home from the hunt, he found her crying in fright over a mouse.

Once I saw her playing at the little girl who runs away from home. She said to me perfectly sober sounding, "Come on, Zora, with your car and let's you and I go on a trip." That was good because I love to go on trips. "Elisabeth Marbury," she went on, "is up at her summer home at Bellglade Lakes, Maine.[6] She wrote me that she wants to see me at once and have a talk. It is something important and I really must go to her. Can we start tomorrow?"

So the next morning bright and soon I had the little Chevrolet all serviced and at the door. Miss Hurst came down with her mouth all set like a Christmas present and got in the car. And in no time at all I was going up the Boston Post Road washing my foot in the gas tank. Early that afternoon we were in Saratoga Springs. Miss Hurst told me to stop.

"We must stop here awhile. For one thing you have never tasted this spring water here and it just does wonders for people. I'd feel guilty if I didn't give you a chance to drink some of this water. Besides, Lummie needs to stretch his legs." (Lummox was Miss Hurst's two pound size Pekinese.)

We stopped in front of the United States Hotel and gave Lummox a brisk walk of about five feet. Then I went over in the park where I saw water that didn't look too private spouting up from a fountain and took a drink. By that time Miss Hurst had come back from the telephone and we all got back in the car. Miss Hurst had that look in her eyes that a child has when it is about to tell its mother that it has seen a fairy.

"Zora, your getting that drink of water reminds me to ask if you have ever seen Niagara Falls?"

"No, I never have, Miss Hurst, but I have always meant to see it some day."

"Oh, well then, you might just as well see it now. Everybody ought to see Niagara Falls as soon as possible. Suppose we go there right now?"

"Fine, but what about Miss Marbury?"

"Oh, she can wait. I couldn't think of letting you go back to New York without seeing the Falls."

So we pointed the nose of the Chevrolet due west with my foot in the gas tank splitting the wind for Buffalo. And next afternoon we were there before sundown.

We parked the car and Miss Hurst stood over there by the car because she had seen the Falls many times. You know those falls is a great big thing, so I went right up to the rail to stand there and look. It is monstropolous. It looks like the Pacific Ocean rushing over the edge of the world. But before I could really conceive of the thing, Miss Hurst called me. Mouth all primped up again.

"Zora, you must see this thing from the Canadian side. Let's go over there and get the view. They light it up at night, you know."

In five minutes we had crossed the international bridge and I had to go into a little building right close there to register the car. When I came out, Miss Hurst was almost dancing up and down like a six-year-old putting something over on its elders.

"Get in the car, quick, Zora. I think we can make Hamilton before dark!"

So we saw Hamilton, and Kitchener, and Gault and many another town in Ontario. It took us two weeks, not a minute of which was dull. I cannot remember in which town it was, but in some town in that part of Canada we saw a big sign that said that Vilhjalmur Stefansson, the great Arctic explorer and lecturer, was speaking on the Chautauqua.[7] So we hunted him up and he gave us free passes to the lectures and they were fine. All about mosquitoes practically eating up dogs beyond the Arctic circle, and how wolves *don't* go in packs.

Then one night Miss Hurst began to make notes and look very studious, and the next morning we were headed for New York state and the city on the Sound. The little girl who had run off to the horizon to hide from her complicated existence was grown up again. The artist was about to birth a book. She had had the fun of running away (who among us has not planned to do that same thing ever since we took off our diapers!) and she had had the fun of fooling me off to go along with her.

Fannie Hurst is a blend of woman and author. You can't separate the two things in her case. Nature must have meant it to be that way. Most career women are different. Their profession is like oil on water. You can see where one stops and the other begins. And then again some women writers are writers and some of them are women. Anytime, day or night, you run across Fannie Hurst you can see and feel her womanhood. And if you read her, you are going to find out that she is an author.

Art and Such[*]

When the scope of American art is viewed as a whole, the contributions of the Negro is found to be small, that is, if we exclude the anonymous folk creations of music, tales and dances.[1] One immediately takes into consideration that only three generations separate the Negro from the muteness of slavery, and recognizes that creation is in its stumbling infancy.

Taking things as time goes we have first the long mute period of slavery during which many undreamed of geniuses must have lived and died. Folk tales and music tell us this much. Then the hurly-burly of the Reconstruction and what followed when the black mouth became vocal.[2] But nothing creative came out of this period because this new man, this first talking black man, was necessarily concerned with his newness. The old world he used to know had been turned upside down and so made new for him, naturally engaged his wonder and attention. Therefore and, in consequence, he had to spend some time, a generation or two, talking out his thoughts and feelings he had during centuries of silence.

He rejoiced with the realization of old dreams and he cried new cries for wounds that had become scars. It was the age of cries. If it seems monotonous one remembers the ex-slave had the pitying ear of the world. He had the encouragement of northern sympathizers.

In spite of the fact that no creative artist who means anything to the

*This essay was drafted as part of Hurston's work on the Florida Writers Project but was not published in her lifetime.

Arts of Florida, the United States nor the world came out of this period, those first twenty-five years are of tremendous importance no matter which way you look at it. What went on inside the Negro was of more importance than the turbulent doings going on external of him. This post-war generation time was a matrix from which certain ideas came that have seriously affected art creation as well as every other form of Negro expression including the economic.

Out of this period of sound and emotion came the Race Man and Race Woman; that great horde of individuals known as "Race Champions." The great Frederick Douglass was the original pattern, no doubt, for these people who went up and down the land making speeches so fixed in type as to become a folk pattern. But Douglass had the combination of a great cause and the propitious moment as a setting for his talents and he became a famous man. These others had the wish to be heard and a set of phrases so they became "Race" Men or Women as the case might be. It was the era of tongue and lung. The "leaders" loved to speak and the new-freed field hands loved gatherings and brave words, so the tribe increased.

It was so easy to become a Race Leader in those days. So few Negroes knew how to read and write that any black man who was proficient in these arts was something to be wondered at. What had been looked upon as something that only the brains of the master-kind could cope with was done by a black person! Astonishing! He must be exceptional to do all that! He was a leader, and went north to his life work of talking the race problem. He could and did teach school like white folks. If he was not "called to preach" he most certainly was made a teacher and either of these positions made him a local leader. The idea grew and traveled. When the first Negroes entered northern colleges even the northern whites were tremendously impressed. It was apparent that while setting the slaves free they had declared the equality of men, they did not actually believe any such thing except as voting power. To see a Negro enter Yale to attempt to master the same courses as the whites was something to marvel over. To see one actually take a degree at Harvard, let us say, was a miracle. The phenomenon was made over and pampered. He was told so often that his mentality stood him alone among his kind and that it was a tragic accident that made him a Negro that he came to believe it himself and struck the tragic pose. Naturally he became a leader. Any Negro who graduated from a white school automatically became a national leader and as such could give opinions on anything at all in which the word Negro occured. But it had to be sad. Any Negro who had all that brains to be taking a degree at a white college was bound to know every thought and feeling of every other Negro in America, however remote from him, and he was bound to feel sad. It was assumed that no Negro brain could ever grasp the curriculum of a white college, so the black man who did had come by some white folk's brain by accident and there was bound to be conflict between his dark body and his white mind. Hence the stultifying doctrine that has not altogether been laughed out of existence at the present. In spite of the thousands and thousands of Negro graduates of good colleges, in spite of hundreds of graduates of New England and Western colleges, there are gray-haired graduates of New England colleges still clutching at the vapors of uniqueness. Despite the fact that Negroes have distinguished themselves in every major field of activity in the nation some of the left-overs still grab at the mantle of Race Leader. Just let them hear that white people have curiosity about some activity among Negroes, and these "leaders" will not let their shirt-tails touch them (i.e. sit down) until they have rushed forward and offered themselves as an authority on the subject whether they have ever heard of it before or not. In the very face of a situation as different from the 1880s as chalk is from cheese, they stand around and mouth the same trite phrases, and try their practised-best to look sad. They call spirituals "Our Sorrow Songs" and other such tom-foolery in an effort to get into the spot light if possible without having ever done anything to improve education, industry, invention, art and never having uttered a quotable line. Though he is being jostled about these days and paid scant attention, the Race Man is still with us—he and his Reconstruction pulings. His job today is to rush around seeking for something he can "resent."

How has this Race attitude affected the Arts in Florida? In Florida as elsewhere in America this background has worked the mind of the creator.

Can the black poet sing a song to the morning? Upsprings the song to his lips but it is fought back. He says to himself, "Ah this is a beautiful song inside me. I feel the morning star in my throat. I will sing of the star and the morning." Then his background thrusts itself between his lips and the star and he mutters, "Ought I not to be singing of our sorrows? That is what is expected of me and I shall be considered forgetful of our past and present. If I do not some will even call me a coward. The one subject for a Negro is the Race and its sufferings and so the song of the morning must be choked back. I will write of a lynching instead." So the same old theme, the same old phrases get done again to the detriment of art. To him no Negro exists as an individual—he exists only as another tragic unit of the Race. This in spite of the obvious fact that Negroes love and hate and fight and play and strive and travel and have a thousand and one interests in life like other humans. When his baby cuts a new tooth he brags as shamelessly as anyone else without once weeping over the prospect of some Klansman knocking it out when and if the child ever gets grown. The Negro artist knows all this but he conceives that a Negro can do nothing but weave something in his particular art form about the Race problem. The writer thinks that he has been brave in following in the groove of the Race champions, when the truth is, it is the line of least resistance and least originality—certain to be approved of by the "champions" who want to hear the same thing over and over again even though they already know it by heart, and certain to be unread by everybody else. It is the same thing as waving the American flag in a poorly constructed play. Anyway, the effect of the whole period has been to fix activities in a mold that precluded originality and denied creation in the arts.

Results:

In painting one artist, O. Richard Reid of Fernandina who at one time created a stir in New York Art Circles with his portraits of Fannie Hurst, John Barrymore and H. L. Mencken.[3] Of his recent works we hear nothing.

In sculpture, Augusta Savage of Green Cove Springs is making greater and greater contributions to what is significant in American Art.[4] Her subjects are Negroid for the most part but any sort of preachment is absent from her art. She seems striving to reach out to the rimbones of nothing and in so doing she touches a responsive chord in the universe and grows in stature. (Names of most important works)[5]

The world of music has been enriched by the talents of J. Rosamond Johnson, a Jacksonville Negro.[6] His range has been from light and frivolous tunes of musical comedy designed to merely entertain to some beautiful arrangements of spirituals which have been sung all over the world in concert halls. His truly great composition is the air which accompanies the words of the so-called "Negro National Anthem." The bitter-sweet poem is by his brother James Weldon Johnson.[7]

Though it is not widely known, there is a house in Fernandina, Florida whose interior is beautifully decorated in original wood-carving. It is the work of the late Brooks Thompson who was born a slave.[8] Without ever having known anything about African Art, he has achieved something very close to African concepts on the walls, doors and ceilings of three rooms. His doors are things of wondrous beauty. The greater part of the work was done after he was in his seventies. "The feeling just came and I did it," is his explanation of how the carpenter turned wood-carver in his old age.

In literature Florida has two names: James Weldon Johnson, of many talents, and Zora Neale Hurston. As a poet Johnson wrote scattered bits of verse and he wrote lyrics for the music of his brother Rosamond. Then he wrote the campaign song for Theodore Roosevelt's campaign, "You're Alright Teddy" which swept the nation.[9] After Theodore Roosevelt was safe in the White House he appointed the poet as consul to Venezuela. The time came when Johnson published volumes of verse and collected a volume of Negro sermons which he published under the title of *God's Trombones*.[10] Among his most noted prose works are *The Autobiography of an Ex-Colored Man*, *Black Manhattan* and his story of his own life, *Along This Way*.[11]

Zora Neale Hurston won critical acclaim for two new things in Negro fiction.[12] The first was an objective point of view. The subjective view was so universal that it had come to be taken for granted. When her first book, *Jonah's Gourd Vine*, a novel, appeared in 1934, the critics announced across the nation, "Here at last is a Negro story without bias. The characters live and move. The story is about Negroes but it could be anybody. It is the first time that a Negro story has been offered without special pleading. The characters in the story are seen in relation to themselves and not in relation to the whites as had been the rule. To watch these people one would conclude that there were no white people in the world. The author is an artist that will go far."

The second element that attracted attention was the telling of the story in the idiom—not the dialect—of the Negro. The Negro's poetical flow of language, his thinking in images and figures was called to the attention of the outside world. It gave verisimilitude to the narrative by stewing the subject in its own juice.

Zora Hurston is the author of three other books, *Mules and Men, Their Eyes Were Watching God* (published also in England; translated into the Italian by Ada Prospero and published in Rome), and *Tell My Horse*.[13]

It is not to be concluded from these meager offerings in the arts that Negro talent is lacking. There has been a cruel waste of genius during the long generations of slavery. There has been a squandering of genius during the three generations since Surrender on Race. So the Negro begins feeling with his fingers to find himself in the plastic arts. He is well established in music, but still a long way to go to overtake his possibilities. In literature the first writings have been little more than the putting into writing the sayings of the Race Men and Women and champions of "Race Consciousness." So that what was produced was a self-conscious document lacking in drama, analysis, characterization and the universal oneness necessary to literature. But the idea was not to produce literature—it was to "champion the Race." The Fourteenth and Fifteenth Amendments got some pretty hard wear and that sentence "You have made the *greatest* progress in so and so many years" was all the art in the literature in the purpose and period.

But one finds on all hands the weakening of race consciousness, impatience with Race Champions and a growing taste for literature as such. The wedge has entered the great inert mass and one may expect some noble things from the Florida Negro in Art in the next decade.

Stories of Conflict

Review of *Uncle Tom's Children* by Richard Wright,
New York: Harper & Bros. (The Story Press), 1938.

This is a book about hatreds.[1] Mr. Wright serves notice by his title that
he speaks of people in revolt, and his stories are so grim that the Dismal
Swamp of race hatred must be where they live.[2] Not one act of understanding and sympathy comes to pass in the entire work.

But some bright new lines to remember come flashing from the author's pen. Some of his sentences have the shocking-power of a forty-four.
That means that he knows his way around among words. With his facil-
ity, one wonders what he would have done had he dealt with plots that
touched the broader and more fundamental phases of Negro life instead of
confining himself to the spectacular. For, though he has handled himself
well, numerous Negro writers, published and unpublished, have written
of this same kind of incident. It is the favorite Negro theme just as how
the stenographer or some other poor girl won the boss or the boss's son is
the favorite white theme. What is new in the four novelettes included in
Mr. Wright's book is the wish-fulfillment theme. In each story the hero
suffers but he gets his man.

In the first story, "Big Boy Leaves Home," the hero, Big Boy, takes the
gun away from a white soldier after he has shot two of his chums and kills
the white man. His chum is lynched, but Big Boy gets away. In the second

story there is a flood on the Mississippi and in a fracas over a stolen row-boat, the hero gets the white owner of the boat and is later shot to death himself. He is a stupid, blundering character, but full of pathos. But then all the characters in this book are elemental and brutish. In the third story, the hero gets the white man most Negro men rail against—the white man who possesses a Negro woman. He gets several of them while he is about the business of choosing to die in a hurricane of bullets and fire because his woman has had a white man. There is a lavish killing here, perhaps enough to satisfy all male black readers. In the fourth story neither the hero nor his adversary is killed, but the white foe bites the dust just the same. And in this story is summed up the conclusions that the other three stories have been moving towards.

In the other three stories the reader sees the picture of the South that the communists have been passing around of late. A dismal, hopeless section ruled by brutish hatred and nothing else. Mr. Wright's author's solution, is the solution of the PARTY—state responsibility for everything and individual responsibility for nothing, not even feeding one's self. And march!

Since the author himself is a Negro, his dialect is a puzzling thing. One wonders how he arrived at it. Certainly he does not write by ear unless he is tone-deaf. But aside from the broken speech of his characters, the book contains some beautiful writing. One hopes that Mr. Wright will find in Negro life a vehicle for his talents.

The Chick with One Hen[*]

In the January issue of *Opportunity* Dr. Alain Leroy Locke published something which he calls a criticism of the Negro books published in 1937.[2] No matter which way you look at it, the piece is an example of rank dishonesty. It has not the innocence of conviction in error. It is a conscious fraud. Dr. Locke knows that he knows nothing about Negroes, and he should know, after *The New Negro*, that he knows nothing about either criticism or editing, both being branches of the same thing. Dr. Locke, conscious of his degrees, pants to be a leader, and in his eagerness to attract attention he rushes at any chance to see his name in print, however foolish his offering.

For instance he says in what he calls a criticism of *Their Eyes Were Watching God*, that the great force of the book was in the folk lore in it as in my other two books. Now, either Dr. Locke does not know Negro folk lore or he lies with malice afore thought. There is not a folk tale in the entire book. A folk *character is mentioned* in one connection that does not affect the story in any way. If Dr. Locke knows nothing about the matter, then he should have kept quiet. If he does know better, then he should have felt the obligation of his many degrees to be honest. To his discomfort I must say that those lines came out of my own head. In the final paragraph he says that the book has "condescension" which is just another instance

*Hurston wrote this essay and mailed it to *Opportunity* magazine after reading Locke's negative review of *Their Eyes Were Watching God* in his annual review of literature from 1937. The essay has never before been published.

of Dr. Locke's trying to bruise the brains of the public with his personal angle.[3] I know just why he said it and I am going to point it out.

Dr. Locke wants to be a leader. He felt sure that his degrees would guarantee that much at least.[4] But the time has passed when Negroes bowed down before mere letters on a piece of paper. And so far, Dr. Locke has offered nothing else to see. Up to now, Dr. Locke has not produced one single idea, or suggestion of an idea, that he can call his own. So far he has not had the courage to even *champion* an idea that belonged to someone else until it was already generally accepted. He waits good to see which way a procession is going then when he is very sure, he rushes up to the head of it by a forced march in the dark and says, "Come on, parade, I'm your leader!" He will approve of *anything* that is already approved of. So lacking in both talent and courage, he has failed to be that which he yearns for—a leader. Now this brings us to the second clause. When he says that there is "condescension" in *Their Eyes Were Watching God*, he is complaining that I do not write like Sterling Brown.[5] And that constitutes a crime in the good doctor's present feeling about Sterling Brown. Not that Dr. Locke has any sympathy with either the down-trodden Negro nor the working man, but Sterling Brown is committed to that political philosophy and that is enough for Locke. His impatience with everyone who does not agree with Sterling Brown could be likened to the fury of a hen with one lone chick, but in this case, the ideas belong to Sterling so we have to say the chick with one lone hen. Sterling, it seems, consults with Locke, so he must be right. Nobody objects to Sterling Brown having his own political view-point. He is free to write volumes and volumes about it. But not between the covers of my books. My great crime is that I did not consult Dr. Locke. Seeing that all of the great critics of America approved of the book Dr. Locke certainly would have approved of it if I had consulted him. And God help those who go ahead and succeed without consulting him!

An instance of Dr. Locke's insincerity. I remember well at Howard University that he was one of the leaders in a hullabaloo against the singing of Negro spirituals.[6] That was before so many people in high places had praised them. Now he tootches his lips all out and shivers with ecstasy when he speaks of "those beautiful and sensitive things." I remember his trembling with emotion over "the faithfulness to Negro religion" in *The Green Pastures*.[7] Which is anything you want to call it but the truth. But nobody was going to catch Dr. Locke not chiming in with anything so popular as that.

Dr. Locke is abstifically a fraud, both as a leader and as a critic. He knows less about Negro life than anyone in America. And if what he did in *The New Negro* is a sample, he does not know anything about editing and criticism either. You can tell by reading what he writes that he intends to be a great *big* fraud. It does succeed in being a fraud, but not a big one. He intends to pass off his "personal" touches as criticism. He is a public turn-coat and ought to be pointed out as such. He has set himself up as an opinion-passer without having the material for the opinions. I will send my toe-nails to debate him on what he knows about Negroes and Negro life, and I will come personally to debate him on what he knows about literature on the subject. This one who lives by quotations trying to criticize people who live by life!!

Jazz Regarded as Social Achievement

Review of *Shining Trumpets: A History of Jazz*
by Rudi Blesh, New York: Alfred A. Knopf, 1946.

Comes now Rudi Blesh to sing the trumpet and the city.[1] The trumpet is the "vocal" trumpet of jazz, and the city was founded by the French and named New Orleans.

"Shining Trumpets" is a fairly large and definitely ambitious book, generously illustrated with photographs of contributors to the history of the art of jazz. There is a meaty appendix to the work and a helpful index. The text at times is plaintive, sighing for the days when Negro dance music was jazz and jazz was king—before the big band was born and swing began its bobby-soxing reign. Then the author waxes indignant at the usurpation and the usurpers of the throne of jazz. At times the book is mighty informative, and the author makes some penetrating observations on persons prominent in the music world and evaluates their contributions.

Rudi Blesh is a prophet and apostle and quite boldly assumes the role of special pleader in this book. For him it is jazz and New Orleans against the world, and he is out there on the battlefield with his sword right in his hand. "Shining Trumpets" is the result of this militant enthusiasm. There was also "the need of a serious book about authentic jazz."

There is no clear-cut definition for the lay reader of what jazz really is other than the statement that it is a method of playing certain instruments. The reader is given a colorful picture of New Orleans in the decade following the Civil War, when jazz is reputed to have originated in that city. He is treated to the spectacle of the marching street bands of the Negro fraternal orders; the dances and social affairs employing these bands in the birthing years of jazz; the wind-voiced instruments, cornets and the like that came to be the foundation of jazz as it was played.

"Jazz emerged. It began not merely as one more form of Negro folk music in America, but as a fusion of all the Negro musics already present here. These, the work-songs, spirituals, ragtimes and blues all stemmed back more or less completely to African spirit and technique . . . took all this music and added elements of American white folk musics. It added, as well, the music and the distinct instrumentation of the marching (brass) bands and the melodies of French dances, (memories even from the French Opera House), the quadrille, polka, waltz, the rhythms and tunes of Spanish America and the Caribbean, and many other musical elements.

"The American Negro poured these rich and varied ingredients into his own musical melting pot and added his own undying memories of his life on the Dark Continent. . . . Under the pot he built the hot fire of creative force and imagination and then, preparatory to a miracle, stirred them all together. For jazz is no musical hybrid. It is a miracle of creative synthesis."

These words, perhaps the nearest in the book to a definition, or explanation of what constitutes jazz, leave the layman hazy. The analytical will certainly wonder how all these elements, available by pure accident at the time, and so unlike in concept and expression, could ever have come to sound so Negroid in the first place, and secondly could have been considered by the author as forming the basis of a permanent art. When Mr. Blesh takes to task the big-name bands that superseded the small jazz bands after 1935 for stealing from the classic composers, is he not accusing them of doing the same thing that he praised in the originators of jazz? If a technique, by his own admission, borrowed from the white military bands along with the instruments, and the material borrowed from so many white sources, could be considered authentic Negro art, then the history of jazz and its dethronement by the big swing bands was inherent in the beginning. Mr. Blesh's complaints against swing and the mores of the times amount to something like that anecdote which was going the rounds some years ago: The wife complains to her husband that the maid is not honest. What is her proof? Why, all those towels which they took from the hotel, the maid had stolen.

Rudi Blesh makes a laudable attempt to present the history of jazz in a self-setting . . . to display his pearl upon the nacre of the oyster shell. But he goes too far afield, and the reader finds his attention directed away from jazz men and the development of jazz to questions of social justice, race, and even anthropology, ethnology, and biology, none of which he deals with in a convincing manner. These rushes of scholarship to the head lend to the book a highly suppositious flavor. The author hands down dicta with the air of being as unoriginated as God.

For example, he tosses off his ideas about the origins of the blues, without seeming ever to have heard of that authentic cradle of the blues, the Jook.[2] He states in one place that the work-songs were brought to this country from Africa. He thinks that the spirituals came from white hymns, the same mistake made by Dr. George Pullen Jackson from examination of too few samples and hasty generalization.[3]

And the author reads out of the lodge any characters that do not meet his personal definition of jazz. In this subjective manner he discounts W. C. Handy, and many others, Negro and white.[4] He completely ignores a magnificent blues singer like Ethel Waters.[5] In the days when she did sing them, she could and did lift a blues song from her well known diaphragm and lam it against the back wall of a theatre with the very best. Yet from "Shining Trumpets" the reader would get the impression that Ethel Waters was as unheard of as God's brother. Mr. Blesh plays down Josh White, who by many is considered a truly great guitarist in the Negro manner, which is the most intricate guitar-picking on earth.[6] Gabriel Brown is not even mentioned, though he too is a great artist on the guitar and in recordings.[7]

Instead Mr. Blesh plays up Lead Belly, who, as a performer, I would swear, is not worthy to unlatch the shoes of either.[8] He is, however, a discovery of Alan Lomax, to whom the author acknowledges his indebtedness.[9] Under the flag of the championship of "pure" jazz, Mr. Blesh either ignores, or takes a swing at, many successful Negro and white exponents of popular music whom he accuses of hybridizing "true" jazz. Louis Armstrong is for him a notable exception.[10]

In spite of these disagreements with Mr. Blesh I find "Shining Trumpets" a valuable contribution to American music. The author has painstakingly assembled an imposing mass of records and analyzed them for the reader. His accounts of jazz-making and jazz-makers from 1870 on makes extremely interesting reading. Without the belaboring which Mr. Blesh gives the point, the intelligent reader can discern in the history of jazz a comforting social achievement. For here is the pageant of the popular music of the nation based on Negro idioms, developed and elaborated and diffused through the nation by white and Negro performers and composers alike, the various elements combining to produce a dynamic voice of the people, by the people, and most certainly for the people—a national art. This mode of expression, certainly more inwardly understood than the austere Declaration of Independence, more nearly typifies the spirit that broods over the continent.

Review of
Voodoo in New Orleans
by Robert Tallant

Review of *Voodoo in New Orleans*
by Robert Tallant, New York: Macmillan, 1946.

Voodoo in New Orleans is offered by Robert Tallant, the author, as the most authoritative work on the subject.[1] The student of folklore therefore expects to find first, properly authenticated material unpublished before, and the answers to the following questions: 1. What is Voodoo? 2. What is its esoteric background? 3. Who were and are the exponents and carriers of this culture in New Orleans? 4. From what sources did the author gather his material? 5. What is the historical background? 6. What inferences have been drawn to add to the sum total of human knowledge?

There is nothing in *Voodoo in New Orleans* concerning Hoodoo (to use the American name) that has not already been published, and in more specific terms:[2]

1. Hoodoo is nowhere defined in the text, except by implication. "Voodoo is an all-embracing term which included not only the god and the sect, but all its rites and practices, its priests and priestesses, and the people who obeyed its teachings" (p.9). No god of that name, nor any of its

possible derivatives has ever been found in Africa by the Herskovits[es], nor any of the other anthropologists who have made studies of African religions. ". . . thousands of these snake-worshippers were sold into slavery in the West Indies" (p. 9). (For this authority, see *Magic Island* by William Seabrook but no scientist has ever found snake-worship predominant in West Africa, an area of ancestor-worship.[3] In the few tribes where it is found, the snake is the symbol of fertility.) ". . . clawing and biting and falling on the ground in embraces of frenzied lust. They were *possessed*." (Again see *Magic Island*. It is apparent that the author defines the functions of Hoodoo as a mere stimulation to sex. He fails, however, to sustain this all the way through.)

2. The author makes no attempt to establish the esoteric background by a comparative study of the African religions, nor the developments in Santo Domingo. He merely quotes several writers of fiction who wrote about occurrences brought to public notice. Thus he follows closely, not what went [on] inside of the Hoodoo establishments, but the popular notions about Hoodoo which include the clichés of all fiction writers when describing a primitive rite anywhere. They dance, go into a frenzy and crawl off into the dark in couples, just as all fictional ghosts rattle chains.

3. The best efforts of the author were in digging out of the newspaper files the names of the best known Hoodoo "doctors" of the past. No light is thrown upon those numerous others, however, who did not for one reason or another attract the notice of the press or the police.

It would have been infinitely more valuable to the work had the author spent less time trying to establish the well-known Marie Leveau as a procurer and a gambler and more upon the aspect of her work as a Hoodooist.[4] The length, too, that he dwelt upon the spurious Marie Leveau II is both worthless and wasted, for numerous women sprang up after Marie Leveau to attempt to profit by her reputation. The New Orleans area is crowded with both men and women to this day who claim to be descended from Leveau in one way or another, and to be following her routines. The effort to establish relationship is just as footless as to comment on the blood relationship between the subsequent Roman Emperors and the family of the

Caesars. An analogous situation exists on the South Carolina and Georgia coasts where so many Hoodoo doctors claim to be a "doctor Buzzard" after the success of the first man of that name, who has long been dead.[5]

A more worthwhile effort would have been a thorough and comprehensive study of the methods of the current Hoodoo doctors of the area. Since in *Mules and Men* I substituted other names for the real ones, Tallant only mentions the name of one of the more successful ones of the present, and that because Rockford Lewis was arrested, and got his name in the papers.[6] There is no study of the current Hoodooists of New Orleans. There are merely some rumors about them. What their methods and procedures are, the author does not tell us. He does favor us with excerpts from the cards that many of the drugstores in the Negro section pass out so freely. However, if the author had witnessed every ceremony of Hoodoo in Louisiana as he claims, the story would have been different. In order to witness the ceremony, he would have had to be a Hoodooist himself. If he had been "in" and become a "Two-Headed Doctor" himself, he would have known about the Hoodoo supply house, been assigned a number and given a list to order from.[7] He would not have had to depend upon the spurious cards from the drugstores, nor the advertisements in Negro newspapers. He would not have fallen into the errors about High John de Conquer, whom he flippantly speaks of as "Johnny."[8]

5. The author offers not one valid source of information for a serious work on folklore. About a matter that is by nature secret, he offers no evidence from the practitioners themselves. Those on whom he relies are: 1. The local press past and present, whose sources of information are alien to the people under study; 2. The stories of a few well-known fiction writers, none of whom were Hoodooists; and 3. The Writers Project of the WPA.[9] The author obviously assumes that being "on relief" automatically qualifies one for the discriminating task of recognizing and collecting material, and endows one with the knowledge and skill to evaluate and winnow it. Not only are all the scientists who have spent years in the African, Santo Domingan, and American fields ignored, but even the Hoodooists

themselves. The author evidently prefers to hear something *about* them, than *from* the practitioners of Hoodoo. It is clearly the popular and sensational rather than the serious and valid approach. Or perhaps he did the best that he could with what he was able to find. Perhaps nothing more definite was available to him.

Tallant would have done well to have been more familiar with even the European background of sympathetic magic. If he had known even that, he would not have thrown in the "inevitable boiling cauldron" (p. 13), wax effigy (p. 14), serpents, frogs and bats (p. 16) which belong to Europe and not to Hoodoo. A casual acquaintance with Shakespeare would have made him more familiar. His Haitian borrowings of the Zombi (*Magic Island* and *Tell My Horse*), and the little coffin of the Sect Rouge (*Tell My Horse*), and the translation of the invocation to Legba (*Tell My Horse*, p. 103) do not escape the notice of the trained eye, though nowhere does the author acknowledge his indebtedness.[10]

6. The historical background is not to be trusted since the author is not even informed about matters that are in the common school histories of the United States. He states (p. 9): "Voodoo came to the Americas a little over two hundred years ago. The raids on the African Slave Coast began about 1724." It is well established that African slavery had been introduced into Santo Domingo by 1520, and soon spread over all of the Spanish colonies, and a little later, the French Antilles. The first Negro slaves were brought to the English Colonies of North America in 1619. And with the coming of the first slaves came their African beliefs, by necessity always secret. There does not now exist, nor has there ever existed any organized cult of Voodoo in the United States. With the disruption of the African background, it has persisted in the Americas, as fragmentary, individualistic efforts. In New Orleans as well as in the rest of the United States, each "doctor" is a law unto him or herself. Marie Leveau nor anyone else has ever been the ruler of a Voodoo cult. From all accounts, she ran the social dances in Congo Square, and she was the best known Voodoo doctor in the United States. It is not to be ignored that the Negro slaves in other parts of the South were

permitted to hold their dances too, and their "white folks" looked on also and enjoyed the singing and the dancing.

7. The author arrives at no conclusion.

Voodoo in New Orleans is totally exterior so far as Hoodoo is concerned. There is no revelation of any new facts, nor any analysis of what is already known. It is rather a collection of the popular beliefs about Hoodoo from the outside. The snake-worship sex-orgies, Greek Pythonesses, and goat-sacrifices, proceeding from false premises, and governed by hasty generalizations. It offers no opportunity for serious study, and should be considered for just what it is, a creative-journalistic appeal to popular fancy.

The late Lyle Saxon says in the foreword, ". . . So much nonsense has been written about Voodoo in New Orleans. . . ." *Voodoo in New Orleans* in no way tends to abate the nuisance.[11]

What White Publishers Won't Print

I have been amazed by the Anglo-Saxon's lack of curiosity about the internal lives and emotions of the Negroes, and for that matter, any non-Anglo-Saxon peoples within our borders, above the class of unskilled labor.

This lack of interest is much more important than it seems at first glance. It is even more important at this time than it was in the past. The internal affairs of the nation have bearings on the international stress and strain, and this gap in the national literature now has tremendous weight in world affairs. National coherence and solidarity is implicit in a thorough understanding of the various groups within a nation, and this lack of knowledge about the internal emotions and behavior of the minorities cannot fail to bar out understanding. Man, like all the other animals, fears and is repelled by that which he does not understand, and mere difference is apt to connote something malign.

The fact that there is no demand for incisive and full-dress stories around Negroes above the servant class is indicative of something of vast importance to this nation. This blank is NOT filled by the fiction built around upper-class Negroes exploiting the race problem. Rather, it tends to point it up. A college-bred Negro still is not a person like other folks, but an interesting problem, more or less. It calls to mind a story of slavery time. In this story, a master with more intellectual curiosity than usual, set out to see how much he could teach a particularly bright slave of his. When he had gotten him up to higher mathematics and to be a fluent reader of Latin, he called in a neighbor to show off his brilliant slave, and to argue that Negroes had brains just like the slave-owners had, and given the same opportunities, would turn out the same.

The visiting master of slaves looked and listened, tried to trap the literate slave in Algebra and Latin, and failing to do so in both, turned to his neighbor and said: "Yes, he certainly knows his higher mathematics, and he can read Latin better than many white men I know, but I cannot bring myself to believe that he understands a thing that he is doing. It is all an aping of our culture. All on the outside. You are crazy if you think that it has changed him inside in the least. Turn him loose, and he will revert at once to the jungle. He is still a savage, and no amount of translating Virgil and Ovid is going to change him. In fact, all you have done is to turn a useful savage into a dangerous beast."

That was in slavery time, yes, and we have come a long, long way since then, but the troubling thing is that there are still too many who refuse to believe in the ingestion and digestion of western culture as yet. Hence the lack of literature about the higher emotions and love life of upper-class Negroes and the minorities in general.

Publishers and producers are cool to the idea. Now, do not leap to the conclusion that editors and producers constitute a special class of un-believers. That is far from true. Publishing houses and theatrical promoters are in business to make money. They will sponsor anything that they believe will sell. They shy away from romantic stories about Negroes and Jews because they feel that they know the public indifference to such works, unless the story or play involves racial tension. It can then be offered as a study in Sociology, with the romantic side subdued. They know the skepticism in general about the complicated emotions in the minorities. The average American just cannot conceive of it, and would be apt to reject the notion, and publishers and producers take the stand that they are not in business to educate, but to make money. Sympathetic as they might be, they cannot afford to be crusaders.

In proof of this, you can note various publishers and producers edging forward a little, and ready to go even further when the trial balloons show that the public is ready for it. This public lack of interest is the nut of the matter.

The question naturally arises as to the why of this indifference, not to say scepticism, to the internal life of educated minorities.

The answer lies in what we may call THE AMERICAN MUSEUM OF UNNATURAL HISTORY. This is an intangible built on folk belief. It is assumed that all non-Anglo-Saxons are uncomplicated stereotypes. Everybody knows all about them. They are lay figures mounted in the museum where all may take them in at a glance. They are made of bent wires without insides at all. So how could anybody write a book about the non-existent?

The American Indian is a contraption of copper wires in an eternal war-bonnet, with no equipment for laughter, expressionless face and that says "How" when spoken to. His only activity is treachery leading to massacres. Who is so dumb as not to know all about Indians, even if they have never seen one, nor talked with anyone who ever knew one?

The American Negro exhibit is a group of two. Both of these mechanical toys are built so that their feet eternally shuffle, and their eyes pop and roll. Shuffling feet and those popping, rolling eyes denote the Negro, and no characterization is genuine without this monotony. One is seated on a stump picking away on his banjo and singing and laughing. The other is a most amoral character before a share-cropper's shack mumbling about injustice. Doing this makes him out to be a Negro "intellectual." It is as simple as all that.

The whole museum is dedicated to the convenient "typical." In there is the "typical" Oriental, Jew, Yankee, Westerner, Southerner, Latin, and even out-of-favor Nordics like the German. The Englishman "I say old chappie" and the gesticulating Frenchman. The least observant American can know them all at a glance. However, the public willingly accepts the untypical in Nordics, but feels cheated if the untypical is portrayed in others. The author of *Scarlet Sister Mary* complained to me that her neighbors objected to her book on the grounds that she had the characters thinking, "and everybody know that Nigras don't think."[1]

But for the national welfare, it is urgent to realize that the minorities do think, and think about something other than the race problem. That they are very human and internally, according to natural endowment, are just like everybody else. So long as this is not conceived, there must remain that feeling of unsurmountable difference, and difference to the average man means something bad. If people were made right, they would be just like him.

The trouble with the purely problem arguments is that they leave too much unknown. Argue all you will or may about injustice, but as long as the majority cannot conceive of a Negro or a Jew feeling and reacting inside just as they do, the majority will keep right on believing that people who do not look like them cannot possibly feel as they do, and conform to the established pattern. It is well known that there must be a body of waived matter, let us say, things accepted and taken for granted by all in a community before there can be that commonality of feeling. The usual phrase is having things in common. Until this is thoroughly established in respect to Negroes in America, as well as of other minorities, it will remain impossible for the majority to conceive of a Negro experiencing a deep and abiding love and not just the passion of sex. That a great mass of Negroes can be stirred by the pageants of Spring and Fall; the extravaganza of summer, and the majesty of winter. That they can and do experience discovery of the numerous subtle faces as a foundation for a great and selfless love, and the diverse nuances that go to destroy that love as with others. As it is now, this capacity, this evidence of high and complicated emotions, is ruled out. Hence the lack of interest in a romance uncomplicated by the race struggle has so little appeal.

This insistence on defeat in a story where upper-class Negroes are portrayed, perhaps says something from the subconscious of the majority. Involved in western culture, the hero or the heroine, or both, must appear frustrated and go down to defeat, somehow. Our literature reeks with it. Is it the same as saying, "You can translate Virgil, and fumble with the differential calculus, but can you really comprehend it? Can you cope with our subtleties?"

That brings us to the folklore of "reversion to type." This curious doctrine has such wide acceptance that it is tragic. One has only to examine the huge literature on it to be convinced. No matter how high we may *seem* to climb, put us under strain and we revert to type, that is, to the bush. Under a superficial layer of western culture, the jungle drums throb in our veins.

This ridiculous notion makes it possible for that majority who accept it to conceive of even a man like the suave and scholarly Dr. Charles S. Johnson to hide a black cat's bone on his person, and indulge in a midnight Voodoo ceremony, complete with leopard skin and drums if threatened with the loss of the presidency of Fisk University, or the love of his wife.[2] "Under the skin . . . better to deal with them in business, etc., but otherwise keep them at a safe distance and under control. I tell you, Carl Van Vechten, think as you like, but they are just not like us."

The extent and extravagance of this notion reaches the ultimate in nonsense in the widespread belief that the Chinese have bizarre genitals, because of that eye-fold that makes their eyes seem to slant. In spite of the fact that no biology has ever mentioned any such difference in reproductive organs makes no matter. Millions of people believe it. "Did you know that a Chinese has. . . ." Consequently, their quiet contemplative manner is interpreted as a sign of slyness and a treacherous inclination.

But the opening wedge for better understanding has been thrust into the crack. Though many Negroes denounced Carl Van Vechten's *Nigger Heaven* because of the title, and without ever reading it, the book, written in the deepest sincerity, revealed Negroes of wealth and culture to the white public.[3] It created curiosity even when it aroused scepticism. It made folks want to know. Worth Tuttle Hedden's *The Other Room* has definitely widened the opening.[4] Neither of these well-written works take a romance of upper-class Negro life as the central theme, but the atmosphere and the background is there. These works should be followed up by some incisive and intimate stories from the inside.

The realistic story around a Negro insurance official, dentist, general practitioner, undertaker and the like would be most revealing. Thinly disguised fiction around the well known Negro names is not the answer, either. The "exceptional" as well as the Ol' Man Rivers has been exploited all out of context already. Everybody is already resigned to the "exceptional" Negro, and willing to be entertained by the "quaint." To grasp the penetration of western civilization in a minority, it is necessary to know how the average behaves and lives. Books that deal with people like in Sinclair Lewis' *Main Street* is the necessary metier.[5] For various reasons, the average, struggling, non-morbid Negro is the best-kept secret in America. His revelation to the public is the thing needed to do away with that feeling of difference which inspires fear, and which ever expresses itself in dislike.

It is inevitable that this knowledge will destroy many illusions and romantic traditions which America probably likes to have around. But then, we have no record of anybody sinking into a lingering death on finding out that there was no Santa Claus. The old world will take it in its stride. The realization that Negroes are no better nor no worse, and at times just as boring as everybody else, will hardly kill off the population of the nation.

Outside of racial attitudes, there is still another reason why this literature should exist. Literature and other arts are supposed to hold up the mirror to nature. With only the fractional "exceptional" and the "quaint" portrayed, a true picture of Negro life in America cannot be. A great principle of national art has been violated.

These are the things that publishers and producers, as the accredited representatives of the American people, have not as yet taken into consideration sufficiently. Let there be light!

Part Three

ON RACE AND GENDER

The Hue and Cry About Howard University

I went to Howard as a Prep in 1918–19.[1] I had met May Miller and she liked me and urged me to transfer from Morgan to Howard.[2] I still have her little letter of friendship and encouragement. I value it too. That was the beginning of a personal and literary friendship that has lasted.

The thrill Hannibal got when he finally crossed the Alps, the feeling of Napoleon when he finally placed upon his head the iron crown of Constantine, were nothing to the ecstasy I felt when I realized I was actually a Howardite.[3]

We used to have "sings" in Chapel every Monday during services and nobody knows how I used to strive to eradicate all pettiness from my nature so that I might be fit to sing "Alma Mater." We always finished the service with that. I used to indulge in searching introspection to root out even those little meannesses that put us far below the class of the magnificent transgressor and leave us merely ridiculous.

It was during the next year (1919) while Howard was not recovered from the S. A. T. C., that Wienstien [sic] came to Howard under government pay to conduct the singing in the "camp."[4] He had a magnificent tenor voice, and wore his khaki well. He had worked with Prof. Wesley, also a tenor, in the war camps of the country and together they had us singing lustily.[5] We liked it. We sang lots of things: "Long, Long Trail a Winding," "K-K-Katy," "Roll Jordan Roll," and "Gointer Study War No Mo'" among other things but we always ended with "Alma Mater."[6]

After Wienstein left, the singing was continued under Wesley. He used to come out before the faculty on the platform and lead the singing daily. The President would arise with beaming face and ask us to sing our songs for him. He said that Negro music began where "white" music left off. We used to respond cheerfully. Then we would select any song from the book we liked. Hymn 245, "God of Our Fathers," and 180, "Immortal Love Forever Full," were our favorites.[7] This went on for weeks and weeks. Spring was approaching.

One day I wrote Dr. Durkee a note and left it on the pulpit as I came into Chapel, asking him to read the 91st Psalm.[8] He has a marvelous speaking voice and I could wish nothing better than hearing him read that beautiful piece of prose poetry. He did not read it. I felt snubbed and disappointed, but the next day he began that beautiful one, "The Heavens Declare the Glory of God."[9] The sun shone in mellow tones through the stained windows, tendrils of the ivy vine crept in the open windows and the sparrows chirped incessantly in the midst of their nest building.

The President knew it perfectly and before he was fairly under way he had his audience on the edge of the seats so that the last tones left us still hanging there. And when we realized that he was really through we sank back tremendously moved.

Howard was unutterably beautiful to me that spring. I would give a great deal to call back my Howard illusion of those days.

Every day after that for a month the President read a psalm. It took a long time to reach the 91st, but I did not care. He never looked in the book—I am certain he knows them all by heart. E. H. Sothern in *Hamlet* has nothing on Dr. Durkee reciting the psalms.[10]

I dwell on these seemingly trifling details to give one a picture of Howard before the storm.

A few days later and the first storm broke. A great number of students but not the entire body of students by any means were holding indignation meetings alleging that they had been forced or commanded by the President to sing "Spirituals." He was denounced as a despot, a tyrant, who was dragging us back into slavery.

Though there were spokesmen among the students, various members of the faculty were credited as the real leaders. Among whom were Miss Childers, Mr. Tibbs and Miss Lewis.[11] Some said Miss Childers didn't like the idea of Wesley leading the singing as she used to "raise" all the songs. The papers printed things down in the city and some members of the Senate denounced us as ingrates and accused us of being ashamed of ourselves and our traditions.

The President held a conference with the students one day after Chapel to find out how he had offended. There were speakers for and against the "Spirituals." John Miles, now of Yale Divinity School, was one of the "Pro's," May Miller and another young lady whose name has slipped me were "Anti's."[12]

The "Pro's" made the usual stand: (*a*) the beauty and workmanship of the songs. (*b*) Only American folk songs. (*c*) Only beauty that came out of slavery. The "Anti's" held: (*a*) They were low and degrading, being the product of slaves and slavery; (*b*) not good grammar; (*c*) they are not sung in white universities.

The thought that any Negro could or would be ashamed of Negro music, had never occurred to Dr. Durkee I am sure, for he seemed pained that he had unwittingly offended and never since has suggested them.

After a few days of bluster this affair died down but not before a perceptible rift had been made in the faculty and student group.

II

A little later that same year, Senator Smoot arose on the floor of the Senate with a book in hand which he informed the Senators was a highly culpable Bolshevistic volume which he had received from the hands of a Howard student.[13] He understood it came from the university library and insinuated that it was in the curriculum. He held forth that a government supported institution that was making bolshevists should be allowed to toddle along without government aid seeing that this was the U. S. and not Red Russia.

Rumors flew thick and fast among the students as to who had engineered the book into the Senator's hands. It is to be remembered that Smoot was head of the Appropriation Committee. Durkee hastened down to the Senate Committee room and explained that the book had been given by the Rand School and it was the policy of the university to accept all gifts.[14] It was neither taught nor recommended. This satisfied the Senator evidently, for finally the appropriation came through. He was denounced by some on the Hill and some off for having cringed before the Senate.[15] He should have informed that body that we could teach what we liked and if the money was withheld we could have the satisfaction of being untrammeled. I even saw a typewritten, unsigned card on the bulletin board on the second floor of The Main Building to the effect: "It is better to lose $250,000 than our manhood."

After the smoke had cleared away, a young man known to be socialistic, a close friend of mine, left Howard forever. I saw him recently in New York. He says he has been around the world twice since 1919 but never feels right to go back to any school.

More than one person was accused of having sent that book to Smoot by the student. Some say that a professor in the law school did it, others that a teacher in the department of history, to embarrass the Administration. Perhaps it will not be known just who, but anyway, Senator Smoot never drew it from the library.

"The University Luncheonette" run by two law students, Dyett and McGhee, was a place where a great deal of discussion went on, Mr. Dyett being known as the anti-administration man.[16]

About this time the "contemptible puppy" rumors began to circulate. Students were beginning to see that there was something wrong somewhere. Some faculty members and the Administration were not so "clubby," so to speak, as they might be. There were stories flying about the campus that certain members were giving certain trusted students "tips" on faculty meeting doings.

Dr. Emmett J. Scott had been made secretary-treasurer of the university, succeeding both Cook and Parks in their respective jobs.[17] This, some felt, was unjust and muttered that an attempt was being made to "Tuskegeeize" Howard.[18] Dr. Scott being the first gun fired. There was no one to whom these rumors could be definitely traced, but the students passing along the complaints always claimed faculty sources. For instance a young lady friend of mine stopped me in the upper corridor of [The] Main Building to tell me that Dr. Durkee should be thrown out. I was astonished and asked her why she thought this.

"Well," she said, "he called Kelly Miller a black dog to his face."[19]

"How did you hear it?"

"A very high member of the faculty—an official told me, and I know he wouldn't lie."

This was the first time I had heard the story, but not the last by any means. I heard it variously repeated. In one story Mr. Miller had been called a "puppy dog"; in another "a black dog," in another a "contemptible puppy." From neither of the principals have I ever heard a syllable on this matter, but whether it is true in any part, it had a tremendous effect upon the students—a Negro professor being called out of his name by a white man—no matter what the provocation, if any.

More and more it came to be that every official act of the faculty must be subject to student scrutiny. In some way or other Z. Alexander Looby, George Brown and Fred Jordan had a pretty thorough knowledge of what went on in the chamber.[20] But Mr. Looby was President of the Student Council and perhaps had a chance to know things that way.

A great many of us took no stock in the hurly-burly feeling that we could not as students act in the capacity of the Administration, but a great number were flattered at these rumored confidences. I discount most of it as being untrue—the figment of persons wishing to enhance their own importance in student eyes by appearing as the confidant of the faculty. One instance I know to be true.

In political science Mr. Tunnell digressed one day from government in general to government in particular and told the class that Dr. Durkee was a joke; that some one (I forget who) had foisted that fisherman on us and that he was being paid a high salary to raise funds, but he was a failure.[21] He then told us the President's salary was $7,000 per year and his house. He then told us that Emmett J. Scott had been brought on from Tuskegee and paid $5,000 ($4,500 salary, $500 incidentals) to divert the golden stream from that school to Howard but he was a white elephant.

Of course I was surprised at such confidences but so much was being said here and there on campus that one could expect about anything. It was evident to me now that the faculty (I mean by that term the entire governing-teaching body) was a Spartan youth concealing a fox under its clothes.[22]

Then there was the instance of the famous note on the desk of Senator Smoot written by Professor Kelly Miller. It had to do with the appropriation rules. The Administration was making a tremendous fight for the $500,000 for the Medical School. A number of Senators were doing battle for and against it, but a strong group had pledged themselves to see it through. President Coolidge in his message to Congress had urged that it be given [to] Howard.[23] Dr. Durkee is a Massachusetts man and his Senators had taken the field openly in his behalf.

In the midst of this came Prof. Miller's note to Senator Smoot asking him not to ask for the half million dollars for fear of losing the regular appropriation of $267,000, I think it was, and threw the Administration friends into confusion. The daily press of Washington accused the professor of attempting to embarrass the Administration since the President stated that Miller's action was unauthorized. I have never seen an authorized version of the affair from Prof. Miller's pen, and shall therefore suspend judgment until I do. There have been a number of stories pro and con, but so far as open statements are concerned, the affair remains where the press left it.

There are those who hold that Prof. Miller aspires to the presidency of the university. No one can deny the urge to ascend in humanity. If we do, we preach stagnation. His ability to bring this about, if it is true, and if so rather to his credit than otherwise, what man is satisfied and his fitness for the job is being hotly debated all over the country at present. Some members of the alumnae claim that all that has happened at Howard in the way of disturbances is part of the ladder up which Dean Miller prepares to ascend. The human mind unexpressed being unreadable, all these things pro and con on the subject are still conjectures. Every one who reads or listens knows how often mole hills of trifling incidents are stretched to mountains and given special significance.

On the other hand there are those who contend that Dr. Durkee is an obstruction in the path of Howard's progress. This calls attention to the accomplishments of his administration. His bitterest enemy cannot but admit that more has been done for the advancement of the university under him than in all the other administrations put together. The following are excerpts from "Facts," a pamphlet issued by the university:

By vote of the Trustees, June 4, 1919, the offices of Secretary and Treasurer were combined, and Dr. Emmett J. Scott elected as Secretary-Treasurer. He began his services July 1, 1919.

The office of Registrar was created as a separate position, and a Howard alumnus, Dwight O. W. Holmes, was elected to that position, and succeeded by Mr. F. D. Wilkinson, upon the former's appointment as Dean of the School of Education.[24]

Both the offices of Secretary-Treasurer and Registrar have been put by these officers on the most modern administrative basis with extensive rooms on the first floor of The Main Building.

The office of Dean of Men was created, and to it elected Dr. Edward L. Parks, former Treasurer.

The office of Dean of Women was created and to it has been elected Miss Lucy D. Slowe, a Howard alumna, formerly principal of the M Street Junior High School of Washington.[25] Miss Slowe is completing her first year most successfully.

For the academic deans has been created a group of offices on the first floor of [The] Main Building, with clerks. The Dean of Men and Dean of Women each also have been given fine offices with clerks.

There is also a University Council, composed of two members of each school of the university, including both undergraduate and graduate schools. The purpose of this Council is for a better understanding between the schools and for a more united purpose. This Council meets three or four times a year.

It has long been felt that an Alumni Secretary was necessary to our greatest success. In June, 1921, the Trustees voted as follows:

> "Authority is granted to the President to secure an Alumni Secretary under conditions which will be of best advantage both to the University and to the alumni, paying such salary as shall be needed, money paid not to exceed $1,000 toward the salary of the person employed."

Mr. Norman L. McGhee, College '19, Law '22, a member of the Secretary-Treasurer's office force, is temporarily heading up this movement for closer affiliation with our alumni.[26]

In February, 1920, the Board of Trustees voted as follows:

> "One Trustee may be elected each year from a number recommended by the Alumni Association of the University, such Trustee to automatically retire at the expiration of his term of office."

Since the report of the Committee, no vacancies on the Board have occurred. It is interesting to note that eight Alumni of Howard University are now serving as members of the Board of Trustees.

Building and Grounds

New buildings erected: The Greenhouse, erected in 1919, at a cost of $8,000, and the Dining Hall Building with class rooms for the Department of Home Economics, erected in 1921, at a cost of $201,000. Plans are now under way for the new gymnasium and stadium. The General Statement, given below, will show numerous renovations made. Howard Hall,

General O. O. Howard's old home, used for so many years as a detention house for incorrigible children, has been reclaimed, the old outbuildings torn away, and the home restored as a dormitory for girls.[27] In The Main Building, a United States post office has been established, thus serving the postal needs of the student body and faculty. In The Main Building, also, has been equipped a Rest Room for girls and also one for women teachers and workers. Both were greatly needed.

The items in the General Statement "Improvement of Grounds" includes the following: Reclaiming of the bank overlooking the Reservoir, formerly a dump for cans and a place for burning rubbish; trees on the campus have been treated twice; large flower beds of rare beauty have been placed; plaza and front of Thirkeld Hall made beautiful and splendid concrete walks and steps to Sixth Street provided; fence surrounding the lower half of main campus; unsightly plot of ground on Georgia Avenue changed into a beautiful little park with paths crossing and steps leading up to Sixth Street; surroundings of Howard Hall graded and granolithic walks and steps placed; grounds surrounding School of Music beautified; underground electric lighting system installed with posts and globes like those used in the District of Columbia—this latter one of the biggest improvements.

It also became necessary for the Trustees to appropriate certain amounts out of general funds so as to complete the improvements and repairs mentioned.

A summary of the amounts spent since July 1, 1919, up to the period ending December 30, 1922:

Repairs to sundry buildings, including The Main Building, Science Hall, the Chapel, President's House, Spaulding Hall and various residence properties of the University	$55,487.34
Repairs to Clark and Miner Halls, dormitories for young men and young women	21,625.08

Improvement of grounds	15,896.98
Repairs to Law School Building	15,530.06
Improvements, Library Building	1,388.61
Repairs to Medical and Dental Schools Buildings	13,745.46
Installation New Electric Feeders, thereby making more efficient the heating and lighting facilities of the University	1,246.20
Repairs to Boarding Hall, while in basement of Miner Hall	478.30
Repairs to Manual Arts Building	732.56
	$126,130.59

Curriculum

At the close of the school year 1918–19, all secondary schools were abolished, leaving a college registration of 1,057. Dire disaster was everywhere prophesied[.] [B]y the following year the college opened with 1,567 college students.

The whole plan of undergraduate work was changed. The four years' college course was divided into two periods of two years each—the first two years named the Junior College, and the second, the Senior Schools. A student entering the undergraduate department will take two years of general college subjects leading to his last two years of specialized work in whatever field he may choose, graduating at the end of four years with his degree from that particular school. The College of Liberal Arts cares for all those students who desire four full years of undergraduate non-professional work.

New courses of study authorized by the Trustees during the present administration:

Architecture

Art

Dramatics

Public Health and Hygiene

Reserve Officers' Training Corps

At the trustee meeting of June, 1919, the old semester system was abolished. Under that system it took the college three weeks to register its students and get to work in its classes. We are now on the quarter system, and register two thousand students and more in two days at the opening of the year, and in one day for the winter and spring quarters, classes beginning recitations the following day.

The General Education Board required as a basis for its help that all finances of the School of Medicine be taken over by the general administration of the university and be handled in one office. When this was done, the Board pledged the university $250,000 as an endowment to the Medical School, providing the university would raise a like sum. This sum, in cash or pledges, must be raised by July 1, 1923, pledges to be redeemed by July 1, 1926. With such an endowment the Medical School may be kept in Class A. Without this endowment the Medical School will lose its Class A rating. Hence, the necessity for every friend of the School to rally to its support now. To show the remarkable spirit among the student body, the President announced that the student body has pledged $24,843. The Trustees, administration and faculty have pledged practically $15,000. The total gifts so far (May 21) amount to about $220,000.

During 1920–21 evening classes were established. The attendance for that year was 46. The registration for 1922–23 is 153. So far we have served 104 teachers from the public schools of Washington.

In 1919 the Trustees, on recommendation of the President, adopted a Faculty Salary Scale, toward which the Administration should work:

Faculty Salary Scale	
Dean	$3,000 to $3,500
Professor	2,500 to 3,000

Associate Professor	2,000 to 2,500
Assistant Professor	1,500 to 2,000
Instructor	750 to 1,500

Over $63,000 have been added to the teachers' salaries alone during this administration. The minimum scale has now been practically reached, and the last two surpassed. Many salaries have been doubled in three years. Average increase of salaries 56 per cent; 26 new teachers have been added.

By recommendation of the President, the Trustees voted that teachers of professorial rank may have the privilege of a sabbatical year of absence on half pay, providing they use that year's leave of absence for advance study in some standard institution of learning, the better to fit themselves for their particular field in teaching.

With the opening of the present administration, 1918–19, total financial income was $220,553.43, of which sum the Federal Government appropriation was $117,937.75.

Our auditors reported for the year 1921–22 a total budget figure of $589,033.87, of which sum the Government appropriation was $363,135.25; $116,000 of the $336,135.25 was appropriated for the new Dining Hall, which has been in use during the school year 1922–23.

We have a School of Public Health and Hygiene with a Director. Under that School comes the Department of Physical Education with a Director in charge; also the Department of Military Education with six officers detailed from the United States Army to care for our Reserve Officers' Training Corps. But the Trustees voted to have student managers of the individual teams, such as football, baseball, track, etc., and also invited the alumni to elect three representatives from the alumni as an Advisory Committee who will meet with the staff of the Department of Physical Education concerning all matters of interest in that Department.

With the opening of this administration there was but one department of the university approved by the rating associations of America. The School of Medicine was Class A.

In the autumn of 1921, the Association of Colleges and Preparatory Schools of the Middle States and Maryland, after most rigid personal investigation, placed our College of Liberal Arts on the Approved List. This means that now our graduates from such college have the same scholastic standing as graduates from any other first-class school in America.

In the spring of 1922, our Dental College was registered in the New York State Board of Regents, thus giving it the highest rating.

Our College of Pharmacy has just been given the highest rating with the Pharmacy Board of the State of Ohio.

Our School of Law is now applying for admission to the Association of American Law Schools, and we are confident of success.

A careful organization of the students has been approved, and under the title of "Student Council," the students have a very large measure of self-government. See table on the following page.

The Board of Trustees in June, 1922, unanimously passed the following vote:

"INASMUCH as the past year has been marked by very substantial progress in the life of Howard University as indicated in the erection of a splendid new Dining Hall building at a cost of $200,000; in the registration of its Dental School in the A grade of dental schools by the Board of Regents of the State of New York; in the acceptance of Howard University on the Approved List of colleges and preparatory schools of the Middle States and Maryland; in the high quality of work done in the various departments and in other respects:

"BE IT RESOLVED, That the Board of Trustees of Howard University hereby expresses its hearty confidence in and its cordial approval of the energy, the sound judgement, and the administrative efficiency of the President and the other administrative officers in the conduct of the life of the institution;

"BE IT FURTHER RESOLVED, That the students, teachers and alumni of the University are cordially invited at any time to confer

Congressional Appropriations for Five (5) Years as Contrasted with First Year of 1918–1919

Name of Appropriation	1918–19	1919–20	1920–21	1921–22	1922–23	1923–24	1924–25	1925–26
Maintenance, salaries, etc.	$72,437.75	$76,437.75	$90,000.00	$90,000.00	$100,000.00	$110,000.00	$125,000.00	$125,000.00
Buildings and grounds	10,000.00	10,000.00	32,500.00	32,500.00	42,500.00	20,000.00	20,000.00	30,000.00
Medical	7,000.00	7,000.00	7,000.00	8,000.00	8,000.00	9,000.00	9,000.00	9,000.00
Medical Addition, New Bldg.	....	....	....	....	....	....	....	370,000.00
Laboratories	2,000.00	2,000.00	2,000.00	2,000.00	3,000.00	5,000.00	5,000.00	5,000.00
Libraries	1,500.00	1,500.00	1,500.00	1,500.00	1,500.00	3,500.00	3,500.00	3,000.00
Fuel and Light	5,000.00	5,000.00	5,000.00	10,000.00	15,000.00	15,000.00	15,000.00	15,000.00
Manual Arts	20,000.00	20,000.00	20,000.00	20,000.00	20,000.00	30,000.00	30,000.00	34,000.00
The New Dining Hall	....	....	85,000.00	116,000.00	....	....	157,500.00	....
*Athletic Field, Gymnasium, Armory and Administrative Headquarters for Department of Health and Hygiene	....	....	....	....	....	40,000.00	....	
Totals	**$117,937.75**	**$121,937.75**	**$243,000.00**	**$280,000.00**	**$190,000.00**	**$232,500.00**	**$365,000.00**	**$591,000.00**

* Note: $157,500 to complete Gymnasium project also authorized, but not yet available.

with the University authorities on matters pertaining to its welfare on the basis of mutual interest."

There is no doubt in the minds of fair people that the President has been maligned. The local alumni is responsible for this to a large extent. Whenever disturbances occur on the Hill, they never investigate but rush to press with bitter denunciations of the President, using that sure-fire gag to arouse the unthinking, "A group of intelligent, high-minded race persons being 'low-rated' and otherwise trampled under foot by a white tyrant." This never fails to stir, for we are yet too close to slavery, evidently to judge a case on its merits rather than on its relation to other groups.

The senseless criticism of the appointment of Dr. Scott has died down, for surely no one can deny that he has made good. No student has been forced to pick cotton as was predicted. The murmurs against Wilkinson as Registrar on the grounds that he was not a college man have proven untimely for he certainly has justified his appointment.

He has been misrepresented in connection with the Curry School at Boston.[28] It is a very small private school founded over forty years ago by Dr. S. S. Curry of Tennessee.[29] There have never been colored students at the school. He received a diploma from the school twenty-three years ago and knows of the great advantages of such training. When after the death of Dr. and Mrs. Curry, Alumni asked him to head up the school for a little while, he did so with the statement in print (he wrote it to Mr. C. Murphy of the *Afro-American*) that he would slowly overcome the practice of forty years and have our colored students admitted.[30] This he feels he could have done.

Notwithstanding the public statements, enemies distorted the whole question and made it seem to the public that he was playing double.

So after seven months as President of the school, Dr. Durkee resigned. Of course his name will appear for a few months, as in all similar cases of school catalogues, but when the board meets in the fall, his name will cease to be carried, as he is no longer connected with the school.

He has been criticized for preaching in the city pulpits of Washington. Some weeks ago he preached in one of the largest pulpits in the city. His services were freely given. At the close of the service a gentleman came up and gave [Dr. Durkee] a check for $500.00 for the School of Religion at Howard. Next week a check for $750.00 came from a good friend who had listened to him preach. He wanted him to place the money in the School of Religion drive. Another friend who listen[ed] to him preach in Washington gave him $400.00 per year for four years for the same drive. A Sunday School class of the city of Washington just sent him $105.00 for the same drive.

These are but a few touches—I could give you scores of such.

He is quoted as drawing a salary of $7,500 and certain perquisites from the school. That is false. His salary is but $7,000 and President's house.

You will note from "Facts" the scholastic standing.[31] The struggle for such recognition may be imagined. Do not fancy that such a rating came to Howard by chance. He could tell you a long story of travel, conferences, writing, etc., with a great internal struggle to get the scholarship of the university up to the point where it could pass the test. Now our graduates go to the graduate schools of the leading universities of America on the strength of their work at Howard and the certification of their deans. There is no longer a question of "colored school." It is simply "Howard University," one of the sisterhood of the great universities. Note our teaching staff (President's Report, page 15).

Let me call your special attention to the close of the President's Report, pages 22–23. Also note Medical Report, page 18. You will see that he has secured $80,000.00 already of the $130,000.00 needed.

All these truths relating to great sums of money raised for

1. New buildings.

2. Endowment.

3. Repairs and improvement of grounds.

4. Teachers' salaries. (We have added to the salaries of teachers in six years, just about $70,000.00. Teachers who were receiving $800.00, $1,000.00 and $1,100.00 per year when he came seven years ago are to-day receiving $2,400.00, $2,600.00 and even $2,900.00. Two or three going over a total of $3,000.00.)

5. Scholarship and scholastic standing.

6. General standing throughout America and the world. (A French scholar visiting Howard this past year, told Dr. Durkee that no group of scholars met for a conference on the European Continent, who did not discuss the growth and achievements of Howard University.)

Now all this vast achievement has been accomplished in the face of bitter opposition from certain persons on the faculty of the university and certain irresponsible persons outside the faculty. The active group on the outside are those who have contributed *scarcely one dollar to all this growth.* Dr. Durkee contributed $200.00 to the Medical Drive and has subscribed $500.00 to the School of Religion Drive. He yearly spends over $300.00 in aiding and helping students and societies of the university. The active opposition all together have not put as much money into the university as he has himself. (See article of Mr. Smith, Alumni and Field Secretary.)[32]

Dr. Durkee often expresses his deepest debt and gratitude to Dr. Scott, without whom the work could not have succeeded; Dean Woodard, scholar, counsellor, loyal lover of Howard; Registrar Wilkinson than whom no university can have a finer, truer officer; Dr. Brady, Professor Coleman, and such like honest, upright, unswerving friends of the right and true.[33] These have made possible what Dr. Leonard has said in his report regarding the wonderful work which the university has, and is, doing.[34]

The following are excerpts from a recent statement of the Trustee Board: Washington, D. C.—The Executive Committee of the Board of Trustees of Howard University, in joint session with the Budget Committee of the Board of Trustees of Howard University, Monday, June 15, 1925, authorized the release of the following statement:

The Statement

In response to requests which have reached the Trustees of Howard University from many loyal alumni and friends, making inquiry as to certain decisions of the Board of Trustees, at its annual meeting, held June 2, 1925, and subsequent meetings of the committees designated by the Board to carry its decisions into effect, the following statement is made:

At a meeting of the Board of Trustees, held October 13, 1924, after a full discussion of the financial and academic condition of the university, it was decided that the general expansion of the university was running far ahead of the university's available resources. An intensive study on the part of officials of the university had led them to feel that there was much duplication of work and considerable expense in the way of overhead.

To relieve the situation of any suggestion of inside pre-judgment, or charge of prejudice, the Board of Trustees, upon recommendation to the Budget Committee, voted that a general survey of the various schools or departments should be made by an expert from the outside. The Board of Trustees considered itself most fortunate in being able to secure the services of Doctor Robert Josselyn Leonard, Director of the School of Education, Teachers' College, Columbia University, New York City, to make its first survey.

It was decided that a survey should be made first of the academic departments to be followed later by a survey of the departments of the School of Medicine. After a thorough-going study of the academic departments of the university, Doctor Leonard appeared in person and submitted a full and complete report, and made certain recommendations at the meeting of the Board of Trustees, held June 2, 1925. These recommendations were given very careful consideration by the members of the Board of Trustees.

Some general idea as to Doctor Leonard's treatment of the situation is indicated by the following extracts from his report:

Retirement Plan for Teachers

Doctor Leonard, in his report, also recommended that the university look toward the adoption of a retirement plan. This whole matter, the retirement of teachers, has been before the Board of Trustees for a number of years. At the meeting of the Board, held February 3, 1925, the Budget Committee was requested to make a report at the June meeting of the Board of Trustees.

Professor Kelly Miller retains his professorship in the department of sociology. It was voted by the Executive Committee that his salary shall rank as $3,500 as dean.

Educational Organization

With reference to the present educational organization of the university, Doctor Leonard states that the present educational organization of the university is sound theoretically; *in fact, it represents the most advanced type*; but, in view of several practical difficulties, a number of important changes are proposed, looking toward consolidation for economy and efficiency, including the merger of the Junior College with the College of Liberal Arts.

"1. The university has neither the staff nor the financial resources to achieve fully the distinctive purpose of the Junior College; namely, to assist young men and women to determine upon the Senior College or professional school best adapted to their interests and abilities.

"2. Doctor Leonard also recommended that the work of the School of Commerce and Finance be merged under the College of Liberal Arts."

At the meeting of the Executive Committee, held June 15, 1925, carrying out the general program adopted by the Board of Trustees, making necessary contractions, eliminations, and the discontinuance of some of the members of the faculty, it was decided to discontinue, as of June 30, 1925, the services of Alain L. Locke, Professor of Philosophy; Alonzo H. Brown, Professor of Mathematics; Metz. T. P. Lochard, Assistant Professor of French; Orlando C. Thornton, Instructor in Finance and Business Organization.[35]

The Executive Committee devoted itself to a very full discussion of the whole matter in all of its phases and decided that the work of the university would not unduly suffer as the work of these professors would otherwise be carried. An expression of appreciation on behalf of the Board of Trustees was voted those discontinued for the services they have rendered the university since they have been in its employ.

Present Registration is 2,123 As Against 1,057 in 1919

A faculty member in criticising the Administration said, "everything that Dr. Durkee has done was needed and will be justified in the further growth of the university. His mistakes are he does the thing suddenly and without enough explanation; so that motives other than the real ones can and are attributed to him. He is rather boyish in his attitude; he will stake all on the word of some one in whom he believes, and a hard vindicative fighter for the things he thinks right. Full of primitive, youthful zeal and deadly serious in his efforts to build up Howard, he has faith in himself and a stomach for hard work. How else could he have done the things he has?"

What or who is behind the efforts to oust Dr. Durkee from the presidency at Howard? I do not know. But certainly this movement is on. In 1919–20 the "Spirituals" demonstration was pulled off. This singing was purely voluntary on the part of each student. Certainly not required of any one. But the President was blamed and denounced—for what?

In this same year (1919–20) the socialistic book found its way into the hands of Senator Smoot, head of the Appropriations Committee, who must decide on funds for Howard. Why? The President's efforts to placate the "watch dog of the treasury" were likewise denounced as "cringing."

A further rift in both teacher and student groups, and the year closed.

The following year when school opened, the third day of Chapel, the President appeared on the platform of the Chapel rather diffidently. Some one began hand-clapping. Instantly the crammed Chapel took it up and gave the President a most stirring reception full of yells, rah! rahs! with "Durkee" on the end.

Having the memory of those last bitter days of the preceding term, he was tremendously moved. He advanced to the pulpit and stood with flushed face and those steel-blue eyes swimming.

"I see that the students realize what a tremendous strain the President has been laboring under for the past months. He—"

He could get no further and bowed his head, signifying dismissal. The students were themselves touched and went forth very pro-Durkee.

But in that same year came the Student Council fight, headed by Looby, George Brown and Fred Jordan. More bitter Chapel sessions, more divisions on the campus, more denunciations from the city for the President.

Later also came the fight on compulsory Chapel attendance by students in the Engineering School led by Priestly, Hardwick and Goins.[36]

This began a long struggle against Chapel attendance by compulsion. I believe few persons exist who do not object to forced spiritual life. The President's contention was that it was his only chance to meet the students. He wanted to keep in touch, to maintain a personal bond, feeling perhaps the need of this in his position. For in the services, he always told of any new acquisitions, any new conquest, our athletes were called up and lauded, in general the life of the university was synchronized there. Then too, he had a chance to "tell his side of things."

In this same year or early in the next there was the dining room strike. The students paid $18.50 per month for board. Some said the food was poor, others more conservative said that the price was too high. The politicians of which there was no dearth accused the Administration and Mr. Scott in particular of gouging. The outcome of it all was that they were given permission to board in the city. But after a brief trial every one was again boarding on the Hill.

The next year, 1921–22, the general protest against Chapel attendance won and no one but freshmen were required to go.

But other strikes have come and gone, more distracting of student attention from class room to problems of administration. Where will it end? We shrug.

No attempt has been made, nor will be made to show that the Administration is perfect or infallible. But their mistakes are made in an effort to arrive at something better than what they have to work with. Their efforts are *constructive*. That cannot be said of the harsh critics throughout the country who neither know what is being done nor wish to know in order that their unfavorable attitude may not change. Indeed, the facts in the case do not alter their opinion at all. Disintegration is the goal toward which they work for the university they profess to love. Never a dollar contributed, never a helpful word, never a constructive criticism from year to year. These are those who tear down in the name of love.

The question arises: Is it best to lend a helping hand to Howard—imperfect as it is, it is our only university—to raise it to our ideal of a university, or by destructive internal warfare, level it to the earth again? This is a world of compromises. Katabolism is easy, growth is hard.

The Emperor Effaces Himself*

I

Eight modest, unassuming brass bands blared away down Lenox Avenue. It was August 1, 1924, and the Emperor Marcus Garvey was sneaking down the Avenue in terrible dread lest he attract attention to himself.[1] He succeeded nobly, for scarcely fifty thousand persons saw his parade file past trying to hide itself behind numerous banners of red, green, and black.[2]

This self-effacement was typical of Mr. Garvey and his organization. He would have no fuss nor bluster—a few thousand pennants strung across the street overhead, eight or nine bands, a regiment or two, a few floats, a dozen or so of titled officials and he was ready for his annual parade. It was pointed out that the pennants were used solely to [indicate] the route as the "Royal African Guards" were mounted and the horses had to [be] shown which way to go.

The uniforms worn by the paraders were so colorless that they gave strength to the rumor that Mr. Garvey's visit to Col. Simmons of the

---

*This satire by Hurston remained unpublished in her lifetime.

K. K. K. had been successful and that he and all of his followers had become members of the "Invisible Empire."[3]

Mr. Garvey himself wore an Admiral's hat hidden under a mass of red and green plumes. Not to appear too partial to the navy, he wore a General's uniform set off by a few fat ropes of gold braid, a sash and a sabre. His men wore black suits with stripes of red braid running hither and yon. Perhaps under anyone else they would have been dressed entirely in scarlet, but Mr. Garvey said 'no.' He was very firm in the matter.

Back in 1920 for his parade Garvey had worn a purple robe with a black hood lined with red and green silk. But he revolted against such gaudiness—hence the plumes.

II

As a military genius he had no faith in himself at all. Tho he was Admiralissimo of the "African Navy," Generalissimo of the "African Legions," he frequently expressed a fearful lack of confidence. But these expressions placed side by side with his mighty accomplishments are proof positive of the man's overwhelming modesty.

"When I get to Africa, with my invincible Black Army, I shall not ask Great Britain what she is doing there, I shall tell her to 'get out!' I shall not ask France what she is doing there, I shall tell her to 'get out!' I shall not ask Belgium what she is doing there, I shall tell her to 'get out!' And so on until I have kicked every white man out of Africa!"

Perhaps he felt this charming reserve because he had never had a day of military experience in his life.

On the walls of his living room in 129th Street, there hung a large picture of Napoleon. On the opposite wall hung one, still larger, of himself. It is evident that he wished no comparisons drawn. If he had, he would have caused them to be hung side by side.

"You already have the governors of Europe trembling," he announced a little further on. "Lloyd George has warned the other statesmen to look out for Marcus Garvey, and you can rest assured that Marcus Garvey is looking out for Lloyd George."[4]

III

With his Negro contemporaries, whom lesser souls might have considered rivals and consequently felt the pangs of jealousy, he was never too busy to pause and pay them compliments. Of W. E. Burghardt Du Bois he said: "Fifty years from now, Du bois will still be sending petitions to Congress. Marcus will be coming up the Hudson Bay (river) with a flotilla of battle-ships, dreadnaughts, cruisers, submarines and aeroplanes to land the first African ambassador in the United States. The next day he will dine at the White House."[5]

The indolence of some of his race bretheren stirred his great spirit. He himself was willing to serve, was eager to save. They not only re-frained from saving themselves, but actively objected to his saving so much. Tearfully he read them out—expelled them bodaciously from the race. Bleached and faded, they go mewing about the limbo of nothing-ness that borders the land of races. Thus passed Du Bois, James Weldon Johnson and William Pickens from among the Negro living, and are seen no more.[6]

But with his officers and others who shared his zeal, he was most gen-erous. Of his wealth of titles, he gave and gave till it hurt them to carry all that he gave them. Behold his "Duke of Uganda!" His "Knight Com-mander of the Sublime Order of the Nile!" "Supreme Deputy Potentates," "High Chancellor," "High Auditor" and "Lords" and "Ladies" aplenty.

For himself he kept almost nothing. He was merely Managing Editor of the *Negro World*, Pres. of the Black Star Steamship and Navigation Line, Pres.-General of the Universal Negro Improvement Association, Supreme

Ruler of the Sublime Order of the Nile, Provisional Pres. of Africa and Commander-in-Chief of the "African Legions."

IV

With rare foresight, he saw that the redeeming of the entire continent of Africa would take time. It would be no easy task to make it safe for the black folk of the world. They must not be too optimistic he told them.

"Ninety days from now (Aug. 1920) we will have an ambassador at the court of France; ninety days from now we will have an ambassador at the court of St. James; ninety days from now we will have an ambassador at the court of St. Petersburg; ninety days from now we will have an ambassador at the court of Moscow; ninety days from now we will have a Black House side by side with the White House in Washington."

He might have demanded the entire site on which the White House stands, but you see, he generously offers to share it. He will permit even the conquered whites to have an executive mansion side by side with his.

V

Democratic soul that he was, he frequently humored the whims of his subordinates. If the military men wished to call themselves "Royal African Guards" and trick themselves out in the uniform of the Jamaica Police, he could not find it in his heart to deny them so simple a pleasure. They probably derived great happiness from the parades in which they figured. He yielded also to the women who wished to call themselves "The Ladies of the Royal Court of Ethiopia."

Modest and reserved himself, he loved these qualities in others. He was most severe with those who endeavored in one way or another to thrust themselves into prominence unduly.

For instance, a stockholder, coming to Garvey's office one day while the guard was absent, actually entered the place unannounced! The President promptly threw him out again. Such methods of advancing one's self in the world cannot be too vigorously squelched.

E. L. Gaines, Capt. of the African Legions, to whom the organization owed a few hundred dollars, went to the office and brazenly demanded his pay.[7] This flagrant example of selfish greed was put down ruthlessly. The Capt. was thrown out also. How else could Mr. Garvey preach to the world the high spiritual aims of the organization if his officers' minds clung to thoughts of pay? Why, he himself would not accept more than five hundred or more dollars per month—scarcely enough to keep a millionaire alive!

Even Sir William H. Ferris, K.C.S.O.N. (Knight Commander of the Sublime Order of the Nile), Vice Pres. of the U.N.I.A, Treasurer of the Negro Factory Corporation, Literary Ed. of the Negro World was so lacking in taste as to lead a group of factory workers seeking pay into the Imperial suite.[8] He was severely rebuked.

"How dare you bring anyone into my office without my consent?" Mr. Garvey asked him. Any amount of abuse heaped upon such vulgar social climbing would not be too much.

One is not surprised to learn that he hated praise. One of his followers who continually shouted "God and Garvey" at every meeting, was silenced by being made Speaker of the Convention.

VI

Mr. Garvey hired several lawyers to advise him at various times. They were evidently men of small calibres. They purported to be lawyers, but invariably he knew more about legal processes than they. Furthermore, a more sensitive, touchy lot never lived. If Mr. Garvey playfully hinted that [they] were useless and need not clutter up the place any longer, they resigned. He was forced more than once to take cases out of their hands and go into court and conduct them himself. He knew no law, but 'his not to reason why, his but to go and try.' Once he was forced to be both lawyer and witness, to ask and answer his own questions.

GARVEY, LAWYER: Do you know Capt. Gaines?

GARVEY, WITNESS: Yes.

GARVEY, LAWYER: How long have you known him?

GARVEY, WITNESS: Four years.

GARVEY, LAWYER: Was he ever employed by you?

GARVEY, WITNESS: Yes.

GARVEY, LAWYER: In what capacity?

GARVEY, WITNESS: I appointed him Captain of the "African Legions."

Later on, he was forced to conduct his own defense before the U.S. Supreme Court. The Govt. either out of fear of Mr. Garvey, or envy of his great conquests, arrested him on the flimsy charge of using the mails to defraud. Fraud? Ridiculous! Of course, he had sold a few trifling thousand dollars worth of Black Star Line stock before he had a ship, he had sold a few passages to Africa on a ship that did not exist, but what's a few little ships among emperors![9] But why the cry of fraud? He had taken the people's money and he was keeping it. That was how he had become the greatest man of his race. Booker T. Washington had achieved some local notice for collecting monies and spending it on a Negro school.[10] It had never occurred to him to keep it. Marcus Garvey was much in advance of the old school of thinkers. Hence he stood in places never dreamed of by Booker T. Washington. There have been some whisperings concerning W. E. B. Du Bois on account of his efforts to lower the violent mortality rate among his people, and advance their interests generally, but he never learned how to keep the people's money and so missed true greatness.[11]

Mr. Garvey sat up one night and learned law. The next day he bravely took the burden of the case upon himself. Even tho he realized he had not a chance in the world against the District Attorney, he assumed the responsibility.

"When I get thru with that little Jew Dist. Atty., he will be so small you will have to hunt for him with a candle."[12]

These are some of the telling points he drove home for the defense in spite of the prosecution.

1. Capt. Gaines, prosecution witness on the stand. (The same who had been dismissed without pay by Mr. Garvey, and thrown out of the office)

   GARVEY: What is your personal opinion of Marcus Garvey—is he honest and sincere?

   CAPT. [GAINES: No. He] spent the money you [sic] got by fraud on race horses and women.

2. Sidney de Bourg: Prosecution witness.[13]

   GARVEY: Did you ever have a conversation with "Lady" [Illegible] concerning Marcus Garvey?

   DE BOURG: Yes.

   GARVEY: What was said during this conversation about Marcus Garvey?

   DE BOURG: She said you were impossible and the only thing to do was to let you fall over a precipice and break your neck.

Judge Julian Mack, before whom the case was being tried, asked both sides to rush because a Zionist convention was to be held soon in Chicago, and he wished to be present.[14] With rare self-effacement Garvey asked, "Would you rush this case to attend a convention when the liberty of Marcus Garvey is at stake?"

His unselfish desire to help is shown by a remark he made to one of his officers out of court.

"I have preached and shown these preachers how to preach, and now I'll show these lawyers how to practice law."

Very Touching! With only one night of law to his name, he was willing to share it with the benighted legal profession.

Tho he had no college training, he was a thirster after knowledge. After his address to the jury, he decided to study law, and asked a law student about entrance requirements.

GARVEY: Do you have to have a college degree?

STUDENT: Yes.

GARVEY: How about a man as famous as me—don't they have any special provisions?

The jury [endangered] his college career by finding him guilty, but Judge Mack was more sympathetic. He urged Mr. Garvey to take a five-year course in Practical Geology as being more helpful than the practice of law.

That instance is not [illegible] of affection for the higher learning. He wanted to be a patron of Letters so he founded or rather created Booker T. Washington University out of a twenty-foot board and nailed it up where all might see. Of course, the alumnae of this university might be only splinters, but even so, it shows the lofty ambition of the man.

VII

He was a fearless seeker after truth. By scientific investigation, he discovered that The Virgin Mary was a black woman, and that Jesus Christ, a mulatto who has been "passing" these two thousand years. So, what could be fairer than showing them in their true colors? [What] could be darker? Nothing, according to the 1924 edition of his modest little parade.

VIII

In five years he will be free. Already he has forgiven us our sins and is willing to stay and help us. But the Government says, 'no.' The Black [Typescript ends here.]

The Ten Commandments
of Charm

"For when a woman is perfectly frank with men, they do not rise up and
bless her. Nay, they rise up and run!"

"And what woman can be happy and likewise neglected?"

1. Be cheerful. Let not thy smile come off. For a joyous damsel is
   a bright light in a dismal world; and a man will pass up long
   eyelashes and a short temper for a damsel that is long on laughter
   and short on complaints.

2. Show not thine hand to any man; neither tell him the truth. For
   a little mystery is better than much wit. And unto a man every
   woman is a brand new puzzle wherein he loseth interest when he
   has found the "answer."

3. Forget not the first law of conversation, which is, Thou shalt
   not talk about thyself; nor the last law, which is, Help every
   man to express himself brilliantly! Thus shalt thou be accounted
   a "fascinating conversationalist," though thou utterest not a
   single word.

4. I charge thee, avoid the telephone as thou wouldst the plague.
   For a woman that pursueth a man with telephone calls affecteth
   him like unto five meals a day. She causes him sentimental
   indigestion. She is as lobster after the dessert.

5. Beware the temptation of the inkwell. For a woman that delugeth a man with letters and perfumed notes shall be called "pest"; and she that singeth the song of her soul in fourteen pages shall be named "Anathenia." Verily it requireth six men all very much in love to write as many letters as one damsel who is only a little in love.

6. Thou shalt not ask questions. For curiosity concerning a man's comings and his goings and his stayings away is a hobble on the feet of Love. And when suspicion en[t]ereth at the door, love flyeth out thru the transom.

7. Remember a man's vanity to keep it nourished. For every criticism from the lips of a woman is an arrow aimed at love; but in the matter of swallowing raw compliments and lumps of flattery, every man (and likewise every woman) is an ostrich! And he that departeth full of self-admiration will come back for more.

8. Hold thyself not cheap. For difficulty is the spice of love; and peradventure the colder the wind thou blowest upon a man's heart, the higher thou fannest the flame of his ardor. A *little* indifference is a wonderful thing.

9. Charm a man if thou canst, comfort him if thou wilt but above all amuse him! For a man who is bored will turn away from Venus herself, to play with a little brown elf who can make him laugh![1] And she who playeth all his game with skill shall be an easy winner in the Game of Hearts.

10. Whatsoever thou doest, forget not thy femininity. For what profiteth it a damsel though she learneth to box, and driveth her own car and carryth her own golf clubs, if she hath forgotten how to blush, to cling and to coquette a little? Yea, be not too independent!

Noses

"All of which goes to prove that there are noses and noses," Mark Antony remarked to Cleopatra one day when it was 99 in the shade, "and even you, perfect one, have a nose." And Mark was right, there are several noses extant, and Cleopatra did have one, too.

In fact, from our first ancestor to the present, the nasal lump has persisted on the human face. The Merchants and Miners' Union declare there are enough noses in existence to supply the present population, in fact they claim that the present supply is inexhaustible. How many men, if they told the truth, would admit they had arrived at success by merely following their noses.

The uses of the noses are as varied as their looks. They are, (1) To separate the eyes; (2) to keep the lips from running up to the forehead; (3) to wear powder—yea Coty's. Rigaud, Hudnut's, ye do well to tremble before the mighty beak; were it not for glistening proboscis, thy traffic ne'er had been[1]—(4) to whiff and locate one's food; (5) to administer the snub; and no snub is so snubbish as the snub administered by the proper organ of snubbing. I observe a great stir among the ladies of l'haut monde for if there were no noses, there would be no snubs and with no snubs there would be no society.

The kinds of noses: (1) Roman nose; like all Gaul is divided into three parts—the start, the bend and the drop; it is a conquering nose, it hastens around the world accompanied by a sword. (2) Anglo-Saxon nose, brother to the Roman nose, is usually wedged in between blue eyes: a snoopy, b[o]ssy nose, a conquest seeking snout, loves to make laws for other people to obey. (3) Greek nose; straight and classic leads directly to the divorce court. Helen of Troy—"The face that launched a thousand ships" (?) [sic]. Nix, the nose. (4) Now pause we before the nasal appendage of Abraham, Isaac and Jacob. This is a refulgent nose, a bounteous nose that "droppeth like the dew from heaven" onto the lips beneath and leads straight to Wall Street. (5) At last we stand before the nose of Ethiopia. No lucre loving beak is this, no conquest seeking snout; a nose that squats calmly upon its face, singing and dreaming and dreaming and singing ad infinitum.

From all these things it can be seen that noses are necessary. I need not plead with you more to keep them, be kind to your nose, make a pal of it, take it everywhere with you. Let us have a rising vote—"Ah, the nose have it!" Let us adjourn.

How It Feels to Be Colored Me

I am colored but I offer nothing in the way of extenuating circumstances except the fact that I am the only Negro in the United States whose grandfather on the mother's side was *not* an Indian chief.

I remember the very day that I became colored. Up to my thirteenth year I lived in the little Negro town of Eatonville, Florida. It is exclusively a colored town. The only white people I knew passed through the town going to or coming from Orlando. The native whites rode dusty horses, the Northern tourists chugged down the sandy village road in automobiles. The town knew the Southerners and never stopped cane chewing when they passed. But the Northerners were something else again. They were peered at cautiously from behind curtains by the timid. The more venturesome would come out on the porch to watch them go past and got just as much pleasure out of the tourists as the tourists got out of the village.

The front porch might seem a daring place for the rest of the town, but it was a gallery seat to me. My favorite place was atop the gate-post. Proscenium box for a born first-nighter. Not only did I enjoy the show, but I didn't mind the actors knowing that I liked it. I usually spoke to them in passing. I'd wave at them and when they returned my salute, I would say something like this: "Howdy-do-well-I-thank-you-where-you-goin'?" Usually automobile or the horse paused at this, and after a queer exchange of compliments, I would probably "go a piece of the way" with them, as we say in farthest Florida. If one of my family happened to come to the front in time to see me, of course negotiations would be rudely broken off. But even so, it is clear that I was the first "welcome-to-our-state" Floridian, and I hope the Miami Chamber of Commerce will please take notice.

During this period, white people differed from colored to me only in that they rode through town and never lived there. They liked to hear me "speak pieces" and sing and wanted to see me dance the parse-me-la, and gave me generously of their small silver for doing these things, which seemed strange to me for I wanted to do them so much that I needed bribing to stop. Only they didn't know it. The colored people gave no dimes. They deplored any joyful tendencies in me, but I was their Zora nevertheless. I belonged to them, to the nearby hotels, to the county—everybody's Zora.

But changes came in the family when I was thirteen, and I was sent to school in Jacksonville. I left Eatonville, the town of the oleanders, as Zora. When I disembarked from the river-boat at Jacksonville, she was no more. It seemed that I had suffered a sea change. I was not Zora of Orange County any more, I was now a little colored girl. I found it out in certain ways. In my heart as well as in the mirror, I became a fast brown—warranted not to rub nor run.

But I am not tragically colored. There is no great sorrow dammed up in my soul, nor lurking behind my eyes. I do not mind at all. I do not belong to the sobbing school of Negrohood who hold that nature somehow has given them a lowdown dirty deal and whose feelings are all hurt about it. Even in the helter-skelter skirmish that is my life, I have seen that the world is to the strong regardless of a little pigmentation more or less. No, I do not weep at the world—I am too busy sharpening my oyster knife.

Someone is always at my elbow reminding me that I am the grand-daughter of slaves. It fails to register depression with me. Slavery is sixty years in the past. The operation was successful and the patient is doing well, thank you. The terrible struggle that made me an American out of a potential slave said, "On the line!" The Reconstruction said, "Get set!";

and the generation before said, "Go!" I am off to a flying start and I must not halt in the stretch to look behind and weep. Slavery is the price I paid for civilization, and the choice was not with me. It is a bully adventure and worth all that I have paid through my ancestors for it. No one on earth ever had a greater chance for glory. The world to be won and nothing to be lost. It is thrilling to think—to know that for any act of mine, I shall get twice as much praise or twice as much blame. It is quite exciting to hold the center of the national stage, with the spectators not knowing whether to laugh or to weep.

The position of my white neighbor is much more difficult. No brown specter pulls up a chair beside me when I sit down to eat. No dark ghost thrusts its leg against mine in bed. The game of keeping what one has is never so exciting as the game of getting.

I do not always feel colored. Even now I often achieve the unconscious Zora of Eatonville before the Hegira.[1] I feel most colored when I am thrown against a sharp white background.

For instance at Barnard.[2] "Beside the waters of the Hudson" I feel my race. Among the thousand white persons, I am a dark rock surged upon, overswept by a creamy sea. I am surged upon and overswept, but through it all, I remain myself. When covered by the waters, I am; and the ebb but reveals me again.

Sometimes it is the other way around. A white person is set down in our midst, but the contrast is just as sharp for me. For instance, when I sit in the drafty basement that is The New World Cabaret with a white person, my color comes. We enter chatting about any little nothing that we have in common and are seated by the jazz waiters. In the abrupt way that jazz orchestras have, this one plunges into a number. It loses no time in circumlocutions, but gets right down to business. It constricts the thorax and splits the heart with its tempo and narcotic harmonies. This orchestra grows rambunctious, rears on its hind legs and attacks the tonal veil with primitive fury, rending it, clawing it until it breaks through to the jungle beyond. I follow those heathen[s]—follow them exultingly. I dance wildly inside myself; I yell within, I whoop; I shake my assegai above my head, I hurl it true to the mark *yeeeeooww!* I am in the jungle and living in the jungle way. My face is painted red and yellow and my body is painted blue. My pulse is throbbing like a war drum. I want to slaughter something— give pain, give death to what, I do not know. But the piece ends. The men of the orchestra wipe their lips and rest their fingers. I creep back slowly to the veneer we call civilization with the last tone and find the white friend sitting motionless in his seat, smoking calmly.

"Good music they have here," he remarks, drumming the table with his fingertips.

Music! The great blobs of purple and red emotion have not touched him. He has only heard what I felt. He is far away and I see him but dimly across the ocean and the continent that have fallen between us. He is so pale with his whiteness then and I am *so* colored.

At certain times I have no race, I am *me.* When I set my hat at a certain angle and saunter down Seventh Avenue, Harlem City, feeling as snooty as the lions in front of the Forty-Second Street Library, for instance. So far as my feelings are concerned, Peggy Hopkins Joyce on the Boule Mich with her gorgeous raiment, stately carriage, knees knocking together in the most aristocratic manner, has nothing on me.[3] The cosmic Zora emerges. I belong to no race nor time. I am the eternal feminine with its string of beads.

I have no separate feeling about being an American citizen and colored. I am merely a fragment of the Great Soul that surges within the boundaries. My country, right or wrong.

Sometimes, I feel discriminated against, but it does not make me angry. It merely astonishes me. How *can* any deny themselves the pleasure of my company! It's beyond me.

But in the main, I feel like a brown bag of miscellany propped against a wall. Against a wall in company with other bags, white, red and yellow. Pour out the contents, and there is discovered a jumble of small things priceless and worthless. A first-water diamond, an empty spool, bits of broken glass, lengths of string, a key to a door long since crumbled away, a rusty knife-blade, old shoes saved for a road that never was and never will be, a nail bent under the weight of things too heavy for any nail, a dried flower or two, still a little fragrant. In your hand is the brown bag. On the ground before you is the jumble it held—so much like the jumble in the bags, could they be emptied, that all might be dumped in a single heap and the bags refilled without altering the content of any greatly. A bit of colored glass more or less would not matter. Perhaps that is how the Great Stuffer of Bags willed them in the first place—who knows?

Race Cannot Become Great Until It Recognizes Its Talent

Thinking about this sun-burst of Negro art that is going to do so much for America and the world in general, reminds me of a story.

It wasn't told to me but I heard that one time a Negro from Longwood, Florida, was over-persuaded to go up to Jacksonville on a trip. He got off the train and went on down to Bay Street to Main. He got there and stood and looked at all the tall houses and the traffic and his mouth fell open and he said, "Well, if the world is any bigger than this, I don't want to see it."

Which means nothing except it be to point out that human beings change their concepts slowly, and the Negroes of America are not exceptional along those lines.

We hear a great deal these days about culture and in that connection the name of William Shakespeare is certain to be mentioned, so I might just as well drag it in. Along in 1066 William, the Conqueror, fell upon England and conquered it. He was a thorough man. He enslaved the British and put iron collars around their necks—the beginning of a serfdom that lasted many centuries. It wasn't that the iron collars were so enduring but the effect of the mastery that the collars represented. The English mind got the habit of looking up to the Norman conquerors. People slaved and starved in English, but dined and took all pleasures strictly in French. It required nearly five hundred years for an Englishman to regain enough self-respect to consider beauty of any kind except in Norman form and Norman terms.

Being English never came into fashion before the reigns of that doughty Welsh family known as the Tudors.

And, of course, nobody dreamed of writing a line of literature in the dialect of the licked and lowly Englishman. Any thought worth its salt had to be embalmed in French or Latin. French because it was the language of conquerors, it being held that the language of such valiant men with swell biting steel must of necessity be beautiful. And there was corporal as well as intellectual power behind that Latin too.

CHAUCER DIFFERENT

Until Chaucer; he saw the beauty of his own language in spite of the scorn in which it was held and he used it as a mold for his immortal "Canterbury Tales." It is a long time between Chaucer and Shakespeare, but in spite of that, in spite of the enhanced prestige of the nation generally, the memory of the iron collar was still strong enough to cause the majority of persons who affected literature in England to prefer the florid, continental elegance of Ben Jonson to the greater genius of the more natural and universal Shakespeare. He was despised for the very qualities that have made him immortal. Shakespeare the man, who quickly grew through his callow, imitative, period and settled down to his labors of shedding so great a radiance upon English history and folk-lore that he must live as long as civilization. Yes, folk-lore. Sprites, fairies, Puck, Caliban, Twelfth Night celebrations, Mid-Summer's Night observances are just as much a part of English folk-lore and folk ways as Hoodoo practices and Brer Rabbit are a part Aframerican folk ways.[1]

Now we stand in America where the English stood in the days of Chaucer, physically but not spiritually free, unable as yet to turn our eyes from the distorted looking glass that goes with the iron collar, thinking like the man from Longwood—if we can reach the Jacksonville of the white man's degrees and language, that the world can hold nothing more. Provincial and fanatic! Who knows [w]hat fabulous cities of artistic concepts lie within the mind and language of some humble Negro boy or girl who has never heard tell of Ibsen, for instance?

WE LYNCH ORIGINAL THOUGHT

We cry out against the ignorance and barbarities in the South that we say bar our way to the heights. But why go so far afield? Stop, at home and consider. The fiendish murder of Claude Neal in Florida was a monstrous thing.[2] A man was ravished out of life in the most heartless manner. But he is one man. And while we are in the chamber of horrors let us take another look. How about the intellectual lynching we perpetrate upon ourselves? Let any Negro do anything but pat himself on the back for being almost just like a white person or if that cannot be achieved, almost like someone who is almost like someone who almost acts like a white person and the cliche who got nothing out of college but a degree seize the rope and faggot and make for the nearest tree, discouraging original thought in thousands of immature Negro minds. And that is the tragedy. The world's most powerful force is intellect. The only reality is thought.

We are standing on the street corner in Jacksonville free as the air except in mind. We can stand there and gape and be hustled about by the busy nation about its business of making and doing. Flattered by our homage but in no way impressed intellectually. We may copy his street corner ever so faithfully and he will say, "A good pupil. Maybe some day you will be able to copy my big street corner in New York." A good pat on the back for the apt pupil. But remember, pupils never stand on equal footing with the master. A little independent thinking is greater than the most colossal copying.

Fawn as you will. Spend an eternity standing awe-struck. Roll your eyes in ecstasy and ape his every move, but until we have placed something upon his street corner that is our own, we are right back where we were when they filed our iron collar off.

Now Take Noses

Now take the nose for instance. Loyal and kind, from the daybreak of creation down to this present moment, it has stood before man's face and remained his best friend and adviser. Art has taken account of its stand, and science has fought battles over its uses and advantages. Noses have remained in fashion longer than any other ornament that man has ever worn, which is proof enough of its popularity. Where goes humanity there goes the faithful nose.

One school of scientists maintains that man came before his nose and developed it to his uses. They point that it was evolved as an instrument by which man found his food and mate. Later on it was used as an implement with which he tilled his fields and attended to the affairs of his neighbors. Still later it was elevated to the position of supporting the eye-glasses of the human race and uplifting the dignity of man. Women found use for it as a powder-base. They also voted it the most suitable rendezvous for lip-meetings, sub-nosa. Hence it became a most haughty social arbiter.

The opposing school of savants pooh-pooh this whole position. This second school asserts and believes that the nose came first and man just sprouted behind it. The nose is the measure of the man, they claim, and further state that it could not be otherwise because a man is as his nose made him. Hence the counsel of all great scholars "follow your nose." And they cite the profound observation of Socrates "as a man nose so is he."

Proceeding on the theory that the nose ante-dated man these savants classify the noses of men and hence the human race.

The Grecian nose goes straight before its face and what happens? It leads straight into love and romance. It was the nose of Helen that launched a thousand ships and every ship was full of men.

Behold the nasal appendage of Abraham, Isaac, and Jacob that droppeth like the dew of heaven upon the lip beneath and leadeth its other parts to the marts of finance. It is the curve of commerce and the man so called and impelled by his nose must follow.

The Roman nose, like all Gaul, is divided into three parts—the start, the bend, and the drop. It starts forthwith to rule the world, bends sharply to seek its means, and proceeds sharply after that to achieve its ends. This leads to conquest and law.

The nose of Africa sits in the shade of its cheek bones and dreams. It points not upward, not downward, not anywhere. It sits and dreams and dreams.

This pre-man-nose school of scholars is divided on one point. Some maintain that the Mongolian nose is a thing to itself. Others state just as firmly that it is the same as the African nose, but that Africa at a very early date took over the Mongolian nose and elaborated upon it.

From all these findings it is evident that people need noses. So be good to your nose. Make a friend and a pal of it. Pet it and take it with you on your vacation.

The nose have it, and so on.

Lawrence of the River

This is about a man of the cow lands of Florida. The heart of cow Florida is the Kissimmee Prairie, which stretches for more than a hundred miles from just south of Orlando to the upper Everglades. It embraces the head-waters of the St. Johns River and it gives Florida high rank among the beef producers of the Union.

Lawrence Silas, a dark brown, stockily built Negro, is in and of the cow lands. He is important, because his story is a sign and a symbol of the strength of the nation. It helps to explain our history, and makes a promise for the future. Lawrence Silas represents the man who could plan and do, the generations who were willing to undertake the hard job—to accept the challenge of the frontiers. And remember, he had one more frontier to conquer than the majority of men in America. He speaks for free enterprise and personal initiative. That is America.

Considering that Florida is in Dixie, it will sound like poker playing at a prayer meeting when you read that Lawrence Silas, Negro, is one of the important men of the cow country. But that is the word with the bark on it. The cattlemen of the state have a name for him, and it is good.

They talk about him readily, and with admiration. They do not tell you about his thousands of head of cattle, his fifty-odd miles of fence, or his chunky bank account. They like more to tell you about his character and his skill, as if to say that you ought to have sense enough to know that a man like that is just bound to have something to put away.

When I called Silas' attention to this recently, he replied quietly: "Well,

I have never had my word doubted in business. My plan is: Treat everybody right and honest; pay your just and honest debts; and tell the truth. Whenever you find a man that ain't right, why, feed him with a long-handled spoon. Pass and repass when you find out he won't deal right."

"But," I said, "you handle a lot of horses. How about horse trading? Don't you have to lie in a horse deal?"

"Some folks do a powerful lot of it, so they must figure they have to. I don't see it that way. If I think somebody is interested in buying, I tell 'em I want to sell the horse. I give so much for him, and I got to have so much for him. He ain't no good to me, but maybe he is all right for you. In that way, nobody can't say I lied to beat him out of his money. Then we don't lose no friendship over the deal."

"As simple as that?"

"Sure. In the first place, I know horses too well to let anybody sell me any crow bait, and then the people know that I know what I'm doing, so any horse I ever own would be good for somebody. They might not suit for what *I* want."

Silas knows horses, their uses and treatment, from nose to fetlock, and cows from horns to tail brush. The other breeders know that he knows. Therefore, the richest dealers and breeders in the business will come to him for expert advice before buying or selling herds.

He buys and sells for Lykes Brothers, one of the biggest outfits in the world. Young Pat Johnson, whose father was one of the Florida pioneers in the game, comes to him for advice as he would to a father. If Lawrence Silas says it is so, then it is so. So be it in the grand lodge.

By repute, his hands are as skilled as his mind and eyes. He can sit on a gap, which is what the cow people call a corral gate, and let the cowboys run—actually run—a herd of cattle past him. No matter how large the herd, amount of dust or the speed, when the last steer has passed he can tell you exactly how many passed the gap. He never misses one—or adds one.

Then take mammying-up, for instance. That is a cow term for matching up every cow mother with her own calf at calving time.

"With hundreds of cows and calves to mammy-up, how can you always tell which calf belongs to which cow?" you ask him.

He smiles tolerantly. "Oh, there is ways to do that. Don't care how many you got of the same color, there is a difference between em if you know how to look. All I need to do is to pen 'em one day, and the next day I can tell you which is which. Supposing you had twenty red calves to mammy-up. If you look at 'em good, one's color is just a teeny bit different. One is got a different set around the shoulder. The hair is curled in a peculiar way on one front leg, maybe, and so on and so forth.

"Things like that will tell you. You can't depend on the calf. He will mammy-up with any cow. Then, too, I think hard about it, and some of them calves come before me in my sleep. It is something you have to get straight, else you'll get the wrong marking brand on 'em. Some breeders who can't do it themselves will get me to do it for 'em, and then set up on the corral fence and wonder can I do it. I always tell 'em, 'Ten dollars for any mistake I make.'"

And when you consider that thousands of cows are calving at the same time, you feel like the man who got down on his knees to ask God for some groceries. He was asking for everything in barrel lots, and finally asked God to send him a barrel of pepper. He caught himself and said, "Hold on there a minute, God. That's a darned lot of pepper when you stop to think about it."

Silas knows what to do for every kind of sickness a cow can have. Then again, cow people insist that the moving, separating and general handling of cows is a highly technical job. Lawrence Silas, with a cigar forever in his mouth, not only handles his own but other owners get his outfit to handle theirs.

It is something out of this world to [a] cow man to see Silas sit on a fence with his dehorner in his powerful hands and point. He can point five hundred pairs of horns in a day.

Lawrence Silas passes that off as all in a day's work. "I ought to be able to do that. I was born to the cow business. I been hunting cows since I was five years old."

"Cow hunting?"

"Oh, that is what we call it here in Florida. When I was a boy that was just what we done. You see, there's been cows down here in this prairie ever since the Spaniards first landed in America. Some of them they brought over got away, and so the Indians had big herds round here on the St. Johns River long before the white folks moved in on 'em. Today you don't hunt wild cows. We got bred cows now with a lot more beef on 'em than they used to have. These imported Brahma and Guizerat bulls make a lot of difference in the beef. Lykes Brothers brought the first ones over here from Texas. B. F. Lester brought up some, and I distributed 'em round here for him.

"Yeah, I been fooling with cows and riding the swamps ever since I was five years old. I was so little, I had to let down the stirrup in order for me to reach it. After I scrambled up on the horse, then I would pull the stirrup up after me. Used to stay out in the woods for months at a time before I was ten years old. I didn't have to, but I always did hate farming, and so I used to stay off to keep from working round the farm.

"My father was a cowman before me. He come down here from somewhere in Georgia, and settled at Whittier (now Kenansville). First he worked for some white folks, while he bought a cow or two at a time as he got hold of the money. Then he would sell his he and buy a she. Afterwhile he got to the place where he could go into business for himself. He owned several thousand acres of land when he died, and over two thousand head of cattle. He left thirteen head of children and no will. Mamma couldn't hold what he left together. It was all gone in no time. I was eighteen years old when he died.

"Well, then, I had to start out fresh. I went to work for some white folks too. I rode the woods, and butchered for men who had big herds. It wasn't long before I was running crews for some of 'em. I have run crews from eight to ten men, and rode herd on ten to twelve thousand cattle and more.

"I learnt to be a good butcher while I was about it. I butchered whole herds at a time for owners who did business with Swift and Company, Cudahy, and other big meat packers like that. I kept the count and the weights myself. They all trusted me to do that. The men that owned the cattle I butchered didn't know how much they had coming to 'em till I turned over the figures to 'em and the money. Nobody ever found me off a penny.

"The way I got my pay was, the cattlemen give me one cent on the pound and the hides for butchering for 'em. I furnished all the help and the equipment. The hides alone would bring me two to three dollars apiece. After it was all done, the owners would come to me for their money.

"I went into the cattle business just like my father did. Bought a cow or two at a time, sold my hes and put the money into shes. Sarah, my wife, she took in sewing and kept the house going while I put all that I could rake and scrape into cows. Finally we come to have quite a few. Yes, indeed.

"Then too, I made some few friends that really meant me good. Mr. Pat Johnson was a big man in the business, and somehow he liked to put me in the way of making something any time he could. Mr. B. L. Lester and Mr. Earl Bronson did the same. It was sort of lucky for me when the Lykes Brothers took a liking to me. From that day on, our friendship ain't ever changed. They was just plain Florida crackers with nothing but grit and git to start off with, but they run the business into the millions. They got cows in Florida, Texas, Cuba, South America and all over. Now they own steamship lines, lumber and I don't know what all. We was friends when they was poor and we're still friends now when they are rich. None of them Lykes boys ain't changed a bit. Mr. Tom and Mr. Howe Lykes let me bring their Brahma bull up to Kissimmee so I could breed 'em when they knowed I couldn't afford to buy bulls. I got some fine ones now, but then it was different."

"You have a mixed crew, white and colored, and they all seem to be enthusiastic about working for you," he was reminded. "How do you manage to keep your men happy like that when so many bosses are having labor troubles?"

"Well," Silas replied, "I try to deal fair and then a little more than fair with my men. When work is plentiful I pay 'em what is right. When work is slack, if any of them come to me in need, I let 'em have the money. I eat what they eat. I sleep where they sleep. I don't ask a one of them to do nothing I wouldn't do my ownself. And then I don't ever use my power of hiring and firing to beat no man out of his manhood. If one of my hands don't like what I got to say about his work, he can invite me out. I'll go out in the swamp with him and give him satisfaction. The rest of the crew can stand around and see fair fighting done. If he is a good cow hand and still wants to work for me after we settle our differences, it is all right with me. 'Course, no man ain't got no business running no cow crews unless he can take care of himself. There ain't no servants in the cow business. Every cow hand figures himself a king."

(Curly, one of the Silas crew, here whispers out of the corner of his mouth that the boss does all right for himself with his dukes, in spite of his fifty-five years.)

"Is cow hunting dangerous? Sure it is, and then again it ain't. You got to know what you doing. Yeah, cows is just as risky as bulls to fool with. She don't fly hot so often, but when she got fight in her she'll hook you quicker than a bull. And you can't dodge her as easy, because she don't take out after you with her eyes shut like a bull. Sometimes a bull will come out of a bunch like a whirlwind, and if you don't watch yourself, you'll be thunder-struck by lightning. Naturally, a cow horse is trained for things like that. He knows how to shift. But even so, it's nip and tuck at times. More than once I have seen my horse doing all he could to get away from a bull charge, but the bull would be so close behind us till the horse's tail would be laying on the bull's horns. One case in particular, it was a long race before the horse outdistanced that bull.

"Another time it looked mighty like I was going to be riding herd on God's big range. I didn't have no flank girth on my horse. I roped a bull, but my throw was short, and I only got the rope on one front foot. That bull whirled and charged, and I mean charged! My horse—he was a good one—kept away from the bull's horns, but I'm telling you, he had to do some fast turning in a close place. Then I felt my saddle slipping. So I had to put the catch dogs on the bull. Them dogs knowed just what to do. They run in and caught the bull and held him. That gave me [a] chance to get the rope off my saddle.

"A lot of times during branding, I've just barely beat a bull to the fence. One of 'em got so close till he hooked my pants clean off me.

"But I reckon a stampede is the worst thing that can happen in the cow business. Stampedes are funny things. When cows are moving around at night and lowing, there ain't no danger of a stampede. But you take a big herd that's all tired out from a long drive, and get all bedded down and quiet, then most anything will stampede 'em. One cow can get up and sniff around, and the whole bunch is up and off like a tornado. If you get caught in front of the bunch, you just got to be fast enough on your feet or horse to keep in front. If you try to cut across the herd, they will run over you. It's a terrible and cruel thing. I have seen 'em run over anchored horses, and them horses would be stomped out as flat as a piece of paper. All the hair off the horse's hide would be tromped off smack and smooth.

"Never will forget one big stampede I saw down round the Everglades at Fort Basinger. Eleven hundred steers went into a stampede. We men heard it in time and run in every direction. The stampede headed for a big swamp. Wasn't a thing we could do. The next morning we followed the trail down to the edge of that swamp. We knowed that the ground was too soft for 'em to get across. They didn't but you couldn't see no cows at all. All you could see was horns—just a whole lake of horns.

"Yes, I done broke plenty wild horses, and I been throwed aplenty. Don't care how good you can ride and how long you been doing it, there's times when you can't stay on.

"Things done changed a lot in my fifty years on the range. The open range is gone. You don't throw cows no more by the chin-and-horn-hold. Now we rope from a horse. The cows been bred up from them stringy-meat wild cows to heavy blooded ones. The Florida cow business done come to be something. It is a good thing too. It's bound to be a lot of help to the Government in this tight spot. I figure we cow men is something like a good catch dog—sort of holding things until the folks up there in Washington can get another rope on things.

"Natural, the future to me looks something different from the past. I don't expect to keep on staying out on the range thirty and forty days at a time like I do now. A human man can't do it. I'm buying more land all the time and extending my fences. When I can't stay on the range so much, I figure on handling smaller herds, but a better grade of cattle. I'll be a cow man as long as I live and always be buying more cows. I might even die out on the range with a cigar in my mouth. Wouldn't be nothing wrong with that."

Biddy, Biddy bend, my story is end,
Turn loose the rooster; and hold the hen.

My Most Humiliating
Jim Crow Experience

My most humiliating Jim Crow experience came in New York instead of
the South as one would have expected. It was in 1931 when Mrs. R. Osgood
Mason was financing my researches in anthropology.[1] I returned to New
York from the Bahama Islands ill with some disturbances of the digestive
tract.

Godmother (Mrs. Mason liked for me to call her Godmother) became
concerned about my condition and suggested a certain white specialist at
her expense. His office was in Brooklyn.

Mr. Paul Chapin called up and made the appointment for me.[2] The
doctor told the wealthy and prominent Paul Chapin that I would get the
best of care.

So two days later I journeyed to Brooklyn to submit myself to the care
of the great specialist.

His reception room was more than swanky, with a magnificent ham-
mered copper door and other decor on the same plane as the door.

But his receptionist was obviously embarrassed when I showed up. I
mentioned the appointment and got inside the door. She went into the private office and stayed a few minutes, then the doctor appeared in the door
all in white, looking very important, and also very unhappy from behind
his rotund stomach.

He did not approach me at all, but told one of his nurses to take me into a private examination room.

The room was private all right, but I would not rate it highly as an examination room. Under any other circumstances, I would have sworn it was a closet where the soiled towels and uniforms were tossed until called for by the laundry. But I will say this for it, there was a chair in there wedged in between the wall and the pile of soiled linen.

The nurse took me in there, closed the door quickly and disappeared. The doctor came in immediately and began in a desultory manner to ask me about symptoms. It was evident he meant to get me off the premises as quickly as possible. Being the sort of objective person I am, I did not get up and sweep out angrily as I was first disposed to do. I stayed to see just what would happen, and further to torture him more. He went through some motions, stuck a tube down my throat to extract some bile from my gall bladder, wrote a prescription and asked for twenty dollars as a fee.

I got up, set my hat at a reckless angle and walked out, telling him that I would send him a check, which I never did. I went away feeling the pathos of Anglo-Saxon civilization.

And I still mean pathos, for I know that anything with such a false foundation cannot last. Whom the gods would destroy, they first make mad.

The Lost Keys of Glory

Lost from sight somewhere in the world, and probably in the United States, is a bunch of three keys. These three small keys were made in Heaven, and once hung from a silver nail driven into the end of God's mantelpiece. The keys are on an exquisite ring of gold, set with rare and very precious stones. The keys themselves are made of some material so rare, that it has never been seen anywhere outside of Heaven. Only God Himself knows the composition of the metal of which these keys were made, and He only knows where it can be found. It is said to be found in a vein of the rocks in the highest hill in Heaven.

Once these three rare and precious keys were known and used on earth. They were sacred to the females among humans, and were in constant use.

How these keys came down from Heaven and into the hands of Woman, is told in an old, old tale known to Negroes. Negroes have always been neighborly with God, and perhaps got the story of the Three Precious Keys directly from Old Maker Himself in a leisurely chat with Him, and this is the way it goes.[1]

In the beginning, The Man and The Woman were made equal. Equal in every way, like muscle and brain strength, and authority over each other. They were equal in property, and lived in a house together.

But being equal in brains did not mean that they had the same thoughts all the time, so Man and Woman used to disagree and have words over various things. When the arguments got too hot, they would stop changing words, and get to changing licks. These fights never ended and settled anything, because neither one could whip the other one, so a great deal of the time, The Man and The Woman lived at variance, and neither one would give in. This went on for a very long time.

Then one day after a big tussle, the Man got a notion how to settle the thing and be done with it. He knocked off from his work in the field, and went on up the stairway to Heaven. When he got up to the Throne, he took off his hat and made his manners, and spoke nice and polite to God.

"What are you doing up around the Throne so soon this morning, Man?" God asked the Man and waited for him to talk.

"God, I hate to be so much trouble, but that Woman! It's no getting along with her no matter what I do and say. She just won't gee nor haw, and that keeps us fighting all the time. There is just no peace around the house. No sooner do I get in from the field, than she is ready to jump me. I can't whip her, Lord, as You know, and she can't whip me. So nothing ever gets settled."

"What do you want me to do about it, Man?"

"I want You to give me more strength than Woman. I want it so I can whip her and make her mind. I want more strength than she has in every way, so somebody can be the boss and have some peace around the house sometime. I'm sick and tired of battling it out with that Woman."

God thought things over for awhile, and then He said to The Man, "Peace is a wonderful thing. I gave you and The Woman the same strength, because I hate bulldozing an[d] oppression. I don't want to see nobody run over, but so that peace can come on earth, I will grant you more strength than Woman in every way, if you promise me just one thing."

"What is it, Lord?" Man asked and wondered.

"That when you are stronger than Woman in every way, that you will not take advantage of Woman, and mistreat her in any form nor fashion. You must be good to her, and kind, and take care of her."

"I promise, God, if You just grant me strength enough to war that Woman down so I can manage her, and have some peace around home. No two can both be boss in any house."

"In the creation of Love between you two, and for the sake of peace in the home, I give you greater strength than Woman. You go and look after her, and mix your strength with her weakness, and from now on [y]ou love her like she was a part of yourself."

The Man took the extra strength and thanked God for it, and came on back down and went to work. He was very happy, but he did not go by the house and tell Woman what had taken place. Time enough to let her know when something went wrong next time. Besides, just as God had said, he began to feel sort of sweet and foolish when he even thought about Woman now.

But it was not very long before Man gave Woman a command to do something, and she flew hot and flew right on him for a fight. The Man whipped her good, and he whipped her quickly. The Woman couldn't figure out how it happened, so after she got her breath, she flew back on him again, and the Man whipped her again. Woman couldn't rest until she had tried him one more time, and she got whipped again, and got told that from now on, Man was Boss around the house, and she was going to do as she was told.

The Woman did not know how this thing had come about, but she was convinced that Man was too strong for her to whip these late days. She was as mad as fire about the change, and made up her mind that she was going to have a talk with God, and find out if Man was telling the truth when he said that God had given him greater strength than she had. So bright and soon the very next morning, Woman flung off her apron and went running up the golden stairs to God.

"Good morning, God," Woman said, and she didn't take time to make too many manners because she was too mad and too outdone for that.

"Good Morning, Woman. What are you doing up around the Throne so soon in the morning?"

"God, that Man has done beat me. Beat me three times running, and it seems like I can't do a thing with him nor about his whipping me. He acts just like he's got more strength than I have."

"He has, Woman. I gave it to him Myself."

"Oh, but there has got to be some changes made. I can't let Man be getting the best of me like that."

"What kind of changes, Woman?"

"Change things back just like they were, or better still, give me more strength than Man has."

"No, Woman. I can't do that. I have given Man more strength than you. No matter how much strength I might give you, Man would still be the stronger. I granted him *more* strength than you, and what I give, I never take back. It must stand that way forever."

Woman was so angry at that, that she never waited to ask God anything else, and she didn't stop to listen. Woman pitched off down the stairs just as fast as she could go, talking out loud to herself and quarreling with God's arrangements. She was so mad that she almost ran over the Devil as she hopped off the bottom step to Heaven.

The Devil and the Woman had always been the best of friends. And being as mad as she was and feeling all outdone, she was glad to have some-body to talk with, and tell her troubles to.

The Devil listened to what the Woman had to say, and then he told her what to do.

"If God said that Man would have more strength than you, it will have to be that way. He never takes back what He gives away. But you don't have to worry about that. Straighten up your face and go fix yourself up, and go on back up to Heaven. When you get there, you rap at the door, and go in nice and polite and act just like you are not angry about a thing. When He asks you what you came for, you ask Him for that little bunch of keys hanging on a silver nail at the end of God's mantelpiece. He will give them to you, then you fetch them on back down here to me, and I will show you what to do with them."

Woman did like the Devil told her, and went on back to God and asked Him for the keys hanging on that silver nail at the end of the mantelpiece, and God smiled and gave the Woman the bunch of keys on the jewelled ring made of gold. She took them in her hand and hurried on back to where the Devil said that he would be waiting for her, and showed him that she had them.

"See these three keys, Woman," the Devil explained. "As long as you have these, you don't need to worry about how much strength the Man has got. You can always make him use his strength to your advantage. Now, this first key here, is the key to the cradle. This next one, is the key to the bedroom, and this other one is the key to the kitchen. All very powerful keys. Now all you need to do is to go on home and lock up all these three places and put the keys in your pocket. Don't go around boasting and bragging about your keys. You don't need to even make mention of them to Man. He can't do without any of these things, so by and by, he will come to you to make some arrangements and he will have to do what you want done. Just go lock up everything, and go set on the front porch and rock."

Towards evening, the Man came in from the field whistling and singing to himself. He paid no attention to the Woman sitting on the front porch and rocking and singing herself a song. The Man headed straight for the kitchen as he had always done. He tried every way that he knew how to get in, but he couldn't, and afterwhile, he went out on the front porch and spoke to the Woman.

"Funny thing, the kitchen is locked up tight and I can't get in."

"I know it, Man. I locked it up myself."

After a long while, the Man was outdone, and had to make some arrangements with the Woman to get inside the kitchen. It was the same way with the bedroom and the cradle. Man found out that he was using his strength for the Woman's benefit nearly all the time.

The new moon came and went several times, and then Man came to the Woman to make a new arrangement. He would willingly give the Woman part of his strength, if she would let him handle the keys just as much as she did.

The Woman was very glad to hear that, and was about to come to an agreement on the thing when the Devil popped up, and shook his head at the Woman.

"Tell him 'No.' You have the best of the bargain. Tell him to keep his strength and you keep your keys. Hold on to them with all your might and main, and you will always have the advantage of Man."

So the Woman kept the keys, and the Man could not even tell just how nor where they were hidden. He got so tired of the that [sic] weakened-down Woman outdoing him all the time, that he made up his mind to go have a talk with Old Maker.

"God," the Man said, "I want to ask a favor of You."

"You are on good and pleading terms with mercy, Man. Go ahead and ask for what you want."

"God, I come to ask You to take those three keys away from that Woman. She's more trouble to me now than she used to be when she was strong. Take back Your keys, so I can have peace and agreement around the house without always having to do what she wants done, in order for me to get things that I want to satisfy myself."

"The keys belong to the Woman, Man. What I give, I give."

"Then please give me some keys just like those the Woman has so I can be independent of her, Old Maker."

"There are no more keys like those three, Man. I never made but one set, and no more will ever be created like that so long as the world shall stand. The keys are in the Woman's hands."

"I was better off the way things were in the beginning," Man muttered in complaint. "I wish now that I had let well enough do me and went along."

"I know, Man, but you asked to have changes made."

"But the way things are now, with all my strength, I'm helpless in her hands. Why, I have no way of knowing whether I'm the father of a child that the Woman bears or not. How can I know that, God?"

"Ask the Woman, Man. That information is something else that goes along with the keys."

Old Maker turned His head, and so Man knew that God was through talking for the day. Man came walking in a slow way back down the golden stairs and went on back to work again.

Man found that he was not unhappy after a while. He got used to the Woman and her keys. Sometimes he would fly hot at the way that Woman generalled things to make up for her weakness, but on the average, he enjoyed showing off his strength before the weakling thing that lived with him in his house. She did all kinds of little things for his delight, and he forgot that he was chained. He came to love the Woman as she was, and was overjoyed with the children that they had. So centuries went by that way. The weakness of the Woman came to be a kind of glory to the Man, and he built a little altar to her and made gifts of ceremony to her weakness and her ways, since it had caused her to do so many things to please his fancy and to humor him that she might continue to use his strength to live. It made Man feel like a divinity to himself, since Woman perfumed herself and smiled and told him how much he favored God.

But the Devil and the Woman had a falling out. They got so that they were not kissing-friends any more. The Woman lost the best friend that she had without knowing what she was doing when she told Old Devil not to foot the premises any more. She got so she wanted equal strength with Man again, and laid aside her keys. She put them down somewhere, and did not use them for such a long time, that after awhile they got mis-placed, and Woman forgot what she had done with them. God did not change His mind and give her equal strength, but Woman didn't notice. She got ahold of some of the tools of Man, and felt that they would give her the same strength that she used to have.

So Woman left her house and her ways that the Devil had taught her, and went on out in the field where Man was working, and challenged Man to a tussle to see which one was boss. She grabbed up every tool that Man had in his tool-box and told him that she would match him row for row in everything that he did.

In 1947, women have entered every profession and skill that men follow for a living. There is no doubt that women are taking themselves quite seriously as the equal of men in all of these pursuits. It is obvious, however, that women are not adequate to the struggle. Like the brief time when Man first got his superior strength, the women have not realized that it is so. In their new freedom, they are still to be convinced that God has given men the upper hand in certain ways.

In proof of this, while women have had a degree of success in all the professions and the Arts, it is nevertheless true that except for the fields of literature and the theatre, the very best women in any particular field have never been able to come up to a standard of excellency equal to the best men in that same profession. In Medicine, Law, Business, Journalism, Statecraft, Chemistry, Physics, Pedagogy, Sculpture, Painting, Economics, and even Dress-Design and Cooking, the men are the top of the field. As has been observed, men even invent the very tools with which women do their housework.

These things are true, but in no way a reflection upon the female kind. Dirt is defined as matter out of place. A woman is a woman no matter where she might be and do. That is an excellent thing in this big, round, world. Our very frailty becomes us. We add that color and the fill-in to Man's bold outline of life. We are the pigments on the palette after men have sketched in the figure. We are that fine indefinable something that goes along with the keys.

And let no woman deceive herself that she ever impresses men by her intellect. We get along in the professions or in public life because some man in power is impressed by our femininity, and for what that compels in him, lets us have our way to an extent. Men pay us the compliment of saying that we are intelligent when we are smart enough to pay him the compliment of his strength. If we accept our success as a grant from him, in other words. The moment that a woman, however capable, attempts to muscle him down by sheer mental strength, we find an implacable opponent in that man. And men can find it in themselves to be much more cruel to a woman opponent than he ever does to another man. It is possible that he decides that it is time to whip her good and put her in her place. We find to our dismay that even groups of men will gang up on us to war us down and chase us back to our knitting. But let us ask that same thing as a favor, and those same men will outdo themselves to help us on our way.

To prove that men always see us as women and not as equals, all we need do is to look at the women whom men will let succeed in public life or a profession. Clare Boothe Luce is a very intelligent and capable woman, but she owes a great deal more to her beauty and her female charm than she does to her brains.[2] No homely, awkward female would ever have gotten as far as she has in this world. The brains are there, but if they had been in an ugly head, the men would have taken pleasure in beating them out and forgetting all about her the next moment. Where men see no possibility of mating in a woman, she holds no interest for them. They may not recognize this consciously, but they work along those lines just the same. The keys compel. It is safe to say that no man has ever admitted any woman as his mental equal since God said, "Let's make Man." Men just say that when the female in question interests him enough.

Nor is it a valid contention that men have control of things and will not let us do. It is only too true that men are in control of the affairs of the nation and the world, and that they are not willing for us to take their places. But the conclusion is inescapable that we have not the same strength as men, or they could not have things their own way. It is begging the question to clamor that men will not let us win. The questions must be answered, first, if we have equal strength, how did those men get control of things in the first place? Second, how have they managed to beat off the challenge of the last century of female freedom? It is true that female students in college make just as high grades as male students do. But let us not forget that for the greater part those text-books have been written by men, and that once out of college, we seem not to be able to apply what we have learned with the same intensity and drive as the men do, and thus fall into second place in the same fields. Nor can we maintain with any degree of conviction that the burden of motherhood is what prevents our measuring arms with the best men in the professions, because it is a notorious fact that most career women do not become mothers, even if they marry.

It is possible that that very thing works against us in the race with men. Nature denied is cruel. The high rate of neurasthenics among modern women may mean something. We have not that same detachment in our emotions, and pretending that we do might be costing us more than we care to admit. Let us ignore the extremely rare woman who has the male detachment, for she is so rare as to be negligible. That known male whom we desire, or the possible one whom we seek is with us even in the midst of the most serious occupation, consciously or sub-consciously. We want some other woman's grown-up boy to pet and mother over, make whatever compensation through a career which we will or may, and suppressed desire is always at our elbow, hunching and nudging us when we should be concentrating on business if we are to be a serious challenge to men. We refuse to admit it in the glory of our new freedom, and the nation is strewn with female neurotics from end to end.

Perhaps in a few more centuries of female freedom, this natural handicap will be overcome, but this will only be true if some way is found to alter biological fact. In the meanwhile, thousands of inter-sexual tragedies clutter up the landscape.

Take the case of the beautiful and utterly feminine girl who decided that it would be an exciting thing to become a woman dentist. Before her freshman year in college was over, she had two opportunities to marry well. She refused both for the stated reason that she did not wish to abandon her career. Men wanted to make a pampered plaything of her, but she wanted it understood that she was going ahead, and prove the case for the women. She seemed not to notice the fact that she got along as well as she did in college because the men in her class were ever ready to help her with her laboratory assignments, and the professors were not indifferent to her beautiful young body and face. They were happy to give her of their strength. She misunderstood and considered herself a whale of a Dentist.

The pay-off came when she opened her office in New York City. First, she found that many men admired her as a woman, but few trusted her as a Dentist. The women stayed away from her office in droves. Then her femininity kept her on the go all the time, and she had little time to devote to her practice, and even less energy for actual work. Still, she refused to marry unless she could find a man who was willing to acknowledge her intellectual equality and take her as a sort of brother-brain. After six years of this sort of thing, she came to realize that she was a failure in her profession, and a bit shop-worn socially. This came about because she had mistakenly considered that as a professional woman, she could adopt the same attitude towards sex-relations as a man. The desirable men who had once held it an honor, and certainly a blessed privilege, to win her as a wife, were either safely married, or now treated her as she had wanted to be treated, that is like a pal. Nothing like marriage in their minds, where she was concerned any more. Dr. Emily F. eventually captured a shipping-clerk in her mid-thirties, shut up her unprofitable office and settled down to be a very humble housewife. A year later, a very frustrated woman, she had a nervous break-down, from which she has never recovered. A divorce followed, and single again, she became a notorious sensation-seeker and publicity fiend. She has made two more marriages, both beneath her intellectually and after short periods, both ended in divorce. She still delights in parading her degree of D.D.S., but otherwise has been working as a Social Worker for years. Always talking about the now-distinguished men who she could have married.

Then there was Winifred B. Another beautiful and feminine creature who came up from the South to study at Columbia U. She met a young and ambitious medical Doctor, and soon they were married. She was introduced to his circle of friends. Among them was a young woman, not nearly so seductive, who was a journalist with a pleasing personality that got her a great deal of admiration from the Doctor and other men. They admired her as a woman and as a good and skillful conversationalist, but the pretty wife of the Doctor did not understand this. She thought that it was the career that made the other so popular. She decided that she herself must have a career. Nothing that the Doctor, who was very much in love with her, could do or say had any weight with her. His income was still small, but out of that he must pay her tuition for two years at Columbia while she took up Sociology, and in addition, he had to pay a housekeeper to look after the home while she attended school. On graduation, she must get her a job and work, which she did. It brought her fifty dollars a week, but the satisfaction which she got out of being a career woman, and rattling off the technical terms of her profession with others in her field, made up for the rest.

This went on for nine years over constant friction in the home. She complained that the Doctor was too old-fashioned, and wanted to crush her personality by confining her to his home. In the meanwhile, and in her absence, the Doctor went in for advanced studies himself. After six years of graduate work, he bloomed forth as a brilliant surgeon, and was offered a place on the staff of a famous hospital in the city of New York. Madam Winifred suddenly came to realize that she was the wife of a very important man. She also came to understand at about the same time that the Doctor was no longer interested in her. He was indifferent as to whether she was at home or at the office where she held down a minor desk. All he wanted of her was a Bill of Separation from her, and not to see her around the place.

What troubled her was that she was a little heavy now through her hips, and the lower part of her face. If the Doctor succeeded in getting rid of her, as he boldly stated that he wished to do, where would she find another man of the eminence of the Doctor? How could she bear to move out of the big, comfortable house where they now lived, and which he owned, to less spacious quarters, which the terms of the Separation demanded? The Doctor won all down the line, and she was forced to accept a monthly stipend, and move into a small apartment. Now she was compelled to hold onto that job which she hated, but which had once been so important to her.

And there was no satisfaction for her anywhere she looked. The journalist had long since married a friend and colleague of the Doctor, and practised her trade only when it did not interfere with her domestic arrangements. *Her* husband got steak and potatoes and apple pie on demand, his slippers brought to him wherever he might choose to flop down and wallow in the house, and his cigar ashes brought no protest from the erstwhile woman of affairs. She had disappointed Winifred in turning out to be her husband's woman as well as his wife, and betrayed the feminist movement in other ways. Winifred came to feel herself led astray and betrayed, and took to sleeping-pills every night so as not to remember what she had once held in the palm of her hand. After a year of bitter reflection, Winifred became a rabid Communist, and took to speaking in Union Square.

Multiply these two cases by thousands, and you have a picture of the numbers of women who have followed the light that failed. Stirred by the example of some more or less successful woman in the public eye, these women miss the point and tear out after the free life of the males. They see in Dorothy Thompson for instance some female figure detached in space from the ordinary human emotions, when in fact, Miss Thompson is dependant on a man for her happiness and recognizes the fact only too well.[3]

There is another aspect of this mirage that is threatening to happiness for both men and women. That is the illusion that women, career women, can enjoy the same sexual freedom as men with the same impunity. It is not so, but let us assume that the world has advanced to the point where this is accepted. Is it not possible that these women contrive a great disappointment for themselves by such a habit? A blasé attitude brought to a new marriage bed is bound to occasion a certain disappointment in the woman. Where is that springtime ecstasy of the intimacies with the one-possible man for her? The adhesive quality that makes of the mate something of a king, if not something only a little short of a divinity and puts up a barrier against divorce is lacking, and the mistaken girl has been cheated out of the greatest emotion of her life. Then too such casual pre-marital affairs kill off the mystery that men used to feel, and make marriage less binding on them too. He feels little of the need to guard and protect. He does not have that instinctive feeling that he cannot turn back, even under the strain of temporary unhappiness that comes in all marriages at times. Oh, she will get along all right. She knows how to take care of herself, he concludes, and lets the marriage slide. What looked like freedom turns out to be a snare for the girl's unhappy feet. More young women, after a brief period of marriage, join the caravan of unattached and unsettled women in the land.

No one can deny that these conditions are. No female careerist can avoid looking at the picture from time to time. And the inevitable question arises inside her, how much is a career worth to a woman anyway? Are not the unknown women, bossing the man of her choice really happier than the career-woman, however famous outside her natural sphere?

There are those three keys of mystery lost to sight somewhere in this modern shambles. Like the legend of the Holy Grail that hid itself from the sight of men after the world became too wicked, perhaps the three keys of the mysterious metal mined only in the veins of the highest hill in Heaven, on the jewelled ring that is made of finest gold, and that once hung on a silver nail at the end of God's mantelpiece, have vanished from the sight and hold of women because they have become too mannish.[4]

And like Sir Percival, and Sir Galahad, of King Arthur's famous court, some modern maidens, wary of the hurly-burly of the times, shall yet clothe themselves in white samite, and after a period of meditation, shall go forth in search of this female and modern equivalent of the Holy Grail, the three keys of feminine mystery, which if found, shall make Woman again the ruler of the earth.[5] The Devil will be re-won as the friend of Woman and from one thing to another, everybody, including God, might come to be happy again.

The South Was Had

The South was had. There is no doubt about it. The matter of segregation and the Eisenhower administration of the South was had.[1] Dixie was sold the Brooklyn Bridge. The only question before the Court is who made the sale? Who did it? Did what? murmurs that out-of-the-corner-of-the-mouth hypnotic phrase, "The South must not forget that Eisenhower was born in Texas," the inference being that he would hold the line on segregation with all the pugnacity and determination of a Brahma bull. Southerners quoted it with a gloating glint in their eyes. How could they lose with the tools they had. Ike was born in Texas.

There are three prime suspects: First, Ike himself, somebody in Ike's machine and the wishful thinking South itself. Ike has an alibi. Eisenhower has the distinction of being the only man who ever won both the nomination and then the high office while being hidden up under a particular bed. This had a two-way stretch. It prevented the highly touted candidate from exhibiting his ignorance of political know-how, and at the same time, prevented the public from getting to Ike so that it could be truthfully, more-or-less, said that Ike knew nothing of the raw deals necessary to edge him past Robert A. Taft and into the nomination. So far as to whom, where and on what occasion Ike muttered the promissory sentence that so bewitched the South, there is no record.

But even if Eisenhower had shouted the reminder where he was born from the house tops, his record in the matter is such that no one in their sober mind should have been persuaded by his words, for:

1. Ike had no choice in where he was born: for his father, having failed in business in Kansas, merely migrated to Texas temporarily to earn a living for his family, and his stay was very short. Ike indeed was born there, but the family stay was so brief that he was still an in arms baby when the family returned to Kansas—the Kansas of "bloody" conflict over the question of Negro slavery, of the raids of John Brown and the like. There the boy became first a toddling Kansas-baby, then a sturdy boy and finally, a man. Never did he show any inclination to return to "The Lone Star State," and did not do so until he was a commissioned officer in the U.S. Army and a married man, when sent there on an army post.

2. While in command of American forces in Europe during World War II, he desegregated his army of his own accord.

3. When he did decide to buy a permanent home, he did not even glance at Texas, but bought a farm overlooking the historical battlefield at Gettysburg, Pa., where the confederate forces had suffered their most crushing defeat and settled down to breed, of all things, black cows—registered black angus cows.

The second suspect is somehow political genius in the Dewey machine, independent of and unknown to Eisenhower who spread this come-over pledge over Dixie. It was very clever in that it said, in effect, pay no attention to what may be said to get the vote of the Northern liberals. A Texan will never betray the South, even if nominated by the Republican Party. You know how Texans stand on segregation. Well, Ike is no different. Naturally this sounded good in Southern ears, and feeling betrayed by their own party under Truman's determined liberalism, they bought into it.

The third suspect is the South itself. Feeling let down by the liberal wing of the Democratic Party led by Truman, the idea of a man born in Texas was blown up into a grand illusion. This was brought about by some of the vocabulary boys like John Temple Graves II who urged the South to break the oath taken during the Reconstruction to one of NEVER NEVER vote the Republican ticket.[2] And why not? They had already been betrayed by their own party. It was desolate and needed a dream, and it is well known that there is nothing that people know that is one-half so precious as what they want to believe.

So from the meager incident of the birth of Eisenhower in Texas, the South dreamed with such violence that it committed what is equal to killing the albatross. It fell upon the political camel and beat it up, in spite of the proverb from the Near East to "Never strike a camel for it will certainly get even sooner or later."

The beating of the camel took the form of thousands of Southern Democrats invading the Republicans' primaries to secure the nomination for Texas-born Ike over Senator Robert A. Taft.[3] Somebody with evil intent evidently held of the constitution-worshipping Ohio Senator as a boogerboo to the South because he had secured a vote of censure for Senator Bilbo of Mississippi for stating on the floor of the Senate his inclinations and intentions of depriving not only the Negroes of his own state, but of the entire nation of their civil rights and shipping them all off to Africa if possible.[4]

Now, not only his colleagues in the Senate, but everybody of importance in public life was aware of Taft's reverence for the Senate and understood that Taft was spanking Bilbo for saying what he did over the floor of the Senate. There was no intent to pursue Bilbo into his own bailiwick and play into the conduct of radical affairs in Mississippi.

There was no provision in the Constitution for such actions, and consequently, Taft would not even have contemplated it. He made himself clear on this point when a group of Negroes inquired of him at Durham, N.C., whether in case he was elected President, would he force the integration of schools in the South. Taft replied frankly that he would not attempt any such thing because there was no provision for such action in the Constitution. The President, he pointed out, was an executive, there to carry out the laws made and provided by the founding fathers or amendments by the legislators, not to make laws of his own. Yet, somebody scared the South into an acre of fits by picturing Taft as a rabid abolitionist and stampeded them towards Texas-born Ike. As a result, they rushed illegally into the Republican primaries in Texas, Louisiana and Georgia to the extent that the will of the actual Republicans was trampled down, and the political maverick, Ike, was nominated. These three states decided the issue. If their illegal delegations had been ignored, Taft would have won.

After the winning ballot and while the Dewey machine was putting over Eisenhower and whooping it up in the aisles, came the tones of the golden voice of Senator Dirksen, commenting bitterly that there was a whole filing cabinet of evidence of fraud performed by the Eisenhower supporters being somehow ignored by the Republicans.[5]

For a long time, this evil appeared to prosper, but now the South screams of the violations of the Constitution by the Eisenhower Republicans which includes the Supreme Court and the Radical or Modern Republicans. The struck camel fights back. Evil never prospers for always. The South is paying for its political sins.

Until 1954 every few blocks one found stenciled on sidewalks or buildings "We Like Ike," but now these expressions are few and hard to find. Southerners no longer wish to remind themselves that Ike was from Dixie, and in a recent column, John Temple Graves inquires plaintively and rhetorically, "Will we ever call him Ike again in the South?"

Take for Instance
Spessard Holland*

The customers in the gallery of the National Theater—those who do not live in the South, as they observe the drama of desegregation now playing, perhaps see the actors on the stage who are cast in the role of segregationists, as purely that and nothing more. Audiences always do—not as performances bringing to life characters created by a drama test but as creators and themselves—which is understandable, but naïve oversimplification to say the least. He is speaking the lines written by the voters of his constituency and his campaign, and though his own sentiments may be identical with that of his voters, not necessarily so. He is out to win an election, to voice views in opposition to the over majority of the constituency would be nothing short of folly—to spend such [an] enormous amount of money, time and effort to win an office, then to do and say what will be certain to offend and lose votes would be the height of foolishness goes without argument.

But in any case, it is not necessarily true, that the candidate has no other sides to himself. Perhaps the claimant of the policeman from the Gilbert and Sullivan light opera *The Pirates of Penzance* says it all in "A Policeman's Lot Is Not A Happy One"—Like the enterprising burglar when not a-

---

*This essay exists only in Hurston's handwriting. The bracketed passages have been burned away.

burgling, the southern politician when not campaigning might like to hear the little brook a-gurgling, and listen to the merry village chimes[.][1] He may be addicted to good music and the finest in literature of all the ages, a fond family man and a lover of justice.

Now take for instance Senator Spessard L. Holland of Florida, who has just won an overwhelming victory over the so-called Liberal, Claude Pepper, by presenting the case of the conservative or segregationist element of Florida, Holland plowed Pepper under by an impressive 86,000 majority which points up which actors follow the script and who got out there and ad libbed.[2]

It would be very easy and plausible-sounding to flip the superior erudition and liberalism of Harvard rejected as too advanced and the acceptance of [Emory]—which would be hasty generalization and utter poppy-cock, for though [Emory] in Atlanta, Georgia, is no Ivy Leaguer among our educational institutions, it is certainly [adequately] equipped and staffed, and has turned out numerous very able men.[3] In addition it is thoroughly accepted that no college can cram a student's head and turn him out a walking encyclopedia of facts. All that a college attempts to do is to teach methods and sources of acquiring knowledge and the graduate goes on from there.

And Spessard L. Holland, like his equally famous class-mate, Senator Russel of Georgia, has really gone on from there.[4] The walls of his library are stacked from floor to ceiling with good books and he has read them, too. Just challenge him and you'll find out how full he is of information on various matters, and how promptly he will go to the wall and pull out a volume of source material on the subject.

As to the subject of the Negro in America he is loaded to the gills. He has perhaps every book ever published on the subject—some are very rare pre–Civil War publications, even. Therefore his stand on segregation is certainly not the result of lack of information. It must be sincere conviction that close mingling of the two ethnological groups can do no good.

Yet Spessard L. Holland is no vituperous despiser of the brother in black, no taking the stand that Negroes are deprived by nature of the power of intellectual advancement. To the contrary, as he thoroughly demonstrated when he was Governor of Florida. His extraordinary efforts [on] behalf of Negro higher education entitles him to be looked upon as the Father of Negro education in Florida. He allocated money to provide facilities and to raise the standard of instructors, and generally gave the push that raised FAMC from a very mediocre and doleful institution to the scholastic rank that it holds today, and the boost went straight across the board to affect even the most obscure elementary school in the piney woods.[5] Teachers with no more than an eighth grade education and the like, who had infested the state system for generations, found themselves being dumped out to be replaced by college-trained men and women. And he sounded the death-knell in Florida to that scourge of public schools in the South the "Professor." These "Fessahs" were principals of schools, and many were as innocent of education as a Georgia mule. They had gotten and were keeping their appointments through the favor of some powerful white individuals who cared little or nothing about the effect upon Negro education, but whether their Negro friend was taken care of. Nor did these "Fessahs" care about anything more than first being called Fessah by the community which was being so defrauded by them, and second getting their salary checks.

Under the Governor Holland drive in Negro education the ground was lined with fallen Fessahs like ripe guavas under a guava tree in season. Young men and women from Columbia, University of Chicago, Northwestern, Howard University, Fisk and Morehouse began to appear as their successors, and an era in Negro education had ended and another begun.

Holland's passion for education called forth screaming protests from certain members of his legislature. The then state Senator from Longwood shouted that Holland had made the state institutions of higher learning sacred cows. "All of the state funds that can't be squandered at Gainesville are being wasted at Tallahassee (FAMC)." Then he asked and was told the salary paid to the late J. R. E. Lee, President of FAMC, and screamed, "Good God, no Nigger on earth is worth that much money."[6]

The spirit of what Holland was doing in Negro education got abroad. A group of instructors at Rollins College at Winter Park ha[d] offered their services to teach in Negro colleges and high schools, but a law placed on the books in the bitter years of the Reconstruction blocked them. In pique at the activities of the Freedmen's Bureau, under which white men and women came south to teach—the law said that no white person could teach Negroes in the State of Florida.[7] There was talk of the state taking over Bethune-Cookman College at Daytona Beach to supplement FAMC, but this too was blocked by a law passed in the same period which said that the state could not maintain more than one institution of higher learning for Negroes.

Holland's drive can be interpreted in this way—instead of hiding the Negro under his coat-tails to sneak him past the enemy lines, he sought to provide him with the weapons to take care of his own advancement—an infinitely more self-respecting attitude.

And that would be typical of the man. He does not indulge in that patronizing habit of entertaining by quoting the sayings of some ignorant [unintelligible], aunt nor uncle. He likes for the person—of whatever race—to be able to hold up their own end of the conversation, or it seems like a waste of time and effort to him, and justly so.

Now, Holland's record in Negro education is well known in Florida—so well known and accepted generally that Claude Pepper saw no point in digging it up and flinging it in Holland's teeth during the recent campaign to brand Holland as a Nigger-lover. The voters of Florida had twice elected him to the U.S. Senate in spite of it, which provides certain inferences. Contrary to outside propaganda, it can be inferred that the Florida voters on the whole do not object to higher education for Negroes, and candidate and governor Holland knew it. Nobody craved a racial ogre as governor. Separate, however equal, was the desire, and still is. That was all required of him.

And here it should be observed that the South has a vocabulary understood by southerners. Certain words and phrases mean one thing to Dixie and something entirely different outside. For example segregation is interpreted as racial hatred by outsiders but not so in the South where it merely means separate social activities and connections—not hatred of Negro individuals. One can understand the fear of interbreeding by the White South if one examines frankly its past history in that connection. All during the centuries of slavery and up to 1900 it went on lavishly, but the white race could maintain its claim of purity by having a law which said that a child belonged to the race of its mother—hence all the products of miscegenation were automatically assigned to the Negro Race, but if the status were upset now, it would be no longer true and the sanction by law would tend to speed up this interbreeding which has been already in practice illegally in the South for so long. Hence the determined and massive resistance in the South to the ruling of the Supreme Court in the integration of public schools. The South is very frank about it. It is not Negrophobia, but a desire to remain unadulterated. Bear in mind that the agricultural South of the past had no need of the European immigrant to do its work, and so missed the waves of immigrants which have influenced the concepts of the North. So the South remains the purest English blood in the nation with all the insularity and self-sufficiency of the "right little, tight little island." It does not welcome change.

And to make confusion more confounded, the ethnic terms of White and Negro do not mean to the South the same thing they mean outside. They are almost abstractions and in both cases need not refer to individuals but denote a relationship. For instance a Negro who tells you he is against White people will correct you if you ask him if he is against Mr. Wilson, Sanders or Thompson. "I don't mean them, I said white folks." Which means [an] unidentified mass of white people at the least in the next county or more likely another state. Faceless masses unconnected with his life. Just white folks—not the people I know. This goes for the other side too. Negroes are just Negroes has nothing to do with the individuals they know. Again, they live somewhere else. The names are really impersonal [in] application. Present Company always excepted. Therefore the politicos white and black can fire away without injuring their relationship with members of the opposite race. It is recognized that it is a campaign.

Another paradox is that there are no secrets in a southern community.

Whatever the Negroes know, some white individuals will be certain to know and vice versa. Somebody is going to tell across the race line. For in spite of all the propaganda to the contrary, friendships of the firmest [and] the most loyal type exist in every community and somebody is going to betray his own race to the other. That is how deep the feeling goes in spite of difference of color, so do not tell to either side what you don't want the other to know. It will surely be told.

This is why Spessard L. Holland, in spite of his vast victory as an advocate of segregation, is not looked upon as an enemy by the majority of Negroes. None have been heard referring to him as white folks, let alone mean old white folks—he is still an individual, the unit that the South deals in. He and his wife are still Mr. [and] Mrs. Senator Spessard L. Holland.[8]

His case is typical of a certain type of men in public life in the South— far from being obsessed by the race issue, they live full cultural lives when not campaigning.

Take for instance Spessard L. Holland. He is devoted to the best in literature. His wife, Mrs. Mary Holland, is an accomplished musician and so their children and grandchildren are exposed to the best in music. The family group gives time to the appreciation of the master works of painters and sculptors, to say nothing of participation [in] outdoor sports. It is a close-knit, affectionate family with Mrs. Mary Holland the sort of center of things or as the Spaniards say the *alma de casa*, the soul of the house. There are some who say that Senator Holland is too withdrawn, but quickly add that Mrs. Holland is available with her warm and lively personality, and in this way serves her husband just as effectively as Martha Taft, wife of the late Robert A. Taft, served her also reticent husband.[9]

Probably most of the southern Senators live a comparable life, wasting no more time and effort on the race issue than is absolutely necessary and feeding their senses on something more agreeable otherwise [End of existing manuscript.]

Part Four

ON POLITICS

The "Pet Negro" System

Brothers and Sisters, I take my text this morning from the Book of Dixie. I take my text and I take my time.

Now it says here, "And every white man shall be allowed to pet himself a Negro. Yea, he shall take a black man unto himself to pet and to cherish, and this same Negro shall be perfect in his sight. Nor shall hatred among the races of men, nor conditions of strife in the walled cities, cause his pride and pleasure in his own Negro to wane."

Now, belov-ed Brothers and Sisters, I see you have all woke up and you can't wait till the service is over to ask me how come? So I will read you further from the sacred word which says here:

"Thus spake the Prophet of Dixie when slavery was yet a young thing, for he saw the yearning in the hearts of men. And the dwellers in the bleak North, they who pass old-made phrases through their mouths, shall cry out and say, 'What are these strange utterances? Is it not written that the hand of every white man in the South is raised against his black brother? Do not the sons of Japheth drive the Hamites before them like beasts?[1] Do they not lodge them in shacks and hovels and force them to share the crops? Is not the condition of black men in the South most horrible? Then how doth this scribe named Hurston speak of pet Negroes? Perchance she hath drunk of new wine, and it has stung her like an adder?'"

Now, my belov-ed, before you explode in fury you might look to see if you know your facts or if you merely know your phrases. It happens that there are more angles to this race-adjustment business than are ever pointed out to the public, white, black or in-between. Well-meaning outsiders make plans that look perfect from where they sit, possibly in some New York office. But these plans get wrecked on hidden snags. John Brown at Harpers Ferry is a notable instance.[2] The simple race-agin-race pattern of those articles and speeches on the subject is not that simple at all. The actual conditions do not jibe with the fulminations of the so-called spokesmen of the white South, nor with the rhetoric of the champions of the Negro cause either.

II

Big men like Bilbo, Heflin and Tillman bellow threats which they know they couldn't carry out even in their own districts.[3] The orators at both extremes may glint and glitter in generalities, but the South lives and thinks in individuals. The North has no interest in the particular Negro, but talks of justice for the whole. The South has no interest, and pretends none, in the mass of Negroes but is very much concerned about the individual. So that brings us to the pet Negro, because to me at least it symbolizes the web of feelings and mutual dependencies spun by generations and generations of living together and natural adjustment. It isn't half as pretty as the ideal adjustment of theorizers, but it's a lot more real and durable, and a lot of black folk, I'm afraid, find it mighty cosy.

The pet Negro, belov-ed, is someone whom a particular white person or persons wants to have and to do all the things forbidden to other Negroes. It can be Aunt Sue, Uncle Stump, or the black man at the head of some Negro organization. Let us call him John Harper. John is the pet of Colonel Cary and his lady, and Colonel Cary swings a lot of weight in his community.

The Colonel will tell you that he opposes higher education for Negroes. It makes them mean and cunning. Bad stuff for Negroes. He is against having lovely, simple blacks turned into rascals by too much schooling. But there are exceptions. Take John, for instance. Worked hard, saved up his money and went up there to Howard University and got his degree in education. Smart as a whip! Seeing that John had such a fine head, of course he helped John out when necessary. Not that he would do such a thing for the average darky, no sir! He is no nigger lover. Strictly unconstructed Southerner, willing to battle for white supremacy! But his John is different.

So naturally when John finished college and came home, Colonel Cary knew he was the very man to be principal of the Negro high school, and John got the post even though someone else had to be eased out. And making a fine job of it. Decent, self-respecting fellow. Built himself a nice home and bought himself a nice car. John's wife is county nurse; the Colonel spoke to a few people about it and she got the job. John's children are smart and have good manners. If all the Negroes were like them he wouldn't mind what advancement they made. But the rest of them, of course, lie like the cross-ties from New York to Key West. They steal things and get drunk. Too bad, but Negroes are like that.

Now there are some prominent white folk who don't see eye to eye with Colonel Cary about this John Harper. They each have a Negro in mind who is far superior to John. They listen to eulogies about John only because they wish to be listened to about their own pets. They pull strings for the Colonel's favorites knowing that they will get the same thing done for theirs.

Now, how can the Colonel make his attitude towards John Harper jibe with his general attitude towards Negroes? Easy enough. He got his general attitude by tradition, and he has no quarrel with it. But he found John truthful and honest, clean, reliable and a faithful friend. He *likes* John and so considers him as white inside as anyone else. The treatment made and provided for Negroes generally is suspended, restrained and done away with. He knows that John is able to learn what white people of similar opportunities learn. Colonel Cary's affection and respect for John, however, in no way extend to black folk in general.

When you understand that, you see why it is so difficult to change certain things in the South. His particular Negroes are not suffering from the strictures, and the rest are no concern of the Colonel's. Let their own white friends do for them. If they are worth the powder and lead it would take to kill them, they have white friends; if not, then they belong in the "stray nigger" class and nobody gives a damn about them. If John should happen to get arrested for anything except assault and murder upon the person of a white man, or rape, the Colonel is going to stand by him and get him out. It would be a hard-up Negro who would work for a man who couldn't get his black friends out of jail.

And mind you, the Negroes have their pet whites, so to speak. It works both ways. Class-consciousness of Negroes is an angle to be reckoned with in the South. They love to be associated with "the quality" and consequently are ashamed to admit that they are working for "strainers." It is amusing to see a Negro servant chasing the madam or the boss back on his or her pedestal when they behave in an unbecoming manner. Thereby he is to a certain extent preserving his own prestige, derived from association with that family.

If ever it came to the kind of violent showdown the orators hint at, you could count on all the Colonel Carys tipping off and protecting their John Harpers; and you could count on all the John Harpers and Aunt Sues to exempt their special white folk. And that means that pretty nearly everybody on both sides would be exempt, except the "pore white trash" [sic] and the "stray niggers," and not all of them.

III

An outsider driving through a street of well-off Negro homes, seeing the great number of high-priced cars, will wonder why he has never heard of this side of Negro life in the South. He has heard about the shacks and the sharecroppers. He has had them before him in literature and editorials and crusading journals. But the other side isn't talked about by the champions of white supremacy, because it makes their stand, and their stated reasons for keeping the Negro down, look a bit foolish. The Negro crusaders and their white adherents can't talk about it because it is obviously bad strategy. The worst aspects must be kept before the public to force action.

It has been so generally accepted that all Negroes in the South are living under horrible conditions that many friends of the Negro up North actually take offense if you don't tell them a tale of horror and suffering. They stroll up to you, cocktail glass in hand, and say, "I am a friend of the Negro, you know, and feel awful about the terrible conditions down there." That's your cue to launch into atrocities amidst murmurs of sympathy. If, on the other hand, just to find out if they really have done some research down there, you ask, "What conditions do you refer to?" you get an injured, and sometimes malicious, look. Why ask foolish questions? Why drag in the many Negroes of opulence and education? Yet these comfortable, contented Negroes are as real as the sharecroppers.

There is, in normal times, a regular stream of high-powered cars driven by Negroes headed North each summer for a few weeks' vacation. These people go, have their fling, and hurry back home. Doctors, teachers, lawyers, businessmen, they are living and working in the South because that is where they want to be. And why not? Economically, they are at ease and more. The professional men do not suffer from the competition of their white colleagues to anything like they do up North. Personal vanity, too, is served. The South makes a sharp distinction between the upper-class and lower-class Negro. Businessmen cater to him. His word is *good* downtown. There is some Mr. Big in the background who is interested in him and will back his fall. All the plums that a Negro can get are dropped in his mouth. He wants no part of the cold, impersonal North. He notes that there is segregation and discrimination up there, too, with none of the human touches of the South.

As I have said, belov-ed, these Negroes who are petted by white friends think just as much of their friends across the line. There is a personal attachment that will ride over practically anything that is liable to happen to either. They have their fingers crossed, too, when they say they don't like white people. "White people" does not mean their particular friends, any more than niggers means John Harper to the Colonel. This is important. For anyone, or any group, counting on a solid black South, or a solid white South in opposition to each other will run into a hornet's nest if he discounts these personal relations. Both sides admit the general principle of opposition, but when it comes to putting it into practice, behold what happens. There is a quibbling, a stalling, a backing and filling that nullifies all the purple oratory.

So well is this underground hook-up established, that it is not possible to keep a secret from either side. Nearly everybody spills the beans to his favorite on the other side of the color line—in strictest confidence, of course. That's how the "petting system" works in the South.

Is it a good thing or a bad thing? Who am I to pass judgement? I am not defending the system, belov-ed, but trying to explain it. The low-down fact is that it weaves a kind of basic fabric that tends to stabilize relations and give something to work from in adjustments. It works to prevent hasty explosions. There are some people in every community who can always talk things over. It may be the proof that this race situation in America is not entirely hopeless and may even be worked out eventually.

There are dangers in the system. Too much depends on the integrity of the Negro so trusted. It cannot be denied that this trust has been abused at times. What was meant for the whole community has been turned to personal profit by the pet. Negroes have long groaned because of this frequent diversion of general favors into the channels of private benefits. Why do we not go to Mr. Big and expose the Negro in question? Sometimes it is because we do not like to let white people know that we have folks of that ilk. Sometimes we make a bad face and console ourselves, "At least one Negro has gotten himself a sinecure not usually dealt out to us." We curse him for a yellow-bellied sea-buzzard, a ground-mole and a woods-pussy, call him a white-folkses nigger, an Uncle Tom, and a handkerchief-head and let it go at that. In all fairness it must be said that these terms are often flung around out of jealousy: somebody else would like the very cinch that the accused has grabbed himself.

But when everything is discounted, it still remains true that white people North and South have promoted Negroes—usually in the capacity of "representing the Negro"—with little thought of the ability of the person promoted but in line with the "pet system." In the South it can be pointed to scornfully as a residue of feudalism; in the North no one says *what* it is. And that, too, is part of the illogical, indefensible but somehow useful "pet system."

IV

The most powerful reason why Negroes do not do more about false "representation" by pets is that they know from experience that the thing is too deep-rooted to be budged. The appointer has his reasons, personal or political. He can always point to the beneficiary and say, "Look, Negroes, you have been taken care of. Didn't I give a member of your group a big job?" White officials assume that the Negro element is satisfied and they do not know what to make of it when later they find that so large a body of Negroes charge indifference and double-dealing. The white friend of the Negroes mumbles about ingratitude and decides that you simply can't understand Negroes . . . just like children.

A case in point is Dr. James E. Shepard, President of the North Carolina State College for Negroes.[4] He has a degree in pharmacy, and no other. For years he ran a one-horse religious school of his own at Durham, North Carolina. But he has always been in politics and has some good friends in power at Raleigh. So the funds for the State College for Negroes were turned over to him, and his little church school became the Negro college so far as that State is concerned. A fine set of new buildings has been erected. With a host of Negro men highly trained as educators within the State, not to mention others who could be brought in, a pharmacist heads up higher education for Negroes in North Carolina. North Carolina can't grasp why Negroes aren't perfectly happy and grateful.

In every community there is some Negro strong man or woman whose word is going to go. In Jacksonville, Florida, for instance, there is Eartha White.[5] You better see Eartha if you want anything from the white powers-that-be. She happens to be tremendously interested in helping the unfortunates of her city and she does get many things for them from the whites.

I have white friends with whom I would, and do, stand when they have need of me, race counting for nothing at all. Just friendship. All the well-known Negroes could honestly make the same statement. I mean that they all have strong attachments across the line whether they intended them in the beginning or not. Carl Van Vechten and Henry Allen Moe could ask little of me that would be refused. Walter White, the best known race champion of our time, is hand and glove with Supreme Court Justice Black, a native of Alabama and an ex-Klansman. So you see how this friendship business makes a sorry mess of all the rules made and provided. James Weldon Johnson, the crusader for Negro rights, was bogged to his neck in white friends whom he loved and who loved him. Dr. William E. Burkhardt Du Bois, the bitterest opponent of the white race that America has ever known, loved Joel Spingarn and was certainly loved in turn by him. The thing doesn't make sense. It just makes beauty.[6]

Friendship, however it comes about, is a beautiful thing. The Negro who loves a white friend is shy in admitting it because he dreads the epithet "white folks' nigger!" The white man is wary of showing too much warmth for his black friends for fear of being called "nigger-lover," so he explains his attachment by extolling the extraordinary merits of his black friend to gain tolerance for it.

This is the inside picture of things, as I see it. Whether you like it or not, is no concern of mine. But it is an important thing to know if you have any plans for racial manipulations in Dixie. You cannot batter in doors down there, and you can save time and trouble, and I do mean trouble, by hunting up the community keys.

In a way, it is a great and heartening tribute to human nature. It will be bound by nothing. The South frankly acknowledged this long ago in its laws against marriage between blacks and whites. If the Southern law-makers were so sure that racial antipathy would take care of racial purity, there would have been no need for the laws.

"And no man shall seek to deprive a man of his Pet Negro. It shall be unwritten-lawful for any to seek to prevent him in his pleasure thereof. Thus spoke the Prophet of Dixie." *Selah.*

Negroes Without Self-Pity

I may be wrong, but it seems to me that what happened at a Negro meeting in Florida the other day is important—important not only for Negroes and not only for Florida. I think that it strikes a new, wholesome note in the black man's relation to his native America.

It was a meeting of the Statewide Negro Defense Committee. G. D. Rogers, President of the Central Life Insurance Company of Tampa, got up and said: "I will answer that question of whether we will be allowed to take part in civic, state and national affairs. The answer is—yes!"[1] Then he explained why and how he had come to take part in the affairs of his city.

"The truth is," he said, "that I am not always asked. Certainly in the beginning I was not. As a citizen, I saw no reason why I should wait for an invitation to interest myself in things that concerned me just as much as they did other residents of Tampa. I went and I asked what I could do. Knowing that I was interested and willing to do my part, the authorities began to notify me ahead of proposed meetings, and invited me to participate. I see no point in hanging back, and then complaining that I have been excluded from civic affairs.

"I know that citizenship implies duties as well as privileges. It is time that we Negroes learn that you can't get something for nothing. Negroes, merely by being Negroes, are not exempted from the natural laws of existence. If we expect to be treated as citizens, and considered in community affairs, we must come forward as citizens and shoulder our part of the load. The only citizens who count are those who give time, effort and money to the support and growth of the community. *Share the burden where you live!*"

And then J. Leonard Lewis, attorney for the Afro-American Life Insurance, had something to say.[2] First he pointed to the growing tension between the races throughout the country. Then he, too, broke tradition. The upper-class Negro, he said, must take the responsibility for the Negro part in these disturbances.

"It is not enough," he said, "for us to sit by and say 'We didn't do it. Those irresponsible, uneducated Negroes bring on all this trouble.' We must not only do nothing to whip up the passions among them, we must go much further. We must abandon our attitude of aloofness to the less educated. We must get in touch with them *and head off these incidents before they happen.*

"How can we do that? There is always some man among them who has great prestige with them. He can do what we cannot do, because he is of them and understands them. If he says fight, they fight. If he says, 'Now put away that gun and be quiet,' they are quiet. We must confer with these people, and cooperate with them to prevent these awful outbreaks that can do no one any good and everybody some harm. Let us give up our attitude of isolation from the less fortunate among us, and do what we can for peace and good-will between the races."

Not anything world-shaking in such speeches, you will say. Yet something profound has happened, of which these speeches are symptoms and proofs. Look back over your shoulder for a minute. Count the years. If you take in the twenty-odd years of intense Abolitionist speaking and writing that preceded the Civil War, the four war years, the Reconstruction period and recent Negro rights agitations, you have at least a hundred years of indoctrination of the Negro that he is an object of pity. Becoming articulate, this was in him and he said it. "We were brought here against our will. We were held as slaves for two hundred and forty-six years. We are in no way responsible for anything. We are dependents. We are due something from the labor of our ancestors. Look upon us with pity and give!" The whole expression was one of self-pity without a sense of belonging to America and what went on here.

Put that against the statements of Rogers and Lewis, and you get the drama of the meeting. The audience agreed and applauded. Tradition was tossed overboard without a sigh. Dr. J. R. E. Lee, president of Florida A and M College for Negroes, got up and elaborated upon the statements: "*Go forward with the nation. We are citizens and have our duties as such.*"[3] Nobody mentioned slavery, Reconstruction, nor any such matter. It was a new and strange kind of Negro meeting—without tears of self-pity. It was a sign and symbol of something in the offing.

The Rise of the Begging Joints

People have been telling me to clap hands, crack jokes, and generally cut Big Jim by the acre. I ought to look and see. Great joy was around me.

When I turned around and asked why I should jump Juba and burn red fire without ceasing, they told me, "Look! No more slavery days, and even the Reconstruction is past and gone. Aunt Hagar's chillun are eating high on the hog."

Now, there is nothing I favor more than clapping hands in drumtime, and dancing all night long, unless it is having something to clap hands and dance over. But I didn't see too clear; so I said, "Pick up your points. Tell me, and then make see, so I can dance it off."

So they told me again and often. Sometimes they said it with arm-gestures. Same thing all over again: the Reconstruction is over. Everything is fresh and new.

But I am not cutting my capers yet awhile, because I don't know for certain.

I do see a great many Negroes with college degrees, fur coats, big houses and long cars. That is just fine, and I like it. That looks really up-to-time. On the other hand, I see some things that look too much like 1875 in the lap of 1944, and they worry me.

Those Begging Joints, for instance. That is not the name they go by, of course. Some folks with their mouths full of flattery call them normal schools, colleges, and even universities. I'm sort of tie-tongued and short-patienced, so I call them a functional name and let it go at that.

The "puhfessahs," principals, presidents and potentates who run these institutions seem to like them mighty fine. They will tell you without your even asking that these are "great works." Further, that they themselves are latter-day martyrs, electing to "carry on this great work for our people, so that our girls and boys may get some sort of an education." They tootch out the mouth to say this so that it oozes out in an unctuous tone of voice. There is also a ceremonial face-making, with eye-gleams, to go along with the sound. If you don't make some fast time away from there, you are going to hear all about how they were born in a log cabin. Think of that! And now, look! They have builded this g-r-e-a-t institution! The mouth-spread would take in the mighty expanse of Columbia University, but you look around and see something that would have been a miracle in 1875, but nothing to speak of in this day and year of our Lord.

The next thing you know, the talk has gotten around to funds. It always gets around to funds. Money is needed to carry on the g-r-e-a-t work. If you don't know any better, you will soon be shaking with apprehension at the prospect of this institution's closing its doors, and never another Negro girl or boy learning her or his ABC's. If you know anything about Negro education, you come out of your spasm quickly when you remember that there are such Class A seats of learning as Howard University, Fisk University, Morgan State College, Atlanta University and affiliates, Tuskegee, Morehouse, Talladega, Hampton, Florida A. and M., Southern, Bennett, Virginia State and Lincoln U. In addition, there are most all the Northern white colleges except Princeton with Negro graduates and students. For those with ambition but less funds, there are the state-supported colleges for Negroes in every Southern state. Quite a number of colored folk have even earned degrees at the leading universities of Europe, from Scandinavia to Spain. But in spite of this, you are asked to shake and shiver over the prospective fate of some puny place without a single gifted person in its meager faculty, with only token laboratories or none, and very little else besides its FOUNDER. The Founder is the thing! And the Founder exists to raise funds.

II

Where can one of these Begging Joints be found? There are two or more left in every state in the South.

With so many good colleges available to Negroes, why do these Begging Joints keep on existing? Because there are so many poor Negroes, and so many rich white people, who don't know very much. They not only do not know, but they are very incurious.

The colored families from which the Begging Joints draw their students know nothing about the importance of curricula for accrediting. They think that just finishing, "schooling out," is all that is necessary. Just so it is a school. Forty years ago, that would have been all right.

But, not only has the responsibility of the Negro population shifted, the concept of education has changed. Competition is keen, and every chick and child has to pull his weight in efficiency or get trampled in the rush. And that is the tragedy of promoting 1880 in 1944.

These institutions have two things to sell to the folks who are still living in the just-after-slavery aura. One of these is "Off" and the other is the equivalent of the medieval virtue belt. And both of these things take us back to the nineteenth century.

In those days, any kind of school education was something for a Negro to have and threw glitter all around the owner, even to distant cousins. To send your son or daughter *Off* to school marked you as a Big Negro, getting more like the white folks every day. So *Off* became the thing to strive for. It then brought the exalted job of teaching school, plus social preference in everything, including marriage.

Off has lost its glamor for the upper-class Negroes, but it still has some attraction for the lower class. Almost never will an upper-class Negro send his children away until after high school, if a fairly good school is at hand.

When he does so, it is either because the local secondary school is not accredited, and will hamper the entrance to some good college, or because of some family reason. It is not because of any prestige of *Off*. But there are still those to whom *Off* is a putting-of-the-family-foot on the ladder, an offsetting of the necessity of labor over the washtub and the cookstove.

The virtue belt factor comes in like this. During slavery, there was no encouragement to continence on the plantations. Quite the contrary. So when freedom came, it is too much to expect that it would have been acquired immediately. There was little immorality really, just a lack of concept about the thing. So in the first generation or so after Emancipation, a mighty lot of girls got "ruint." Hence the first boarding schools that got started were more like reformatories with instruction on the side. You sent your daughter *Off*, if you could afford it, more to keep her from getting "ruint" than to get her educated. Naturally, even those ugly-faced, chilly, warden-like matrons were no more capable of thwarting a boy-crazy girl than her parents were. Daughter often brought her "diploma" home in her arms. Love will find a way.

The instances of this are fewer these days, but the idea is still being sold to parents—"We are not like that Howard University and Fisk. We don't allow our girls to go walking across the campus with boys. And no dancing together. No, indeed! Of course we are not so big as those places. But the students at those schools don't learn a thing! All they do is socialize."

It is true that schools like Howard and Fisk place no ban on ordinary social contacts. It is true that there is some dancing. But what the Begging Joint Puhfessahs neglect to mention is that such colleges as Fisk and Howard run small chance of getting any under-privileged, unrestrained girls among the students. They have the pick of the nation. The instances in them of the kinds of troubles the Puhfessahs suggest fearsomely are negligible.

But the simple woman toiling over a washtub or cookstove does not understand that. So she sends her daughter *Off* to get her virtue guarded at the instigation of the Puhfessah. It may be a locking of the stable after the horse is stolen, but that is not important to the Puhfessah. If daughter is an indifferent student, that is not important either. The "Great Work" needs students.

No students, no school. No school, no excuse to seek funds. No funds, the principal is faced with a major change in his or her way of living. A terrible loss of prestige, plus going to work with the hands. No more of giving his life and life's blood to the "doing of something for my people." The Puhfessah is in no position to pick and choose among prospective students. He must take what he can get or fold. And God forbid the folding!

So the doer of good works begs a good living for himself in the name of his people. And that is all right for him. But what about the poor black men and women bowed down over mops or standing over the white folks' cookstoves to send daughter or son to the Begging Joint to dress the stage for the educational Fagin?[1] And what of the students?

After two to four years, the janitor's or cook's son or daughter has a piece of paper tied up with a snatch of ribbon. He has had several years of considering himself above the commonality of Negro existence. He thinks sweat and overalls are not for him. But after many trials he finds that he can get no better employment than the boy he used to know next door. He is Jim Jones again, instead of Mr. James Jones, prospective principal of a school. He cannot meet the requirements of the state board of education. He may have learned a smattering of some trade at his school, but not enough to help him much when skilled labor is demanded. Time and money have been wasted, and in addition, he feels a bitter loss of face.

Some, pressed by necessity, adjust themselves and do what must be done. Others, just as pressed, never do. They cannot fit in where they think they belong, but will not adjust themselves to the level of their fitness. They go through life scornful of those who work with their hands, and resentful of the better-prepared Negroes. I know one man of this type who will not wear overalls, no matter how dirty the job that he has to do, because to him overalls are the symbol of the common Negro, and he fears to be so classified. The highest "position" that he has ever held since he left college more than twenty years ago has been a minor one in a 5- and 10-cent store, but he clings to his delusion of grandeur still.

But the Begging Joint does a greater disservice still to the individual, the race and the nation. That is its perpetuation of the double standard in education. Even Southern legislators have come to realize that there must be only one, and the requirements of the Negro state and county schools are being sharply raised year by year. Beginning about 1930, the State of Florida began displacing its old-fashioned Negro teachers with more highly trained ones, preferably from the big Northern colleges. Other states have similar policies. But the little private schools are out of the jurisdiction of the state boards of education. All the state boards can do is to refuse to accredit the backward institution.

The raising of the Negro educational standard is an obvious necessity, since the inefficient are a drag on all. The city, the state and the nation need all their useful people, regardless of race and kind. There is no longer any place for the black man who "does very well for a Negro." He has got to be good these days, and I said *good*. But the Begging Joints are still doing nothing but trying to put exclamation points behind what was considered good away back in 1880 when the majority of white people thought that all Negroes were something less than human.

III

As for the white donors to these Begging Joints, they need their heads examined. I am going to give them full credit for being friends of the Negro. But even so, why can't they be intelligent about it? If they want to do something for Negro education, why not look into things and give where it will do some good?

Mind you, I am not seeking funds. I am not a Founder—and not consequently, a member of the Order of Higher Mendicants, as George Schuyler so aptly puts it.[2] You were the one who said that you wanted to do something for Negro education. All I say is, if you really mean what you say, come out of the backwoods of your mind and do some figuring up. You must know that this is the greatest industrial nation on earth, and that during the last two generations the colleges have followed the trend of the nation's needs. Industry calls for scientists. They must be trained in laboratories, so colleges of any account must have these. Not only must they have them, but they must be able to make the constant replacements necessary to keep up with the latest developments. What would have equipped a whole college fifty years ago is required now for a comprehensive laboratory in chemistry, physics or biology.

Now, why give five thousand dollars a year to Chitterling Switch "college" in the backwoods of Mississippi, let us say, when it is some one-cylinder outfit with perhaps two hundred students? Five thousand is not going to be any help, not in the fix it is in. Five million is more like what it would take to bring it up to our time. Five thousand will just about pay for prexy's new Cadillac, with possibly enough over to put some chairs in a classroom.

Since you really want to do something for Negro education, give to one of the colleges which already has something to build on. You might, for instance, give to Howard University's great medical school. It is turning out doctors and dentists who compare favorably with the best in the nation. The money could be used for laboratory replacements, or to sponsor some serious young medico who yearns to make further research.

Or if you are inclined towards literature, give to the library fund of Morgan State College, at Baltimore. It has a fine library, and aims to make it the finest possible. Or give to the school of Social Sciences at Fisk which, under Dr. Charles S. Johnson, has increased the sum total of knowledge in that field considerably in the last decade. Or send the check to Atlanta University. Its department of sociology is getting grand results.

I wouldn't put it past some of you to tell me that it is your own money, and you can do with it as you please. You are free, white and twenty-one. But you see, I'm free, black and twenty-one; and if you tell me that, I will know and understand that you have no genuine interest in Negro education. Or if you think you have, you haven't taken a sounding in the last forty years. You are way behind the times.

I have made it my business to talk with patrons of the Chitterling Switch kind of a school in the last three years, and I have been astonished at the number of persons giving money to a school without even inquiring into the curriculum or looking into the training of the faculty. They were under the delusion that the school to which they gave was at the top of Negro education. Indeed, the majority had the idea that there was no other kind for Negroes to go to. Some did not even know that there were colleges provided by the state. They fluttered about raising little sums impelled by the fear that if the little place were not kept in condition to keep on crippling young black folks, there would be no other chance for Negroes to get hamstrung for life.

Instead of being a help, many donors have been giving aid and assistance to the defeat of those who need a chance more than anyone else in America.

What is the general history behind these little knowledge-traps? First, there were those little piney-woods schools opened by the Abolitionist church groups immediately after the Civil War. Fired by the Cause, hundreds, perhaps thousands, of pious Northerners came South and gave themselves to teach the freemen how to read and write. But the greatest emphasis was on the Bible. Those who were first taught, were urged to go forth and spread what they had learned to others.

Then the "do something for my people" era came in on the trailing clouds of Booker T. Washington. He was responsible, but he didn't mean any harm. Tuskegee was a success, with tycoons of industry and finance rolling down in their private cars to see and be awed. They saw Booker T. Washington as the Moses of his race. They donated millions to give body to the idea of Booker T. So in the black world the man was magical. He sat with Presidents. He went abroad and stood with kings. This school-founding thing was something!

Without the genius of his idea and the surprise of its newness, hundreds of other Negroes deserted pulpits, plows, washtubs and cookpots and went out founding schools. The outer offices of financiers began to be haunted by people "doing something for my people." There remained only one Tuskegee. But then, there was only one Booker T. Washington.

These Begging Joints were a natural part of the times when they were started and in a way they were all right for then. But they are unburied corpses in 1944. We can bury the carcasses any time we will or may. All it takes is a made-up mind.

Crazy for This Democracy

They tell me this democracy form of government is a wonderful thing. It has freedom, equality, justice, in short, everything! Since 1937 nobody has talked about anything else.

The late Franklin D. Roosevelt sort of re-decorated it, and called these United States the boastful name of "The Arsenal of Democracy."[1]

The radio, the newspapers, and the columnists inside the newspapers have said how lovely it was.

All this talk and praise-giving has got me in the notion to try some of the stuff. All I want to do is to get hold of a sample of the thing, and I declare I sure will try it. I don't know for myself, but I have been told that it is really wonderful.

Like the late Will Rogers, all I know is what I see by the papers.[2] It seems like now, I do not know geography as well as I ought to, or I would not get the wrong idea about so many things. I heard so much about "global" "world-freedom" and things like that, that I must have gotten mixed up about oceans.

I thought that when they said Atlantic Charter, that meant me and everybody in Africa and Asia and everywhere.[3] But it seems like the Atlantic is an ocean that does not touch anywhere but North America and Europe.

Just the other day, seeing how things were going in Asia, I went out and bought myself an atlas and found out how narrow this Atlantic ocean was. No wonder those Four Freedoms couldn't get no further than they did![4]

Why, that poor little ocean can't even wash up some things right here in America, let alone places like India, Burma, Indo-China, and the Netherlands East Indies. We need two more whole oceans for that.

Maybe I need to go out and buy me a dictionary, too. Or perhaps a spelling-book would help me out a lot. Or it could be that I just mistook the words. Maybe I mistook a British pronunciation for a plain American word. Did F. D. R., aristocrat from Groton and Harvard, using the British language say "arse-and-all" of Democracy when I thought he said plain arsenal? Maybe he did, and I have been mistaken all this time. From what is going on, I think that is what he must have said.

That must be what he said, for from what is happening over on that other, unmentioned ocean, we look like the Ass-and-All of Democracy. Our weapons, money, and the blood of millions of our men have been used to carry the English, French and Dutch and lead them back on the millions of unwilling Asiatics. The Ass-and-all-he-has has been very useful.

The Indo-Chinese are fighting the French now in Indo-China to keep the freedom that they have enjoyed for five or six years now. The Indonesians are trying to stay free from the Dutch, and the Burmese and Malayans from the British.

But American soldiers and sailors are fighting along with the French, Dutch and English to rivet these chains back on their former slaves. How can we so admire the fire and determination of Toussaint Louverture to resist the orders of Napoleon to "rip the gold braids off those Haitian slaves and put them back to work" after four years of freedom, and be indifferent to these Asiatics for the same feelings under the same circumstances?[5]

Have we not noted that not one word has been uttered about the freedom of the Africans? On the contrary, there have been mutterings in undertones about being fair and giving different nations sources of raw materials there? The Ass-and-All of Democracy has shouldered the load of subjugating the dark world completely.

The only Asiatic power able to offer any effective resistance has been double-teamed by the combined powers of the Occident and rendered incapable of offering or encouraging resistance, and likewise removed as an example to the dark people of the world.

The inference is, that God has restated the superiority of the West. God always does like that when a thousand white people surround one dark one. Dark people are always "bad" when they do not admit the Divine Plan like that. A certain Javanese man who sticks up for Indonesian Independence is very low-down by the papers, and suspected of being a Japanese puppet. Wanting the Dutch to go back to Holland and go to work for themselves! The very idea! A very, very bad man, that Javanese.[6]

As for me, I am just as skeptical as this contrary Javanese. I accept this idea of democracy. I am all for trying it out. It must be a good thing if everybody praises it like that. If our government has been willing to go to war and sacrifice billions of dollars and millions of men for the idea, I think that I ought to give the thing a trial.

The only thing that keeps me from pitching headlong into the thing is the presence of numerous Jim Crow laws on the statute books of the nation. I am crazy about the idea of this democracy. I want to see how it feels. Therefore, I am all for the repeal of every Jim Crow law in the nation here and now. Not in another generation or so. The Hurstons have already been waiting eighty years for that. I want it here and now.

And why not? A lot of people in these United States have been saying all this time that things ought to be equal. Numerous instances of inequality have been pointed out, and fought over in the courts and in the newspapers. That seems like a waste of time to me.

The patient has the smallpox. Segregation and things like that are the bumps and blisters on the skin, and not the disease, but evidence and symptoms of the sickness. The doctors around the bedside of the patient, are desperately picking bumps. Some assume that the opening of one blister will cure the case. Some strangely assert that a change of climate is all that is needed to kill the virus in the blood!

But why this sentimental oversimplification in diagnosis? Do the doctors not know anything about the widespread occurrence of this disease? It is NOT peculiar to the South. Canada, once the refuge of escaping slaves, has now its denomination of second-class citizens, and they are the Japanese and other non-Caucasians. The war cannot explain it, because enemy Germans are not put in that second class.

Jim Crow is the rule in South Africa, and is even more extensive than in America. More rigid and grinding. No East Indian may ride first-class in the trains of British-held India. Jim Crow is common in all colonial Africa, Asia and the Netherlands East Indies. There, too, a Javanese male is punished for flirting back at a white female. So why this stupid assumption that "moving North" will do away with social smallpox? Events in the northern cities do not bear out this juvenile contention.

So why the waste of good time and energy, and further delay the recovery of the patient by picking him over bump by bump and blister to blister? Why not the shot of serum that will kill the thing in the blood? The bumps are symptoms. The symptoms cannot disappear until the cause is cured.

These Jim Crow laws have been put on the books for a purpose, and that purpose is psychological. It has two edges to the thing. By physical evidence, back seats in trains, back-doors of houses, exclusion from certain places and activities, to promote in the mind of the smallest white child the conviction of First by Birth, eternal and irrevocable like the place assigned to the Levites by Moses over the other tribes of the Hebrews. Talent, capabilities, nothing has anything to do with the case. Just FIRST BY BIRTH.

No one of darker skin can ever be considered an equal. Seeing the daily humiliations of the darker people confirm the child in its superiority, so that it comes to feel it the arrangement of God. By the same means, the smallest dark child is to be convinced of its inferiority, so that it is to be convinced that competition is out of the question, and against all nature and God.

All physical and emotional things flow from this premise. It perpetuates itself. The unnatural exaltation of one ego, and the equally unnatural grinding down of the other. The business of some whites to help pick a bump or so is even part of the pattern. Not a human right, but a concession from the throne has been made. Otherwise why do they not take the attitude of Robert Ingersoll that all of it is wrong?[7] Why the necessity for the little concession? Why not go for the under-skin injection? Is it a bargaining with a detail to save the whole intact? It is something to think about.

As for me, I am committed to the hypodermic and the serum. I see no point in the picking of a bump. Others can erupt too easily. That same one can burst out again. Witness the easy scrapping of FEPC.[8] No, I give my hand, my heart and my head to the total struggle. I am for complete repeal of All Jim Crow Laws in the United States once and for all, and right now. For the benefit of this nation and as a precedent to the world.

I have been made to believe in this democracy thing, and I am all for tasting this democracy out. The flavor must be good. If the Occident is so intent in keeping the taste out of darker mouths that it spends all those billions and expends all those millions of lives, colored ones too, to keep it among themselves, then it must be something good. I crave to sample this gorgeous thing. So I cannot say anything different from repeal of all Jim Crow laws! Not in some future generation, but repeal *now* and forever!!

I Saw Negro Votes Peddled

Millions of Americans no doubt harbor the illusion that the Period of the Reconstruction ended in the 1870s, and in dying out took all of its symptoms with it.[1] No more herding of the Negro vote by greedy Carpet-baggers and their allies, the opportunist-minded southerner who came to be known as the Scalawag. No more prostitution of the purposes of free election by packing the polls with Negro voters who balloted as they were told without understanding what any of the commotion was about. Those days were gone forever.

Those, like myself, who held that delusion, were never so mistaken. In the Florida Primary election of May 2, 1950, I saw Negro votes being peddled on a big scale. Single-shotting was the order of the day.

To those who might not be familiar with the term, single-shotting in an election means to go into the booth and pull down a single lever, ignoring everything else offered for public consideration. This erratic behavior on the part of a voter is a dead giveaway. First, it signifies that the voter is unprepared in his own mind to comprehend, even vaguely, the contested issues. Second, it betrays the fact that the ballot-wielder has been coached. The instructor does not trust the voter's mind to retain but so much, so no risk is taken by trying to teach the whole ticket. Just go in and pull down lever Number 2, for instance, then come on out. Lever Two, you know a 2 when you see it, don't you? Pull 2, t-w-o, and come on out and get your pay. That or these, are the mechanics and the explanation of single-shotting.[2]

It was while registration was going on that a murmur reached me that this was going to be a hotly-contested senatorial fight between the incumbent, Senator Claude Pepper, and his challenger, Representative George Smathers, and that an organization from the North was going to come into Florida to organize and deliver the Negro vote in a lump. From historical background, I did not believe that an outside agency would dare to interfere in a southern election, nor did I believe that the Negro vote could be handled as a dark, amorphous lump. Then and there I made up my mind to be in Florida for this struggle at all costs, and to be in Miami, Florida's largest city and the seat of Dade County, which from rumor was slated to be the hottest battleground.

So I planted myself there and saw the intense and well-organized drive to get the mass of Negro voters registered for the polls. I heard about the payment of a dollar to each prospective voter, because you cannot keep secret what thousands of people know. It was a dollar for each person who registered to vote, and twenty-five cents a head for the bush-beaters who rounded up the people and delivered them to the registration centers. By all accounts, this was the heaviest registration of Negro voters in the history of the State, and perhaps many years will pass before it happens that way again.

Whether there was an organization behind this intense drive was answered for me positively by a Negro schoolteacher who was part of the movement.

"You were correctly informed," she told me with assurance, and even a touch of pride in her voice. "They are really here. That is no rumor at all. It is a positive fact. They are our friends from up North here helping us out and they are doing a wonderful job. The Negro vote holds the balance of power, and the way we are organized now, I can't see any way in the world for our candidate to lose. Not with the help we've got. It's in the bag!"

That was the way it was. Under the promise of gain, if you can call a dollar bill gain, the inert section of the Negro voters were needled into action and registration. The Negro bush-beaters herding the prospective voters in, and the organizers in the shadows directing the bush-beaters.

But all of the Negroes did not hurry to register for the sake of a dollar.

There was a sector of the socially-conscious who already knew something about the organization from the inside, and saw in the election a beautiful Trojan horse. They were on hand to open the door and let out the warriors on the State. The long-delayed capture of the South by the left-wing was at hand.

Estimating that at least 50 percent of those Negroes who had been pressured to register would forget all about the whole thing in a few days, I watched to see the inertia set in. But the organizing experts had thought about that too. For that section of the voters who could be appealed to through their desire for political power, the FEPC issue was kept at white heat. It was going to do everything for them, down to frying the breakfast bacon and hanging out the wash on the line. For those indifferent to such things, a kind of voter's soup kitchen was provided. With the proper credentials, those pleading poverty could go to certain addresses and draw groceries. Here they were exhorted to hold firm and to spread the good gospel wherever they went. Vote right, and there would be a lot more free things besides groceries. That was the kind of government that poor people needed, and that was the kind that they would get if they voted right.

Then there were those post-primary promises. They gave unthinking voters visions and made them dream dreams. One young woman worker told me with a confident smile that the day after the election she would be driving a Cadillac. Just like that! A drab middle-aged woman told me about the groceries that she had already got, and how she had been promised plenty of sheets and towels for her house, which she needed badly. She only wished that she had thought to ask for some new wool blankets too. Her next door neighbor had put in for some. Nice, new, pink-colored blankets and all. Still another woman was glowing over the promise of two new inner-spring mattresses for her beds. She was very excited about the prospect. The wishes of many years coming true at last. Oh, no, she answered my doubts. She was not going to be fooled. She had told the white woman whom she did day-work for twice a week about it and asked her please to go down to the party headquarters the day after election and pick out her two mattresses for her and have them delivered right away.

This voting for what-you-could-get was making me feel sick and sad. The right to vote, to express the will of the individual in the affairs of the community, the commonwealth and the nation, was nowhere to be found in this general talk that was going on all around me. The concept was entirely missing. The exercise of the franchise, the most potent, the most sacred thing that man has conceived and strived for since humans began to live in communities, was counted as practically nothing.

I am only too aware that corrupt politicians buy white votes, and that unthinking white voters sell them, and often very cheaply. But this right ought to be held in higher regard by Negroes than any other citizens in the United States. For us, this prized symbol of citizenship has a long and somber history. It commences with the very inception of the United States. There have been more public debates, more moral preachments, more laws proposed and passed, more contention, and yes, more human bloodshed to bring us to the place where we can cast a ballot, and thereby express our will in the affairs of our country, than anybody else in America. It is positively astounding that any adult Negro could look upon the right to vote as a small thing, let alone regard this highest right in civilization in such a way as to put a price upon it. In the establishment of human rights and the importance of the individual from Greece to Rome to England and to its elevation in the United States by fundamental law, it has been a long, terrible and bloody road. Americans held it so sacred that they laid down their lives on the field of battle that we too might share this right. It struck me as ironical that what others thought worth giving their lives to gain for us, some among us could think so little of that we could sell it for a dollar and think that we had gained something.

The polls opened at seven o'clock on the morning of May 2nd, and I was on hand to see what would happen. Observing as carefully as I could, I went through the colored neighborhoods from one polling precinct to the other. Except for minor human-interest details, the scene was the same everywhere. The organizers were on hand at every place, carefully setting up their pitches the required three hundred feet from the polls. There was somebody seated at a little table. The turnout in the early hours was tremendous, for the most part arriving in trucks, forty to fifty to a load.

The voters tumbled out of the trucks, made their way to the little table, got each a little piece of paper and formed in line to the polls. The lines moved in and out of the polling-place with astonishing speed and smoothness. They knew exactly what they were to do and they did it fast. Pull down one lever and come out of there so that as many as possible could do the same thing before they had to report to their jobs of work.

When these voters came out of the booth, they all did the same thing. They went back to the little table and were handed another piece of paper. With it in hand, they hurried back to the truck and it sped away to bring in another load.

Even if somebody had not told me, my intelligence would have informed me that those pieces of paper passed out after the vote could mean but one thing: That these men and women were being paid for their votes. I further learned that the piece of paper was worth two dollars to the holder when presented at the proper place. But just to pile things up, I got around among the party workers who were handing out literature to voters headed for the polls. One snarlingly refused to give me any answer to my question. But several others gave me an answer, and their phrasing was so identical in every case that it could not possibly be by accident.

"That is a lie. We are not paying anybody to vote. These are poor working people, and we are *giving* them two dollars apiece to pay them for their *time* to take off long enough from their jobs to cast their votes. You can't make nothing illegal out of that. No law against it at all." Then they grinned in a way to let me know that they had all the answers.

Standing around in the warm Miami sun, I suffered from a number of thoughts. During slavery, a healthy Negro slave brought anywhere from seven hundred to two or three thousand dollars on the block. That was trading in Negro bodies. Now, here Negroes were selling something supposed to be infinitely more precious than our bodies at two bucks a throw. My mind flashed to that big scene in *Uncle Tom's Cabin*, where Uncle Tom declaims, "My body might belong to you, but my soul belong to God!" Then I smiled. This election certainly was a big joke on poor old Uncle Tom.[3]

From *Uncle Tom's Cabin*, I passed easily to the Reconstruction, what it had meant to Negroes; what it had meant to southern whites, and what it portended in later years for both races.

The over-simplified accounts of those days, heard in my childhood, made it an age of splendor for Negroes. Nobody ever connected up the awful decades that followed for the Negros with those ten years. I was a freshman in college before I came upon any details of the Reconstruction. Then an elderly and very scholarly Congressman from Virginia described for me the political mechanics of that time in the South.[4]

The Carpet-baggers were in power in the South. Not being citizens in southern states, they could not vote. The newly freed Negroes were citizens, and had the vote. Some southern opportunists hurriedly took the oath of allegiance, and between them and the Carpet-baggers, the Negro voters were exploited to political advantage. An election went something like this: The candidates and their supporters contracted for Negro votes. Naturally, few could read or write, but there were no voting-machines in those days. Come the day of election, anywhere from a hundred to five hundred Negro voters could be found locked in a barn, with plenty of corn liquor promised to them as soon as the voting was over. Also, they would get a whole dollar apiece to spend.

At the first call, a sort of foreman would unlock the door and march out his herd of voters in their shirt-sleeves and to the polls. After voting, they would be taken back to the barn to put on coats, then back again to the polls to vote again. Later they put on hats and the candidate "voted their hats." If necessary, they were scrambled up and brought back to the polls from another direction and voted again. All that a white man needed who yearned for a place and power under that system was a few hundred dollars and a tough conscience, and he was in. With the long and bloody struggle for Negro freedom in mind, many of those Scalawags must have laughed a-plenty to themselves. At a dollar a head, and voting each man three times, a Negro that during slavery would have brought at least seven hundred dollars on the block, he could now buy for thirty cents. And the Scalawag could profit infinitely more by the cheap vote than he could from the voter's sweaty labor, and he did not even have to feed and shelter him.

The measures of Rutherford B. Hayes brought to an end the golden picnic of the Carpet-baggers and Scalawags.[5] Before the fury of the re-enfranchised South, they scattered in every direction. The men who had reaped the harvest from the conquered and prostrate South were gone, but their naive tools, those Negro voters, were still around. And the scars of those Reconstruction years remained. Negroes had repeatedly voted their shirts, their coats and their hats, but had nothing to show for it but empty hands and eyes to cry with. Yet they were called upon to pay for what their exploiters had done. In those dark after-decades arose the Knights of the Ku Klux Klan, disfranchisement of the Negro through the three disabling clauses, the Grandfather Clause, the Property Clause and the Literacy Clause.[6] There came the habit of lynching, and later segregation in every state in the South.

Those Negroes, fresh out of slavery, cannot fairly be held responsible for what went on during the Reconstruction. They were illiterate. They had no background for making decisions, even to small ones that concerned their daily lives. The scoundrels who took advantage of them are the most reprehensible in all history. And to make it worse, they had brought off this monstrous villainy under the cloak of "friends of the Negro." But where, oh, where were these "friends" while the terrible decades rolled in? Unlike the turtle, their voices were not heard in the land.[7]

But this is A.D. 1950. I am standing on the corner in Miami, Florida. It is a southern city with hundreds of very modern and comfortable Negro homes, lived in by Negroes of many professions, from the finest colleges and universities in the United States. Among them are physicians, dentists, lawyers, morticians, pharmacists, teachers, registered nurses, ministers, journalists and the like. A Negro judge presides over a municipal court, and Negro policemen patrol the streets. Free public schools available for Negro children are ably staffed by Negroes. Negro business men control into the millions in wealth. Yet, here is the saddening picture of hundreds of Negros, no thousands, being herded to the polls just as in 1870, and paid two dollars for votes.

To me, it represents an incalculable loss of prestige to the American Negro. There is something ironical about peddling your vote for two dollars, then calling yourself a "Race champion" fighting for more Civil Rights. There has to be an overload of self-pity and insufficient self-confidence and respect to cause a Negro with a ballot, the most potent weapon in a republic, to make him feel that he needs to be led to the polls to express his convictions on public affairs. It has to be a lack of something to cause him to sell his vote, then look for some "friend of the Negro" to look out for his advancement. It is like a man in a jungle facing a tiger and throwing his high-powered rifle away, then calling for some friend to come help him.

Why so many of our so-called leaders spend so much time and energy hunting up "friends of the Negro" is more than I can understand in this day and age. It is self-evident that these persons who talk so loudly and so much about rights and things like that have no appreciation of their present status. They have not yet conceived of themselves nor the mass of Negroes as American citizens, with the same responsibilities towards the nation as others here. To them, Negroes are still wards of the nation, to be done for, but with no responsibilities for the welfare of the United States. We are just here like tourists. Therefore, it is not up to us to fight for able, impartial executives, legislators and jurists. We get carried away by anybody who comes along and claims to be a "friend of the Negro."

Like voracious bluefish swarming around a school of menhaden, this type of politician has been fattening off of us since 1865, and most of them have done very well for themselves by their insincerity and our credulity. But even so, these political craftsmen cannot claim credit for originality. This "taking the heathen" gambit is only a variation of the old missionary game. For nearly three hundred years the English boasted that the flag followed the missionary. The routine is, finding the competition too keen among your own kind, you fit out a clipper and set sail for the "heathen." Appoint yourself his pitying partisan until you can land enough force to take him. Variations of this racket have been worked out and succeeded all over the world, even right here in these United States. Sell dope to the heathen. The dope may be beads, lengths of calico, whiskey, opium, friendship, or some other stupefying stuff. It is a good way to make a big man out of yourself in a hurry. The old game is still good as long as you can buy their votes for two dollars and put them to single-shotting.

But no fairly intelligent Negro has any right to be deceived by any political "friend" who offers to buy his vote. The fact that he offers to buy it tells you what he thinks about your character, and the petty amount gives you his estimate of your intelligence. Lumped together, you are two dollars worth of integrity and brains.

Nor need the Negro leaders of the vote-selling, single-shotting Negro electorate hope for legislation in our favor. They do not seem to realize that when the candidate has paid them off at the polls, he has no further obligations. Nor has he any cause to wish to further our interests out of respect. Under our Constitution, there is no royal ruler. That quality is distributed among the citizens of the United States. Every American is part of the king that rules over this nation. To sell your vote is to abdicate your part of the throne, and that is that.

And how can the Negro leaders who hailed these outside organizers so loudly, and the voters who so slavishly followed their counsel, reconcile their "friendship" with the fact that they neglected the twelve-and-a-half-million-dollar school bond issue at the polls? A generous slice of this was earmarked for the improvement of Negro schools in Dade County. If these organizers had really been our friends, they would have stressed the improvement of Negro schools over the senatorial race. But this was certainly not the case. I sampled 164 voters as they left the polls and asked if they had voted for the Bond Issue. Many of them behaved as if they were hearing about it for the first time. Thirteen stopped and told me that they had voted for it. Two of the others told me that it was not important. What they needed to do was to get the right man in the Senate. That school business could be looked after later on. What can be clearer proof that, no matter what they said about being "friends of the Negro," it was not true.

The Negro vote was thought necessary to elect the candidate they were here to put over, and that was all.

Negro participation in the southern primaries has only just now been restored after generations of being outlawed. So the Smathers-Pepper race is, or was, of the greatest importance for Negroes. It does not gain its significance for us and the nation half so much from what the candidates said about Negroes, and how they said it, as from our own concept of the value of the franchise as expressed by our behavior at the polls. Evaluations of Negroes as participating citizens are certainly being made. Serious and analytical minds will search out whether we see it as our responsibility to serve the common good by supporting men of high caliber for important offices, or whether, ignoring such qualifications, we rally around the "good masters" from the Negro point of view. That will determine whether we are slave-minded mobs or reliable citizens.

One very successful professional Negro man observed, "This FEPC is not the big bonus that our people are taking it to be. In the first place, it is unworkable, and if it could be made to work, it would be a two-edged sword. These Negroes don't seem to realize that. If it could be enforced, what would hinder white office workers, insurance agents and executives, morticians and the like, from penetrating Negro business and throwing thousands of us now gainfully employed out of work? Don't fool yourself that none of them wouldn't do it if there is a good living to be made at it, and there is. White teachers could then man our schoolrooms. We had better learn to think before we yell so much."

"Man!" a pastor of a large Baptist church shouted. "You've got something there. You've got cold chills running up and down my back. Supposing this law passed and a white preacher who has been struggling along for years with a little charge of around a hundred members made up his mind to take over my five-thousand-members-church! And that mob is out there single-shotting for him right now. I'm watching out for these 'friends of the Negro' from now on."

Whatever the issues might be at any given time, it is certainly high time that the Negro voter took his responsibility seriously. It is time for us to cease to be the single-shotting herd. Let us vote our heads instead of our hats and shirts. Each voter approaching the polls fired by his own well-considered convictions and performing this most sacred civil duty in the way that it was intended. It is time for us to cease to allow ourselves to be delivered as a mob by persuasive "friends" and become individual citizens. In other words, turn our backs upon the concepts of the Reconstruction and not keep turning back the clock. To take a look at the calendar and read it right. Find out that this is the Year of our Lord 1950, and *not* 1865.

THE END

Mourner's Bench

Why the Negro Won't
Buy Communism

The American communist party held a convention in New York recently. Henry Winston, a Negro, and organizational secretary, got up and fervently preached a crusade to sweep the American Negro into the party wholesale.[1]

Then in recent days our attention is called to The Peace Information Center and its Dr. W. E. B. Du Bois, a Negro, indicted by the Dept. of Justice, and now changed into the American Peace Crusade, called the most important pro-Soviet offensive in America.[2]

This dove-tails right in with other observations. The experts who watch communist strategy point out that the reds are now beginning a rather important drive to build up their Negro membership. They give two reasons for this. First, the commies hope to lump the American Negro in with all the other colored peoples of the world, so that we will feel that if we fight against the North Koreans, or Mao's hordes, we will be acting against our own best interest. That all colored people of the world must hang together against the whites.

Second stanza: Since Negroes, like all other workers, will be increasingly important in [the] defense industry, the communists hope and pray to use us to do their dirty work in the way of sabotage and espionage. It can easily be seen that they do not think very highly of us and our character by that.

So, the current party line is to muss us up in every way. Even to observing Negro History Week. If, as and when the eleven red leaders go on to jail, it is reported that four Negroes will be among those who will succeed them. Not long ago, Howard Fast had to eat crow for four columns in the *Daily Worker*.[3] It seems that he had made a slighting remark about Lt. Gilbert, the Negro who was found guilty by court-martial in Korea.[4] Fast had said that Gilbert had no business to be fighting in Korea in the first place, but for fear that might frighten off some Negroes, even that much was counter to the new party line. So he had to beat his breast resoundingly and whine that he was guilty of "white chauvinism." The American Negro's feelings simply must *not* be hurt-ed.

This present hassling over the American Negro is just some more of an old soup-bone warmed over. Common meter, Brother Peter. It has been around twenty-five years since certain Negroes of my acquaintance picked up their doll-rags and headed for Russia. The very first I heard of was Wayland Rudd, a minor actor. Then William Patterson, a Harlem lawyer, the Goode brothers, and later Paul Robeson, who at that time was the idol of the American public.[5] At the time, Russia seemed like an odd kind of a place for a pleasure visit, but otherwise, I paid the matter no mind.

My active curiosity was aroused when around 1930, Langston Hughes and Louise Thompson led a group of some twenty-odd Negroes to this same Russia.[6] It was beginning to look like a trend. The rumor was that these people had been selected to produce Negro plays in the Soviet Union. But among them, there was no director, no playwright, no nothing theatrical. Just an oddment of young Negroes at loose ends. So when I asked questions, I was told that the Kremlin was extremely interested in the American Negro. The communists wanted to be our kissing-friends.

I was very interested to know just why they were grinning up to our faces. The press of the world was reporting actual starvation and nakedness in parts of Russia. So I knew that there was some kind of a bug under that chip when I was told that the "People" of the Soviet Union were terribly distressed over the "horrible conditions" existing among the American Negroes. That just did not sound natural to me. People who are hungry and cold just do not worry about things like that thousands of miles away.

Yet and still, my informants gleamed and glowed as they told me how the Russians fairly vominated a thing like race prejudice, and meant to come to our rescue. In fact, Russia was the sworn champion of *all* the darker peoples of the world. And in particular, we American Negroes were so downtrodden, they deeply pitied our case.

Right then and there they lost one black sheep. I was poor, but I certainly did not feel pitiful. But anyhow, I wanted information, so I asked just what Russia *could* do, even if our condition had been as they claimed. I just could not conceive of Uncle Sam letting Stalin sit in on, say, a Cabinet session, nor presiding over the Senate and swinging votes. I tried hard to visualize armed Russians invading our Georgia and dealing with a mob that had been a little hasty with a brother in black.

So what the hen-fire could Russia do for us? And why did we look so valuable to Stalin? Numerically we were a scanty tenth of the population of the United States. We did not sit in on the policy-making bodies of government. We had no control whatsoever over the Armed Forces of the nation. Compared to the vast wealth of the nation, economically, we did not weigh too much. Nor were we overcrowded with technicians and scholars. So why did the communists want us so badly?

From reading, listening closely in silence, and watching things, I discovered our peculiar value to Soviet Russia. I soon saw that they did not love us just because our skins were black. The USSR was bent on world conquest through Asia. They saw in us a shoe-string with which they hoped to win a tan-yard. A dumb, but useful tool.

In spite of the world brotherhood propaganda, it was obvious that Soviet Russia was bent on carrying out the Czarist Russian plans to be masters of Asia. Once they had had a toe-hold in China, but had been expelled from there early around 1904 by the more alert and ingenious Nipponese. Now, while pretending to feel for the little peoples of the whole world, meanwhile issuing hot denials of imperialistic intentions, the Soviet was bent and bound to continue the march to the East. And that was right where we American Negroes could come in handy. With the war-like and determined Nipponese standing across their way of empire, plus Western influences, we were badly needed. The Asiatic millions must be led to fear and thoroughly hate the sight of a white skin. To rise up against their leaders, place their dependence on Russia, swamp Japan, and throw the "white oppressors" out. It must be repeated and kept in mind that this passionate love of the non-whites did not apply to Japan, for obvious reasons.

So the brains in the Kremlin eagerly seized upon the race propaganda of the United States, feeling sure and certain that they really had something that they could use. It could be dusted over Asia to good effect. A horrible example of white rule over darker races. A most frightful scarecrow to shake at the peoples of Asia, and thus hasten them into the arms of the Soviets.

With the then twelve million Negroes in the United States won and done, we could be filed away for the day of revolution here. The dumb black brutes to bear the actual burden of physical combat. Highly expendable. One white zealot discoursed to me at length on the glory we would win under the party, come that day, and millions of us would fall out in the streets behind the barricades to win freedom for the oppressed masses from our "masters."

When the man kept on mentioning Negroes, Negroes and nothing but Negroes "out there," I was moved to enquire:

"While we're out there tussling with the might and power of the Armed Forces of these United States, just where will *you* be?"

"Oh, for God's sake! If we are willing to do the *thinking* for you, *you* ought to be glad and *proud* to do the fighting."

Accented just like that.

"I see," I murmured, and I did. I said it calmly enough, but inside I had jumped as salty as a mackerel. This gang looked down upon us and despised us. They discounted our abilities and integrity infinitely more than those southerners from whom they were pretending to defend us. On top of that, their raw flattery and insulting patronage was intended to hide cold and ruthless hearts. The plan was obviously to herd in the dumb black fools, and when the time arrived to use us up like so many worn-out undershirts and think nothing about it. I thought some more, and by then there was nothing in the drugstore that would kill them all quicker than I, come that day. Nobody has ever yet celebrated being taken for a chump, even by a smart man, and when it is tried by a dumb chuckle-head, that puts knobs on it.

While they waited for the day of revolution, the third important use the communists planned for the Negro masses was to lie down and act as the mud-sills of the proposed American peasant party. This was to be maneuvered in a way to carry out unknowingly, the program of the communists.

The party felt a deep need for such a stratum. The Kremlin had launched out on the conquest of the world by analogy. Then they began to see that what worked among the peasants in Russia did not work so well here. This country was too rich, the working man too well fed, clothed and housed. There was no grinding poverty to make men bitter and desperate. The place was much too juicy and jumpy. Poor folks went up the ladder and rich men tumbled down. What was needed was a permanent bottom-class. Somebody who could be made to feel at the scratchy bottom, and no chance to get up from there without the violent overthrow of their "masters."

They found nothing like that on hand. As one rich and well-born matron said to me, "We do not employ Americans of any color as domestic help. White or black, there are no American servants. They are all millionaires, temporarily short of funds. Instead of being content where they are, they plan to be the boss themselves next year."

So, without putting a name to it, the commies went about creating a permanent lower class by dialectic persuasion. Wealthy persons *per se* were born vipers. There was a great weeping and wailing over share-croppers and the like. All unskilled labor was glorified in words, but bedded down as far as possible to form a foundation for this peasant class. The pleasures of peasantry were lauded to the skies. To make it appear inevitable, the nation was flooded with propaganda about there being no more frontiers; no more chances at all for free enterprise; not a prayer for a lone individual to rise by his own efforts. No more *nothing* but collectivism. It was like a rotting fog hovering over the land. It was as if from a vigorous youth, the United States had arrived overnight at a decaying old age. It was a case of don't try anything. If you could barely keep alive then you were spying noble. The trade unions were invaded and the line peddled that the members were really serfs. No more individuals at all. Their case was really pitiful. Nothing to do but hate bosses and work toward the day when they could do away with their hated oppressors. So labor disorders of an unheard-of intensity and violence swept the nation. It has taken years for many to come out of this fog and return to the American tradition.

The proposed peasant party failed to come off. Mostly, it failed because the Negro, the intended mud-sill, refused to hold still so that he could be built upon. What the party overlooked is the fact that the Negro is the most class-conscious individual in the United States. The biggest snob in America, bar none, is a Negro house servant. It works in varying degrees up and down the line. Kings and potentates, yes! Good groceries, fast cars and fancy shoes, yes indeed! Draped down in raiments of needlework, the average American Negro would much rather call in ten doctors to tell him how near he is dressed to death, than to have one commissar come around to tell him how near dead he will be before he is allowed a change of clothes. The party, misinformed, grabbed the wrong sow by the ear. The dear peasant in the Soviet Union, in his shapeless felt boots and slurping his cabbage soup, meant exactly nothing to us. Just the thing we are striving to get away from. For us to long for that would call for much more persuasion than the party has been able to deliver.

How dead the permanent bottom-man is in the United States was pointed up by last November's elections. The huge majorities piled up by Taft and others who opposed regimentation of the working man said a mouthful.[7] The average American still sees himself as a yeasty man. Why kill the boss? He might be the big boss himself next year. It has been done time after time and again. Every man a king when he gets his break.

Having decided to mount their world rule on black American backs, it is interesting to note how the reds went about the important business of capturing the American Negro.

For a blueprint, they took an ancient and long-discarded folk piece. The analogy of the "white mare." It got to be said during the Reconstruction that the highest ambition of every Negro man was to have a white woman. While one of their faces was spouting about how deeply they resented, and would die defending us from, white slanders the other face was patronizing us insultingly with a revival of these old notions. As a supreme inducement to join up, prospective party members were grinningly offered white mates.

Facts are facts, and it cannot be denied that some of us were influenced. This explains why so many of the Negroes high in party councils have white wives, or husbands, as the case may be. But it also explains one reason for so few party members. When you look at the thousands who could have and did not, it tells you something. The vaunted foundation for sweeping the whole Negro body into the party was laid on sandy land. The structure went shackly even before the winds began to blow.

All the way along, there has been entirely too much dependence placed on sex. Very few of us felt the need of help in a case like that. It offended the thoughtful among us because it amounts to a tacit belief that we are a people totally under the sway of sexual pleasures, a sure and certain way to get us.

The "white mare" apparatus failed to pay off. Yes, it is true that mules will unhesitatingly follow a white mare anywhere and at any time. But it is known there's danger in arriving at conclusions by analogy. It is possible, and even probable, that we might not be mules. But the reds evidently thought so. That is why Harlem swarmed with party-sent white women during the pressure drive of the Thirties. Even white girls of high school age were up there under party orders and doing their level best to "persuade" Negro prospects, and then bring them on through "religion."

But it is still to be observed that Negro membership is slack and scanty. There is constant turnover in membership from backsliding. What happened to the misled little girls is another story. Perhaps there is some connection between this "pig-meat" crusade and the later dismissal of numerous teachers from the New York City school system.

By such whoopdedoo was the Lincoln Brigade recruited to go to Spain in a vain attempt to place the Russian Bear at Gibraltar.[8] But believe it or not, even we can learn a lesson. The disillusionment lingers on.

Another Bear trap was the one polished over by Winston in a recent red convention. That is, that the party must infiltrate into Negro protest organizations, and generally seize upon Negro causes, and otherwise come to be looked upon as our saviors, champions and friends.

This is a very old soup-bone to be warming over. God knows that we have had the experience of communist help, and it sure has been a lesson to us. The notorious Scottsboro Case is a horrible example of how they "help things out."[9] The case was wormed out of the hands of the National Association for the Advancement of Colored People, who had taken steps for a quick and quiet settlement. The party wanted nothing of the kind, and they got it. The party stirred up years of world-wide publicity for themselves as the defenders of darker peoples. The boys got life-time in jail and other unhappy bonuses. There have been other and less-heard-of cases where they "helped us out." Like the joke about the corn salve. The toe is gone, but the corn is still there. We have come to be real shy of party help, so Henry Winston, speaking for Moscow, can save his breath.

And their shaking of the North Koreans and the Chinese communists in our faces is less than intelligent. There never has been any bond between us and the Chinese. So once more and again, the commies show off their estimate of our intelligence. When numerous Negro homes are mourning the death of their sons, husbands and brothers, and boiling over with rage at the knowledge of butchery and inhuman torture of their loved ones at the hands of these same yellow skins, to now be exhorted to treasure them and take sides with them is too much to expect of us, even though we are supposed not to be able to remember nor feel resentment at a thing like that.[10] The sense that God gave a billy goat would have prevented this insult.

In a recent issue of the *Pittsburgh Courier*, a GI in Korea publishes an earnest letter to American Negroes, warning us against the falsity of communist propaganda. In the letter, he strove to make us realize how horrible the commies are in fact. He goes on to wonder how any American Negro can think of joining the party, or even stoop to read the *Daily Worker*.

In addition to jumping into causes as our defenders, another device to supplement that is to set up false enemies of the Negro and then make a great to-do about knocking them over. It is always interesting to note that these "enemy" individuals whom the party discovers to be *our* enemies, have always been less than enthusiastic about communism. In every case, they are and were bedded in Americanism. So we cannot dodge the suspicion that these "enemies" stand in the way of the change in form of government here. They must be done away with as popular public figures if the revolution, the dictatorship of the proletariat, is to come to pass. Our dark hands must be used to pull them down and out of the way of the coming revolution. These "white chauvinists" must go! All done for Negro benefit, you understand.

Another important "defense of the Negro peoples" is by literature and art. The Negro press and protest organizations were not near enough. Besides, they were not telling the thing right. Becoming disgusted and all put out about this condition, the party got the bright idea of their "literary defense" hoping thereby to make a clean sweep of us into the party ranks. And it did not matter, whether despite our mulishness, we saw the light, and joined up in any great numbers or not. The material was what they wanted for anti-American propaganda abroad. So they were going to do it for our benefit, whether we liked it or not.

They acknowledged that the Negro press and protest organizations soapboxed a gracious plenty about racial grievances fancied or real, but there was no real meat to the thing. No hopelessness, no despair, no suggestion of scrapping the Constitution, no mention of revolution. Just lawing and jawing for a better adjustment into the framework as is. And mingled in was offensive material concerning fine cars, big houses, wealth and education among Negroes here. It indicated a "black bourgeoisie, black chauvinists" no different from the white capitalists, and utterly detestable. Lying counter-revolutionists and all that. The party decided to ignore them and create its own Negro literature.

Established Negro writers were approached to produce the kind of fiction that the party could use and approve of. The formula was, you can't win, Negro, you can't win! Expanded, the poor, dear colored character starts off to be something in the world, but he or she gets trapped by our form of government, and down he goes into the lowest depths like buttonless britches. Pity the poor, black brute! Rotted away morally and in every direction, but not his fault at all. It lies at the door of the people of these "United Stinks." In other words, the formula of "The American Tragedy." The Negro characters could not get too low and revolting. The lower and more despicable the better. The sop to the Negro public was, poor thing, what could he do under this American way of life? Negroes here are doomed from birth!

The reward to the compl[ia]nt author was pre-arranged critical support, plus sales boosting and handling. For those who drew back from representing a whole race thus falsely, vile slander and abuse.

It was brushed off as chauvinism that it was not just a matter of race pride, but utterly against fact. Like everybody else in the nation, a Negro can take his choice. The thousands on thousands of very successful Negroes in numerous fields could be offered in proof, so it was and is obvious that you *can* win.

But the party had orders that this evidence of Negro success under the American system must be suppressed. The outside world must see us as a low, degraded mass, and impossible to be otherwise under constitutional government. Poor things! They will take us by our hands and lead us away from all this, and back to the Middle Ages with them. From where we stand, that is just like Mrs. Astor battling to free herself from her enslaving Cadillac to win her way into a Russian droshky.[11]

Now with their intense efforts for at least a full generation, why have the commies gained such a comparatively few Negro adherents?

The party's first and foremost failure was under-rating our intelligence and self-esteem. I have no way of knowing whether they just scraped up any old-fogy notions that they found lying around, or whether they were briefed up by the earlier Negro sycophants that they got hold of. Certainly, the high-sitting black comrades do not object to the insulting program. The rest of what the party has to offer us as a way of life is as morbid and ugly as the devil's doll-baby, when we are on the hunt, like everybody else, for something pretty; tasty-like, to make our side-meat taste more like ham.

I reiterate, it is amazing how commies can hang on to a mere notion in the face of facts. They try to change the whole world, but refuse to let anything change *them*. They simply will *not* see us as Americans, nourished on the same ideals as other Americans, and so headed in the same direction. So why would we want to swap freedom for bondage? Why wouldn't we like this freedom-feeling as well as the next one?

I will not contend that we Negroes are more religious than other Americans, but certainly we are more ceremonial. Negroes own more church property per capita than any other group in the United States. There has to be a reason for that. We *must* like it. So how expect us to turn godless in a lump? Like a lot of other Americans, many of us do not attend church regularly, but we have no thought nor intention of doing away with God. We like Our Maker, and feel better to think He is somewhere around on the premises.

The anti-white program was another mistake. We do *not* hate white people as the commies are determined to believe. As fellow-citizens, it is our privilege to give each other skull-draggings on occasion, but laying all jokes aside, we certainly have no wish and desire to kill off the pink-toed rascals. Even if they were not useful as they are, we'd keep 'em for pets. Where is the kick in being an American if you don't call everybody out of their names now and then from the President on down? Just as natural as the Fourth of July. Are these commies so blind through the eyes that they have not seen us always in there fighting just as hard as anybody else in a common cause? From the revolution on down.

Go against our own country because the Chinese are yellowish in color, indeed! We would fight them just as hard, just as fiercely, if they were lam and damn black. They are not Americans. It has been proved too many times and by different countries, that nationality is stronger than race.

The party got misput on the road again when it fell for that old "leader" foolishness. They have proceeded from the premise that all they needed to do was to capture, or buy in, a few well known Negro names to have the whole tribe of A'nt Hagar's children come tumbling in behind them like a passel of sheep. There is no such of a person among us. Not since Booker T. Washington has there been any "Moses of his race."[12] Like the rest of Americans, we use our privilege of acting contrary and doing our own picking and choosing.

Instead of running like a fool at a funeral after the commie captives, however popular and prominent they might have been before they were taken in the raid, once they are shackled and begin to spout the trite jargon, somehow they seem to repel, rather than to attract. They give off a funny kind of smell. We stand off and look at them, remembering the flash and shine they had in their former existence, then shrink away from the morbid spectacle of their commie state. We shake our heads and mumble, "What happened, what could have happened to make so-and-so like that? They've come to be significant!"

It could be that feeling of strangeness is the inside key to the failure of the party to attract Negroes in any numbers. It feels ghosty, and too much like marrying a zombie. Death on the breath, and something feeling corpse-like to the hand.

Now take it to pieces, and everything is old and mouldy. What they call new and progressive is nothing but momucked-up dialectics. Just like children talking hog-Latin. What it is about is at least a thousand years old. The social devices of the Middle Ages, when the serf was bound to his master, which they have just found out about.

Their touted "significant, socially-conscious" literature is a steal from the old morality plays. Authors and other artists must cater to the Kremlin as they used to do to the Medicis.[13] Their labor arguments pre-date the machine age. The worker must own his tools in this highly mechanized age indeed! If that is kosher, then the very next time I go on the air, I'm carrying off the [microphone]. They are still waging a war against "the masters," somebody who has been dead and gone too long to talk about. All in all, the commies carry on exactly like they have been in a trance like the Sleeping Beauty since the days of Genghis Khan.[14] Awakened by the smell of blood from World War I, they sprang to life like the sons of the dragon's teeth, to continue their bloody march across the face of the world.

So this Russian philosophy does not take with us. We are not that morbid by nature. You need a huge inferiority complex to be a commie, something for hate to feed on. The reservoir of party thought is too much like the Dead Sea. You can and will get gassed to death just trying to fly over it. We are too American to fit in. Our idea of top dog is one who can muscle it out from the shoulder. Russia claims a great victory from World War II, when in fact, it was something like Max Schmeling lying flat on the canvas yelling "Owoooo"—then demanding the championship of the world.[15]

The majority of American Negroes indignantly refuse the role that the party has assigned to us. That is, to go around the world like Paul Robeson, W. E. B. Du Bois and a few others, and be the "horrible example," the pitiful object, the face on the bawl-room floor.[16] On top of that, we are loyal Americans. To paraphrase Benjamin Franklin's definition of a Tory, to us an American commie is a person with his so-called head in Moscow, his trashy body over here, and whose neck ought to be stretched.

Or better still, somebody ought to take and re-bury the dead. By behavior and flavor, they are zombies. Something oozes out like a viscous seepage from a morgue. It is poisoning the air of the world.

So Henry Winston and the whole party in convention assembled, may call upon us like the priests of Baal, but I predict that few will answer.[17] Not even for Chinese hides will we come, not being in the luggage business. Years ago, they had us at the mourner's bench. If they failed to bring us through religion then, they certainly will not make it now. Their latest cry can well up from nothing but desperation. The last stroke of exhausted nature.

From the very beginning, in dealing with us, the party has been led astray by the illusion of color. It has been tested and proven that we feel closer to the American white man than to any foreign Negro. The differences between us and foreigners are deep and fundamental.

We are Americans, and so, a wing-footed people. We are confirmed in our springing impulses by the spirit that hovers this continent. The party is a Society of the Dead, soulless zombies, but even the living-dead have no place among the living people. They, and their way of life, died on the plains of Asia centuries ago. May they now return to their graves and remain there in decent death.

A Negro Voter Sizes Up Taft

"Mister Republican," otherwise known as Robert A. Taft, of Ohio, has an over-average chance to be chosen Elephant Boy of 1952.[1] At a characteristic Taft gait, the senator has been moving up the field since 1936, until now he is in position to make his bid for the collar of roses.

To gain the Presidency of the United States, Candidate Taft must win a considerable percentage of the Negro vote. The question before the house is: Can Taft call the backsliding Negro vote back to the Republican fold? I believe he can. Negroes in Ohio are already enthusiastic about him, and politically conscious Negroes all over the nation are talking about him and sizing him up.

It was not ever thus, even in Ohio, nor yet in his native city of Cincinnati. I was in Cincinnati for three months during the time his first campaign for the Senate nomination was making up.

"What are the Republicans running that Taft for?" was the prevalent query among Negroes. "He might be all right, but he's so standoffish that we don't know too much about him. Why in the world didn't they put up Charles P.? Now that's our Taft, the swellest white man in Ohio or anywhere else, as far as that goes. Charles P. could get every one of our votes."[2]

It has been said that the only sure way to keep from agreeing with Robert A. Taft is not to let him get to you with his reasoning and that big brief case stuffed with facts. Once he gets to pulling facts out of that case and reasoning with you, you're sunk.

The majority of Negroes decided to keep Taft out of reasoning range.

They were going to pay him no attention whatsoever. But they did not count on his wife, Martha Bowers Taft.[3] She is said to have crashed a meeting staged by a group of Negro women in Columbus, Ohio, and asked to be allowed to say a few words. Out of politeness, they told her that she might speak briefly, but her subject had to be Abraham Lincoln.[4]

Now, the charm and wit of Mrs. Robert A. Taft are famous. She got in there and began to talk, the next thing you know, she was making speeches to colored groups all over Ohio. Her husband got in his licks and won the nomination by a majority of 75,000, and went on to win his Senate seat by 170,000.

Taft has strong Negro backing in Ohio now. They no longer mention him politely as "Mister Robert A." He is "Senator Bob" now, and they take great pride in him. They admit now that his quiet manner fooled them.

"That Taft is a fighting piece of plunder!" The speaker might have been looking with admiration at a jet plane or an A-bomb. "A tough piece of leather, and well put together. A regular catawampus on the Senate floor. He's something!"

Another said the same thing in another way. Exhibiting a full-length picture of Senator Taft, he paid tribute to the senator's gameness. "There never has been such a brainy and nervy fighter in the Senate since Daniel Webster.[5] Don't tackle Taft unless you really mean to do some fighting. See that?" He pointed to Taft's paunchy middle. "Don't kid yourself that that's fat. That's his craw. It's so full of grit that he just has to carry it a little low. And then to allow more room for all those brains he's got. In fact, all the rest of him is made out of brains."

Still another clapped hands for Taft. "Nobody else but Senator Bob could have thrown that Bilbo clean out of the Senate."[6] With the gestures he made, you got the idea that Senator Taft had grabbed hold of his colleague from Mississippi and flung him bodily from the Senate chamber. "Don't let his quiet way fool you. That Senator Taft is a killer."

It is to be remembered that Senator Bilbo, "The Man," was re-elected to the Senate from Mississippi. For years he had amused himself by saying things from the Senate floor infuriating to Negroes. He was the sign and symbol of all that Negroes in general hated in American politics. Senator Taft, chairman of the Republican Policy Committee, personally took the lead, stood on the floor of the Senate and asked for a vote against Bilbo's taking the oath. The Senate upheld Taft's motion and their vote thus excluded Bilbo from his seat in the Senate. Needless to say, the Negroes all over the country rejoiced at this.

The Negroes of Ohio point with pride to Taft's long record on other measures affecting Negroes: anti-poll tax legislation, FEPC, consideration of the civil-rights bills, housing and rent programs, discrimination in selecting displaced persons for admission to the United States, withholding Federal education funds wherever racial discrimination was practiced, cloture, attempts to limit debate and break up filibustering, anti-lynch bills and, under the Taft-Hartley Law, a clause to protect Negroes' right to work regardless of the discriminatory union rules. In August, 1942, he voted to exempt all servicemen from poll taxes in national or congressional elections. This enabled many Negroes to vote for the first time in the South.

"Our Senator Taft's record is wonderful," an educated woman active in Ohio politics observed, "but I'm not sure that all of our people understand his motives. Senator Taft is not pro-Negro. He is not prowhite. He is not prolabor, nor promanagement. The man has some strange passion for justice. He would work just as hard to stop us tomorrow if he believed that we were oppressing anybody. And he will tell you so if you ask him. That gives what he does more weight with me. He is not trying to win our votes so much as he is trying to do what is right."

Others see Taft in that light also.

In appearance, Taft is a tall man, over six feet, and with a big build on him. His hair is sandy and thinning. You couldn't exactly call him handsome in the face, but he certainly does not look anything like the devil's doll baby either. He has something more impressive than a Hollywood handsomeness, that look that few men are favored with—obvious intelligence coupled with equally obvious self-assurance, indicating an ego well nourished. It is as plain as day that the man is so used to things that he need make no gestures to impress you.

And Robert A. Taft came by that look naturally. His folks had been good livers for a hundred years before he was born. He has been bred to physical comfort, good manners and public responsibility. His grandfather, Alphonso Taft, was Secretary of War under President Grant.[7] His own father, William Howard Taft, held various important public offices before he became the twenty-seventh President of the United States, and after that became Chief Justice of the United States Supreme Court.[8] Senator Taft's mother was a somebody.[9] He married a somebody. Martha Bowers Taft's father was Solicitor General of the United States.[10] She is a well-educated, witty and charming woman.

Everything around the Taft home is bred up. You ought not to be surprised to hear that the Taft dog's folks had belonged to Thomas Jefferson and his cat's great-grandmamma to the twentieth power had frisked around the billowy skirts of Martha Washington.[11] The Tafts, like the Adamses, the Cabots and the Lodges, have done quite some descending.[12] They could line out their "begats" if they wanted to.

This being used to things is always serviceable to white people dealing with Negroes. Raw manners quickly repel us. "Maybe he means all right," we murmur behind his back, "but he doesn't know how to talk to people." We do love manners and behavior in people in high places. Otherwise, we don't believe they should be there. That is one of the many reasons why American Negroes did not go in any numbers for communism. For the most part, the Reds were much too crude and common.

But going back to descending: Senator Taft's political attitude has descended, too, but not from any public figures of recent years. He is not, and makes it plain that he does not aspire to be, a people's man in the popular concept of that term. He is never stampeded by public clamor. Loaded down with information and readied with conclusions drawn from what he has learned, his move is to direct the enthusiasm of the populace rather than to be led by it. To him, a leader is supposed to lead, and more than that, be prepared to do so.

This attitude makes Senator Taft something right out of the Federalist.[13] This was the position of the men who held high office in this republic during the period brought to a close by the advent of Jacksonian democracy.[14] Beginning with him, the mob took over. It is easily possible that the men who governed this country from the end of the Revolutionary War down to the election of Jackson were not all giants, but no one can honestly maintain that the Spoils System, that looking after party stalwarts with high-paying and responsible public offices, regardless of ability and fitness, has improved the quality of American statesmanship.[15] Robert A. Taft is a political throwback. Like Alexander Hamilton, Taft just will insist on principles, which makes him a lone wolf at times, but also makes him remarkable in this day and age.[16]

Senator Taft is without doubt a mental man. Not even his worst enemies have ever accused him of being dumb to the fact. His public utterances emerge from massed information. Start something, and he immediately goes hunting down the facts. This habit and inclination are valuable in a public official. He is supposed to know more than his constituents about public affairs. That is what he is sent to Washington for. Only by knowing all that can be found out can he ward off all hurt and danger from those who have trusted him and sent him to look after the public welfare.

The institution of the relief program in the early '30's is a case in point. The excuse given by the New Deal Administration was that they heard that somebody said they would start a revolution against the government if it was not done.[17] A man like Taft would not have rushed to appease such people the moment that he heard such a rumor. If the program was necessary, he would have instituted it without a doubt, but not for that reason. The very first thing Taft would have undertaken would be to find out just who were the people thus threatening the security of the nation, and in exactly what tone of voice. As soon as possible after that, the entire country would have known who those individuals were by name.

You may well ask what good that would do? Just this: That we would have known in the early '30's, and not in the late '40's, that communists were working in our midst. We would have learned something of their habits of boring from within, of their penchant for fishing in muddy water and taking advantage of unhappy conditions in a nation which they meant to knock over. If somebody in authority had only sought to know who was talking revolution, and why, a show of hands would have been forced, and the opportunities to worm their way into the very government, to do the evil here that they have, would have been lost to them then and there.

It is significant that nobody asked a thing, or if they did and were told, they carefully kept the information from the American people. The recent trials and convictions of persons once high in New Deal positions and other New Deal followers, point out to us the high price that the United States has paid for ignorance. If it is denied that it was ignorance, then the implications are even more serious. Anyway, it is well known that in those days anybody who endorsed the Constitution was a "capitalistic reactionary," and to admit patriotism was to be classed as a "dirty chauvinist." Anybody worth a samovar of tea was a "liberal," was known as an "intellectual," and went about talking about "directives" instead of plain orders.

Negroes suffered from this like everybody else. Only, if you were a Negro, you were even more detestable. You were a "black reactionary," or a "black chauvinist," or belonged to the hated "black bourgeois." It came to be known that the way to a government appointment was to make with the dialectics.

Therefore, Taft's open-faced Americanism will be welcome to great hordes of Negroes, who, like their white compatriots, have been part of the American underground for so long. Yes, there has been an American resistance army for a number of years, a sort of guerilla hand doing what they could to restore constitutional government. So Senator Taft, should he become the candidate, is going to have to deal with that slippery word "liberal." This egg of leftist dialectics has been sneaked into the mental nest of Negroes as well as whites, and with the same purpose and confusion. It is just another instance of the sly and cynical prostitution of the dictionary by the leftists for their own ends. The word "liberal" is now an unstable and devious thing in connotation. For example, card-carrying members of the Communist Party describe themselves as liberals to hide their party affiliation. Pinkos and other degrees of fellow travelers boast of being liberals.[18] Led astray by leftists, who do not, however, admit they are pro-Kremlin, great numbers of uninformed persons believe that the proper interpretation of the term "liberal" is a person who desires greater government control and Federal handouts.[19]

In the same way and for the same purpose the word has been debauched for Negroes to mean "pro-Negro." So persons like Henry Wallace, Frances Perkins, Eleanor Roosevelt and Vito Marcantonio have all been classified in our minds as liberals.[20] And that is why Negroes themselves are not spoken of as liberals, since, naturally, the relaxing of racial lines is something that must come from the other side of the race line.

So, when Negro voters shall ask of Candidate Taft whether he is a liberal or not, there will have to be a definition of terms.

Senator Taft speaks of himself as a liberal, but his concept is far from the new interpretations. As he sees it, "A liberal is a man who believes in freedom of thought, who is not a worshipper of dogma. . . . My objection to many of the self-confessed liberals today is that they are constantly advocating measures to reduce freedom."

"Senator Taft," said the *Indianapolis Star*, "is a liberal in the traditional sense that Jefferson and Washington were liberal. He believes the main purpose of government is to set men free and to keep them free, to give equal opportunity to all and special favor to none."

This brings to our minds the picture of Alexander Hamilton leaving his sick bed to travel to Rhode Island to influence a vote to break the deadlock for the Presidency between Jefferson, with whom he had violently disagreed on many occasions and who belonged to the opposing party, and Aaron Burr.[21] Hamilton persuaded the Rhode Island voter to cast his vote for Jefferson, "because he is the better man."[22] That is the kind of liberalism that Taft is talking about.

So it can easily be seen, as the saying goes, that some changes are going to be made. We, as Negroes, will not find Taft a liberal in the sense that we have been taking the word to mean, for Taft, I repeat, is not pro-anybody. Neither pro-Negro, prowhite, prolabor, promanagement or anything else. He is for the cause and the occasion when he believes it to be right. He states this carefully in four points in his My Political Credo:

1. The people and the individual retain true liberty.

2. All citizens are assured equal justice under law, that they may have life and liberty.

3. All citizens have equality of opportunity, particularly in their youth.

4. All citizens have a standard of living which will make happiness possible.

Therefore, it is as plain as white on rice that we, as Negroes, are included, provided we can disregard the intense political appeals to racial antagonism of the last few administrations, and see ourselves once more simply as citizens. You heard the man say "All." And from his record, when Taft says "all" he means all.

For this reason, we will be disappointed if we expect a continuation of the BFSUAKDOFSM, which being interpreted, means, the Bureau For the Setting Up and Knocking Down of Straw Men, to gain the Negro vote.[23] Taft, from his record, will not attempt to whip up either racial or sectional antagonisms to make himself look like our valiant protector. If his record of seeking to do away with inequalities where we are concerned does not impress us, we will have to go unimpressed. He will not create any false situations and then "straighten them out" for our votes. I will try to show you what is meant by that.

In company with four friends, I sat in the office of a prominent Negro politician in Miami in 1950. The Smathers-Pepper campaign had just come to an end, so, naturally, we talked about politics.[24] At one point I got to "A-Man-Who"-ing—a peculiar cry heard around convention halls in political-nesting time—at a great rate. I was touting a prominent man who, if elected governor, would initiate certain measures bound to be beneficial to the Negroes of Florida.

A doctor politician flagged me down. "Stop right there, Zora. You have wasted a lot of time telling us all about what the man is for. Now get down to practical, vote-getting politics and let us know whom and what your man is against."

He saw the astonishment, if not the dismay, in my face and smiled sardonically.

"Negroes, Zora, don't vote for things. You have to show us something we can vote against."

He went on to elaborate, and illustrated his point by cases in recent political history. The perfect example, he told me, was the Marian Anderson–Daughters of the American Revolution incident of 1939.[25] It will be remembered that the manager of Marian Anderson sought a concert hall for her appearance in Washington, D.C. It had to be large, in view of the tremendous drawing power of the great artist. They applied to two places—Constitution Hall, the property of the D.A.R., and the auditorium of the White High School, the property of the District of Columbia.

Now, there is no one so blind in the eye or deaf about the ears as not to know how tight is racial discrimination in the national capital. The D.A.R. refused their hall to Miss Anderson, allegedly on racial grounds. The school board likewise refused the auditorium of the high school, but, after some pressure and dickering, agreed to allow her to appear there, but only under the condition that no Negroes be allowed in the audience.

The BFSUAKDOFSM saw its duty and "went and done it." Mrs. Eleanor Roosevelt, Harold Ickes, et al., rushed into the fray and gave with a howl against narrow-minded racial discrimination that could be heard in Addis Ababa, and probably was.[26] But it was all directed against the D.A.R. No loud screaming against the school authorities. Mrs. Roosevelt, in her protest against racial discrimination, even resigned from the D.A.R. The Straw Man Knockers-Down raised the biggest crop of political corn outside the champion state of Iowa. They foamed up the biggest cloud that has hung under heaven since the Flood, and all whipped up out of political egg-white.[27]

This statement is not unjust, and certainly not meant to be unkind. Mrs. Roosevelt is a very charming woman, but I like a little fact mixed in with what is handed to me as gospel truth. What happened cannot be interpreted entirely as a battle against racial discrimination, because if Mrs. Roosevelt took out after everything like that that went on in Washington, D.C., she would be hollering night and day, and every day. Restaurants, theaters—everything downtown in Washington practices Jim Crow. I can't recall that Mrs. Roosevelt campaigned mightily against these places, as she did against the D.A.R. Therefore, there is but one conclusion to be drawn. Marian Anderson is a world-famous artist, and deservedly so. Her name was bound to be known to Negroes all over the nation. "Defending" her would attract notice and bring in the Negro vote.

As far as the high-school auditorium is concerned, to jump the people responsible for racial bias would be to accuse and expose the accusers themselves. The District of Columbia has no home rule; it is controlled by congressional committees, and Congress at the time was overwhelmingly Democratic. It was controlled by the very people who were screaming so loudly against the D.A.R. To my way of thinking, both places should have been denounced, or neither.

Throughout the New Deal era the relief program was the biggest weapon ever placed in the hands of those who sought power and votes. If the average American had been asked flatly to abandon his rights as a citizen and to submit to a personal rule, he would have chewed tobacco and spit white lime.[28] But under relief, dependent upon the Government for their daily bread, men gradually relaxed their watchfulness and submitted to the will of the "Little White Father," more or less.[29] Once they had weakened that far, it was easy to go on and on voting for more relief, and leaving Government affairs in the hands of a few. The change from a republic to a dictatorship was imperceptibly pushed ahead.

Senator Taft has taken notice of this. "What concerns me is that people gradually come to accept limitations on freedom as the necessary incidents of government. They hear of outrageous treatment given to others with a kind of dull interest, instead of the fiery resentment such incidents would have aroused in the past."

We Negroes will look in vain if we expect Taft to set up imaginary enemies and situations to defend us from. Look for him to do nothing but stick to the facts. He will not make a great to-do of saving us from Bilbo's ghost, for example. He will not dramatically call "The Man" back from his grave and challenge him to mortal combat. It is hardly likely that Senator Taft will introduce a bill in the Senate around election time to bar striped tigers from entering the United States on passport to eat up Negro children. In short, he will not go into the straw-man business at all. He will give us credit for average intelligence as he has the Negroes of Ohio.

Can Robert A. Taft win the Negro vote of the United States? Again I express optimism. I am only one person, and do not pretend to represent the 15,000,000; but then, nobody else can either. Since the passing of the late Booker T. Washington, there has been no single Negro figure who could influence too much the national Negro vote.[30] We have become as individualistic as the white electorate. In these days, candidates must appear before Negro audiences to state their case. They can no longer arrange with a leader to have the mass of Negro votes delivered. That is, except in some cases where votes are bought among the whites also—among the fringe which does not care what the issue is or who wins. All they want is a little handout.

About all a Negro leader can do now for his candidate is to arrange for the most profitable spots for him to appear and get out the crowd. When a Negro subleader begins to tout his candidate to the average voter, he is told, "I have never seen the man. If he wants my vote, let him come and ask me for it. I'm a man just like everybody else. I'm not going to vote for anybody who iggs me."

"Igg" is short for "ignore." The Negro voter is likely to be more sensitive than a white one. He fears being played down. The candidate must certainly not give that impression. He must, as far as possible, spread his platform before the Negro voter in person. Igg him, and he is sure to vote against you. My people will say, "He thought himself too good to come among us. To hell with him!"

This far from November, 1952, thoughtful Negroes are already exhibiting tremendous interest in Robert A. Taft. "The man doesn't do any bragging on himself," a leading Negro minister told me, "so I haven't paid too much attention to him until here lately. But when you come to study it, his record on Negro affairs is unbeatable. He reminds me of Teddy Roosevelt, and he was some man.[31] I think I'll get up a sermon on Taft."

Then the divine extended his remarks, "It's about time this country went back to moral living. The trouble we are having with our young men and women is too distressing. Taft sounds like he stands for what we need right along in here. Yes, I think I'll have to preach about him. Some morals have just got to come from somewhere or else we're headed for hell."

The oft-repeated statement that Taft has no color disputes itself. Color does not necessarily mean exhibitionism, and that was just what Taft meant by his sarcasm when he told the reporters, "You won't find any color in me. I'm too darned normal." Taft has to have strong appeal to keep on winning and winning as he does. His very sincerity and truthfulness are a kind of color. People can feel sincerity. As an old woman told me, "What comes from the heart, honey, will go to the heart again." Deceit may have flash, but truth has color.

Say what they will about theatricals, despite his lack of those tricks Taft has aroused more genuine interest among us than any other Republican since Theodore Roosevelt. He has shown himself an old-line statesman with the affairs of today at his finger tips. He has fought the good fight and shown his integrity.

Why assume that we are impervious to reason and indifferent to decency? There is the example of the Negroes of Ohio to look at.

Never mind that we have run off after strange gods like the rest of the nation. We, too, can see the light and be reasoned with. Let Taft come among us with his blunt truthfulness, his bag of facts and his reasoning, and tell us what he has to say.

Court Order Can't Make Races Mix

Editor: I promised God and some other responsible characters, including a bench of bishops, that I was not going to part my lips concerning the U.S. Supreme Court decision on ending segregation in the public schools of the South.[1] But since a lot of time has passed and no one seems to touch on what to me appears to be the most important point in the hassle, I break silence just this once. Consider me as just thinking out loud.

The whole matter revolves around the self-respect of my people. How much satisfaction can I get from a court order for somebody to associate with me who does not wish me near them? The American Indian has never been spoken of as a minority and chiefly because there is no whine in the Indian. Certainly he fought, and valiantly for his lands, and rightfully so, but it is inconceivable of an Indian to seek forcible association with anyone. His well known pride and self-respect would save him from that. I take the Indian position.

Now a great clamor will arise in certain quarters that I seek to deny the Negro children of the South their rights, and therefore I am one of those "handkerchief-head niggers" who bow low before the white man and sell out my own people out of cowardice. However an analytical glance will show that this is not the case.

If there are not adequate Negro schools in Florida, and there is some residual, some inherent and unchangeable quality in white schools, impossible to duplicate anywhere else, then I am the first to insist that Negro children of Florida be allowed to share this boon. But if there are adequate Negro schools and prepared instructors and instructions, then there is nothing different except the presence of white people.

For this reason, I regard the ruling of the U.S. Supreme Court as insulting rather than honoring my race. Since the days of the never-to-be-sufficiently-deplored Reconstruction, there has been current the belief that there is no greater delight to Negroes than physical association with whites. The doctrine of the white mare. Those familiar with the habits of mules are aware that any mule, if not restrained will automatically follow a white mare. Dishonest mule-traders made money out of this knowledge in the old days.

Lead a white mare along a country road and slyly open the gate and the mules in the lot would run out and follow this mare. This ruling being conceived and brought forth in a sly political medium with eyes on '56, and brought forth in the same spirit and for the same purpose, it is clear that they have taken the old notion to heart and acted upon it. It is a cunning opening of the barnyard gate with the white mare ambling past. We are expected to hasten pell-mell after her.

It is most astonishing that this should be tried just when the nation is exerting itself to shake off the evils of Communist penetration. It is to be recalled that Moscow, being made aware of this folk belief, made it the main plank in their campaign to win the American Negro from the 1920s on. It was the come-on stuff. Join the party and get yourself a white wife or husband. To supply the expected demand, the party had scraped up this-and-that off of park benches and skid rows and held them in stock for us. The highest types of Negroes were held to be just panting to get hold of one of these objects. Seeing how flat the program fell, it is astonishing that it would be so soon revived. Politics does indeed make strange bedfellows.

But the South had better beware in another direction. While it is being frantic over the segregation ruling, it had better keep its eyes open for more important things. One instance of Govt by fiat has been rammed down its throat. It is possible that the end of segregation is not here and never meant to be here at present, but the attention of the South directed on what was calculated to keep us busy while more ominous things were brought to pass. The stubborn South and the Midwest kept this nation from being dragged farther to the left than it was during the New Deal.

But what if it is contemplated to do away with the two-party system and arrive at Govt by administrative decree? No questions allowed and no information given out from the administrative dept? We could get more rulings on the same subject and more far-reaching any day. It pays to weigh every saving and action, however trivial, as indicating a trend.

In the ruling on segregation, the unsuspecting nation might have witnessed a trial-balloon. A relatively safe one, since it is sectional and on a matter not likely to arouse other sections of the nation to the support of the South. If it goes off fairly well, a precedent has been established. Govt by fiat can replace the Constitution. You don't have to credit me with too much intelligence and penetration, just so you watch carefully and think.

Meanwhile, personally, I am not delighted. I am not persuaded and elevated by the white mare technique. Negro schools in the state are in very good shape and on the improve. We are fortunate in having Dr. D. E. Williams as head and driving force of Negro instruction.[2] Dr. Williams is relentless in his drive to improve both physical equipment and teacher-quality. He has accomplished wonders in the 20 years past and it is to be expected that he will double that in the future.

It is well known that I have no sympathy nor respect for the "tragedy of color" school of thought among us, whose fountain-head is the pressure group concerned in this court ruling. I can see no tragedy in being too dark to be invited to a white school social affair. The Supreme Court would have pleased me more if they had concerned themselves about enforcing the compulsory education provisions for Negroes in the South as is done for white children. The next 10 years would be better spent in appointing truant officers and looking after conditions in the homes from which the children come. Use to the limit what we already have.

Thems my sentiments and I am sticking by them. Growth from within. Ethical and cultural desegregation. It is a contradiction in terms to scream race pride and equality while at the same time spurning Negro teachers and self-association. That old white mare business can go racking on down the road for all I care.

EAU GALLIE

ZORA NEALE HURSTON

Which Way the NAACP?

Beneath the sound and fury of the drive of the National Association for the Advancement of Colored People to integrate the schools of the South, what is the real intent? Is it conceived that a Negro child is "advanced" by sitting in the same class-room with White children, and if so, how and why? Or is the push a determined attempted jail-break from the imaginary cage of race on a national scale?

Since in every state in the South the identical text-books are issued to White and Negro schools, then the implied inferiority must rest on the Negro faculty. That assumption is strange in this day and age, for now there are Negro instructors who are graduates of the finest schools of education in the nation, and in some instances, graduate degrees from foreign universities. Naturally, some teachers are more efficient by native abilities and preparation, as among the Whites. This is a very important point, for there is no denying that intelligence is an individual matter, rather than one of race.

So it has to be believed that mere physical contact is advancement in itself. The inferiority complex which brought forth the "Tragedy of Color" doctrine has to enter in. One has to be persuaded that a Negro suffers enormously by being deprived of physical contact with the Whites and be willing to pay a terrible price to gain it. In other words, the Negro masses are expendable, and the end justifies the means.

For example, it is reported from a reliable source that the reason for the troops entering the Girls' restroom at Central High in Little Rock was that a group of the White girls had ambushed one of the Negro girls in the rest-room and was forcing her head down into the toilet-bowl.[1] The troops were summoned to rescue the struggling, screaming girl. Can such a degrading experience be gloated over as "advancement"? Yes, if more physical contact can be regarded as the ultimate goal and ideal. No, and emphatically no, if my own dignity and self-respect are taken into account.

One has to be willfully blind and deaf to assume that forced physical contact will bring about anything but just that. Physical contact has al-ways been bountiful in the South without bringing about social mingling. There has never been *legal* segregation in the North, because of historical circumstances too well known to go into here, yet social mingling is con-spicuously absent. And there is determined segregation in residential areas. Nor do Northern Negroes, products of non-segregated schools, make a better showing in the Northern Universities than those from the South. To the contrary, the majority of entrants are from the South, and on the whole, come off very well indeed. Therefore, what is the interpretation of "advancement" in the minds of the NAACP?

Then let us examine another word in the title of the NAACP. The choice of "Colored" instead of the more universal term Negro is significant. To non-Negro, and even to the uninformed among us, Colored and Negro are synonymous, but that is a kind of slurring over. It brings up that ancient line drawn during slavery between the usual classification of house servants and field hands. When the plantation owner had children by a slave girl, and most of them did, the master did not put his left-handed off-spring out into the field to work. They were the house servants—the valet or ladies-maid, the butler, the coachman, etcetera. They ate from the kitchen of the "Big House" and were given the cast-off clothing of the master, mistress and their sons and daughters. So far was as possible during slavery, they were privileged characters. Naturally, they were proud of their lighter color. The cleavage was as wide as their enslaved condition permitted. Indeed, some masters went so far as to free these mulattos and give them land. The records show that many of them, though held at a distance by the Whites, held black slaves, and fought in the Confederate armies to preserve slavery of the blacks. There existed the same difference in the South as the Tidewater and the Piedmont among the White. These mixed-blooded folk did not speak of themselves as Negroes, but "gentlemen of color." It is very possible that this historical meaning of "colored" was taken into consideration when the organization was named. Very, very possible and probable, since being a Negro was looked upon as such a tragedy by the founders.

Dr. William E. Burghardt Du Bois is the originator of the NAACP.[2] However he did not call it that in the beginning. The idea struck him while he was at Niagara Falls, [and so] he called it the Niagara Movement. Later, it was reorganized with White people in high places in the set-up, [and] was re-named the National Association for the Advancement of Colored People.

The early years of the NAACP were devoted almost exclusively to attempts to destroy Booker T. Washington, founder of the Tuskegee Institute in Alabama.[3] Other goals were announced, but the real drive was a "Get Washington" movement. Why this was so, provides interesting observations on the Association, and throws light on its meaning.

Du Bois, of Great Barrington, Massachusetts, is a graduate of Harvard, and took his doctorate at the University of Berlin. He is in his late eighties at this writing. In his youth, indeed up to the time of World War I, the matriculation of a Negro at a White college or University was so rare that on graduation, he became a national celebrity on the face of it. He also became a Race Champion—the great profession of the time—automatically. It was assumed by both White and Black that such a Negro was a phenomenon for having been able to learn what the average White man could learn, college curricula being tempered to the average intelligence instead of genius. It was [d]ue to the fact that only recently had the Negro masses emerged from slavery and were overwhelmingly illiterate, but burdened by poverty that a scant handful of the 4,000,000 and Northerners who were free before 1865 got there.

Imagine the chagrin of Du Bois when he returned from Germany to find that Booker T. Washington, a graduate of Hampton Institute, had the world by the ears because of the success of Tuskegee, and his "Dignity of labor" slogan. This was unfitting [o]n more than one count from the view-point and expectations of Du Bois. In addition, Washington had emerged from extreme poverty and [was] a Southerner. And though Du Bois was definitely brown of skin, and Washington the son of a White man, it is obvious that Du Bois could [not] conceive of such a person as rating higher than a Dr. of Philosophy from Berlin, to say nothing of ivied Harvard. The situation had no name. So, though Booker T. was a definite mulatto, and his accomplishments, it is obvious that the cultured Du Bois could not think of Washington as anything but a field hand out of place. To worsen matters from the point of view of Du Bois, when he landed in New York, a telegram awaited him from Washington, offering him a position at Tuskegee. He promptly refused it.

Then began the bitter, even scurrilous attacks upon Washington, Tuskegee, and industrial education for Negroes. He and the organization lost no opportunity to ridicule skilled labor for Negroes. The implication was that these 4,000,000 recently freed people ought to take off for Harvard and Berlin. How this was to be done by this largely illiterate mass was never explained. The traditional cleavage between house servant and field hand was prominent, though lofty terms were used to name it, but it was the same old thing.

Against Washington's overwhelming popularity which amounted to idolatry among the Negroes, Du Bois and the NAACP could muster a small band who were encouraged to speak of themselves as the "Better Thinkers" although what this meant was necessarily vague. It was clear that they were merely "Think-we-are-better" without any excuse for the posture. Anyway, the wicked attacks continued until the death of Booker T. Washington in 1915. The founder of the famed Tuskegee fat with the success of his school and floating on a cloud of world-wide fame, never even dignified the attacks by a single answer, unless we take for an answer a statement he made on one occasion when he said, "I will not allow any man to drag me down so low as to make me hate him."

During these years, the inventor of the NAACP was spreading the drear gospel of "the tragedy of color." In his autobiography, *From Dawn to Dusk*, he laments in deep crepe of his not being invited to the Junior Prom of his high school in Great Barrington, Mass.[4] He was the only Negro in attendance. His works follow this theme continually. He is convinced that the Negro race was betrayed by the North because the Reconstruction was not prolonged sufficiently to beat the South to its knees in the matter of race acceptance, and that the nation would not have any problem of race in the 20th Century if this had been done.

Such were the early years of the NAACP. It is obvious that the drive since 1954 is to resurrect the Reconstruction era in our time. And there is at hand a weapon which was not available in the decade that followed Emancipation—the vast number of Negro voters. All office-seekers love votes. And even though the number of Negroes who are paid members of the organization is small compared to the 16,000,000 Negroes in the United States, office-seekers cannot be sure and are not inclined to take chances. Hence the present power of the NAACP in Washington.

But even so, the "advancing" has its drawbacks and stumbling-stones. It is the reluctance of the Negro masses to come right forward and be advanced. This apathy creates the need for martyrs to the cause. Even martyrs or block-busters are in short supply for inappropriate individuals are too often selected. Too many are those self-conscious sample who are hungry for "note," (Notoriety) get the call. Or perhaps hungry for extra spending-change. For the charge that these martyrs to the cause of advancement are generously paid by the NAACP to suffer publicly has never been refuted. Unsuitable candidates make the testing hard for the organization.

For example, Virgil Hawkins, a teacher in the public schools of Florida for some years before his love of studying law, and studying law at the University of Florida, came down upon him.[5] He is now fifty years old, and allegedly paid for his persistency by the NAACP, insists that his ardor is undiminished, and this despite a shabby showing in the entrance exam. He thus holds a high place in the martyrology of Advancement.

Another sample [indicating] the hunger for candidates is Autherine Lucy, the Tactless.[6] She had also been teaching before her overwhelming passion to be a librarian came down upon her. And though the School of Library Science of Hampton Institute in Virginia has an excellent reputation, to say nothing of the splendid Library Schools at Howard University, Washington, D.C., and at Morgan State, Baltimore, nowhere suited her taste like the University of Alabama. Mere learning was beside the point. She was a crusader for advancement, and seemingly, you cannot advance seated among Negroes. By a court ruling, she was admitted, but promptly talked herself out of the place. Her humiliating experiences for the few days that she was enrolled at the University must have left scars that will be there to the end of her days. Neither the stipend she is said to have received from the NAACP nor her brief notoriety can erase them. She was expendable.

As to the experiences of the nine Negro students who were chosen to block-bust at Central High School in Little Rock, no matter how generous the stipend they allegedly received, it was worth it. One degradation after another. What price glory? What price for your head thrust down into a toilet stool? Or is the price regarded as cheap for the pleasure of bringing the White students and their parents to the place where they will eventually tolerate you in the same class-room? They too were expendable. The end, perhaps, justifies the means. What price proximity?

The implications flowing from the present drive of the NAACP are tremendous and revealing. First, it is quite obvious that the drive had been planned by the organization before the Supreme Court ruling in 1954. The price of political support by the Negro voters of the large Northern cities in particular, and of the nation in general was the mingling in the class-room of both races during the tender, formative years. But why so picayune? Why not bargain for the end of all segregation at a whack? This is where the revelation comes in. The Reconstruction-mindedness; the glory of the house servant Heaven. Rather than new and progressive, it is a lapping back to the foggy past. In close to a century of education and progress by the American Negro, self-consciousness of race and an inferiority complex stemming out of the past, we should have come to the place where notice by the Whites and the bolstering of proximity as a sign of tolerance would be utterly unnecessary. Not what counts with the majority in the nation, but what counts within ourselves should have arrived by now.

Why fumble around with old fogeyism in exalted phrases when the whole new world is before us? It has been thoroughly demonstrated that the old shibboleths of the early days of freedom never were anything more than meaningless sounds. Race Champion, Race Solidarity, Race-Consciousness—nothing but collections of syllables.

There is no use in smothering the fact that Negroes hold Negroes in contempt. That fact cancels out all the full-mouthed phrases. It is universal among high and low. Negroes have never made a celebrity among themselves, but a Negro rated as a celebrity by Whites is immediately given applause. "Old Cuffy" an African word that means Negro is always looked upon as ridiculous. "The Buckra" (Whites) has the prestige even when complained of. This attitude is so general that it is deeply imbedded in Negro folklore.

"This was a common-clad Negro, and one time they called on him to lead the church in prayer. He was not used to being called upon in public—him being too poor to have a lot of friends—so he felt bashful-like. So he hung back and begin to pray.

'O Lord, I, I er, I got something to . . .'

The church begin to bear him up and waited for him to go ahead. 'Amen, Brother Stumpy, go on and ast Him.'

'Lord, I got something to ask You but . . .'

'Unh-hunhhhh, Brother Stump, go right ahead.'

'O my high-riding, my conquering God, I got something to ask You, but I know You can't do it. I . . .'

'Aw, go ahead and ask Him. You know He can do anything, even to leading a butt-headed cow by the horns. Don't hold up the church so long. Go right ahead and ask Him.'

'Well then, Lord, I want to ask You to get the Negroes together, but I know You can't do it. Amen, Lord.'"

Folklore says that even before this prayer, another humble, but observant Negro was called upon to petition the Throne. His prayer went like this:

"O Lord, here I is, knee-bent and body-bowed this morning before Your mighty throne, depending upon Your tender mercy. O Lord, I ain't nothing; my wife she ain't nothing, and my children, they ain't nothing neither. And if You fool around with us, You won't be nothing neither. Amen."

This refusal to be led by Negroes is well known to those Whites who have dealt with Negroes for any length of time. Through no fault of their own, Negro foremen and even military officers can not get the results from a Negro group that a White foreman or officer can. General Eisenhower found that out in Europe during the late War. Hence his desegregation of his troops.[7] Nor does high education and competence in a Negro wipe out this attitude in either educated Negroes nor the masses.

"The higher the monkey climb, the more he show his behind," says it succinctly. Perhaps this attitude was taken into account when the Niagara Movement was reorganized into the National Association for the Advancement of Colored People, when a White man became the head, and this has been true ever since. Yet, never has the organization found a home in the heart of the Negro population. Never has the membership reached anywhere near the proportion commensurate with the Negro population of the nation.

This brings up the query as to who is the *real* directive power in the NAACP? The Secretary, editor of *The Crisis*, and attorney are no more than fronts. This is obvious because, with the exception of the late James Weldon Johnson and Dr. Du Bois, they became known by reason of their position in the NAACP, not because of any accomplishments which attracted national attention previously. Numerous Jews of wealth and influence hold membership in the organization. Many Gentile[s] in the same category are members. It is certain that the vast sums of money which have been spent and are still being spent since 1954, were not contributed by Negroes.

The charge so frequently flung that the NAACP is a Communist front outfit has not been proven. There is some kind of an affinity, however. The methods of going about things, common friends, and the like. "We are not Communists," several members of the NAACP have explained to me, "but we cannot be their enemies, for they have taught us how to get what we are after. We cooperate to an extent." And certainly [neither] the organization nor its house-organ, *The Crisis*, has exhibited any distaste for such well-known Party members as Paul Robeson and Langston Hughes.[8] The father of the NAACP, Dr. Du Bois has now been an outright Communist for many years. But it must be made clear that he had quarreled with the Association and left it before that, but still is looked upon as friendly there.

That brings us to the type of cases which the NAACP has defended. The organization has a notable record against lynching. This crime of passion against established law took hold of the South after the Reconstruction—in reality was brought forth out of that unfortunate period. The spotlight thrown upon the practice by the NAACP definitely hastened its extinction. From the record, it appears that Negro men accused of rape against White women have been most interesting to the Association. Defense of Negro women for any cause has been rare except in the recent era of integration of schools in the South.

One case that seemed to fit into the category of defense by the Association, but which was not handled was the case of Mrs. Spaulding, widow of the late C. C. Spaulding, founder of the North Carolina Mutual Life Insurance Company of Durham, N.C.[9] This company is the largest among American Negro insurance ventures. Mrs. Spaulding is an intelligent, well-informed woman of the highest moral probity.

Well, once upon a time, being a high-ranking Republican, Mrs. Spaulding was given an appointment under Oveta Culp Hobby in the newest Cabinet department, which is Welfare, Health and that sort of thing.[10] Days and weeks shuffled along and Mrs. Spaulding took note of something which the NAACP has been doing quite some fighting over. It seems that a government-supported hospital in Texas discriminated against Negro patients. Or did not admit them. Mrs. Spaulding complained to Mrs. Hobby that this hospital should positively NOT receive government money if Negroes were excluded. Mrs. Hobby, a good Texan, not only sent the money, but fired Mrs. Spaulding.[11] Naturally, it was expected that the NAACP would highly approve of her stand, and fight for her re-instatement from Hell to breakfast. But it did *not*.

When Drew Pearson reported the case in his column, this writer did not query the head office, but asked a member why there was no protest to Eisenhower in the matter.[12] The first statement was that Mrs. Spaulding was a Republican, and this was a party friction. It was pointed out that Mrs. Spaulding was indeed a Republican, but that was beside the point.[13] The patients who might want to be cared for at the hospital could not possibly be all Republicans. The issue was not based on party. Then it was added that Mrs. Spaulding did not wish to be defended, and so the NAACP had taken no action.

But Josephine Baker, an American Negro chorus girl, who had gone to France a generation ago and had become an over-night sensation in the French musical comedy theatre by crawling along a trunk of a tree on the stage dressed only in a string of bananas about her hips, and from this monkey-like start, became an international celebrity, fared infinitely better than Mrs. Spaulding.[14] Miss Baker's reputation was enhanced by her marriages to several European White men.

Like the late Will Rogers, all that I know is what I see by the papers.[15] It was stated that Miss Baker sashayed into the Stork Club in New York and ordered herself a steak. Walter Winchell, the famous columnist and radio commentator, was there, and seated a few tables from where Miss Baker and her party settled down.[16] Now, the Stork Club servitors were accused by Miss Baker of acting down right trashy about fetching along the steak. The affair bloomed into a real fracas in words. It was a celebrated newspaper cause. Now, Walter Winchell, a close friend of Walter White, then Secretary of the NAACP, and a sympathizer with the program of the Association, was bitterly assailed by the NAACP as a traitor because he did not come to the rescue of Miss Baker in the fracas. The Communists joined with the NAACP to give Winchell a thorough working over. The violent and scurrilous attacks upon Winchell by these elements became national in scope. But it must not be overlooked that WW looked down his celebrated nose at the Party and all its works and members and continually spoke of them as "Scummunists." That was the way he thought about them. Yet, he had ever been as helpful as he knew how to the NAACP. The much-married Miss Baker was elevated to the niche of a sacred icon.

Why Mrs. Spaulding was made out to be corn-bread and Miss Baker to be high-class cake is hard to fathom. The only explanation is that Miss Baker has ever worked along the line of integration, while Mrs. Spaulding has been sort of neglectful about the thing.

Which way the National Association for the Advancement of Colored People? It is reported that membership has slumped since the federal troops were planed into Little Rock. Be that as it may, it will remain a self-constituted dictatorship so long as it does not ask and receive a mandate from the entire Negro population of the United States. From the slow gains in *Negro* membership, it should be inferred that the organization has not found a home in the hearts of the people. It is possible that the people get the scent of the outsider around the place. And no Non-Negro can feel with Negroes, however hard he or she might try and wish to. It is something like the beautiful but dumb Marie Antoinette out of lack of comprehension saying, "Let them eat cake."

The term "advancement" needs a more precise definition, for the Association has not concerned itself with Negro promotion except along certain lines. It has certainly broadened its base in the last two decades—approved what it bitterly denounced when the late Booker T. Washington was advocating it.

There must be some recognition of reality. One that cannot be denied is the fragmentation, the seeming distaste for cohesion for a common cause among Negroes. The eminent author, Richard Wright reports this stumbling-block from West Africa.[17] The tribal masses do not want the same thing that the British-educated chiefs propose. In Ghana, formerly the Gold Coast, Kwami Nkrumah has succeeded to an extent with his Convention People's Party because he discerned this and took advantage of it.[18] Nkrumah was himself educated in the United States.

And if the passionate drive for integration of the schools of the South is intended to speed up the absorption of the dark minority by the majority

Whites, it is useless, the haste is. Nobody doubts that this will come in some distant time, but there are factors outside of White resistance to it to be gotten around. First that integrated schools have had no effect upon social acceptance in the North. Second, the open bid by the Communist Party in the thirties for the Negro population of America to come right forward and mingle and be mingled, was such a dismal failure. The Party itself has admitted it. Now would it have been anything permanent had the inter-marriages come off [sic]. Negroes are aggravatingly, but persistently, some-timey and faddistic. They make over something new with great enthusiasm for a short time, then forget all about it. When the "new" had worn off of a marriage tha[t] had been for so long denied them, (say a month or so) he would remember that he had divorce in his heels. Then back to the life and ways to which he was accustomed. There were numerous cases of this during the Communist membership drive. And it should have been noted the dizzying speed—say a mile a month—with which Negro men rushed forward to take advantage of the offer in the first place. Again, Richard Wright points out (*Black Power*, by Richard Wright, Harper & Brothers, $4.00) Nkrumah was smart enough to leave the people to their customs instead of trying to turn them into black Englishmen. They yearned for "free-dooom!" from colonialism and that was what he set out to gain for them without interfering too much with ancient custom.

The undisclosed master-minds who direct the program of the NAACP must take the realities into account if the organization expects to be of, for, and by the Negro people and cease to drive at what they feel Negroes *ought* to want or resent.

Where are you bound, NAACP?

Part Five

ON THE TRIAL
OF RUBY
McCOLLUM

Zora's Revealing Story of Ruby's 1st Day in Court!

"Way down upon the Suwanee River; far, far away—
There's where my heart is turning over . . ."[1]

. . .

SUWANEE COUNTY COURTHOUSE, LIVE OAK, Fla.—Stephen G. Foster could not have envisioned such a scene and situation when he wrote the lines that immortalized the river that flows not too far from this little courthouse . . . and the Negro musical idiom that he loved so well. Foster did lament in another of his world-famous songs that "The head must bow, and the back will have to bend . . . wherever the darky may go. . . ."[2]

. . .

Thus it is obvious that he did not contemplate a Negro woman on trial for her *life* for shooting to death a prominent white physician.

FIGHT ON TO SAVE RUBY FROM CHAIR

But in the Year of Our Lord 1952 . . . we see Mrs. Ruby McCollum seated at the table beside her defense counsel, under indictment of first degree murder . . . for shooting Dr. C. LeRoy Adams five times.[3]

Ruby McCollum is not yet on actual trial for her life. The hearing here in the Suwanee County courthouse was announced as a sanity hearing.

. . .

IT IS HER attorney's opening gambit in the struggle to save her from the electric chair. As the minutes drag on, it develops that it is not really a sanity hearing. It can be put down as a crafty move on the part of her counsel, P. Guy Crews, to disqualify two local white physicians as mental experts, and thus bar them from testifying as to his client's sanity at the time of the slaying.

Under oath, both Dr. Workman and Dr. Price admitted that they had NOT been trained in psychiatry, though Dr. Workman stated that he had some experience along that line for a year in Cincinnati, Ohio.

With the way thus cleared, P. Guy Crews then besought the court to authorize the prosecution to allow the defense to call in a specialist, at the expense of the defendant, to establish the mental responsibility of Ruby McCollum prior to, and at the time of Aug. 3, 1952, the date of the homicide.

SHE IS NOT GOING TO BE MOVED ANYWHERE.

THERE WAS a pause during which Judge Adams chewed choppily and rapidly on his tobacco! Then he said:—"It is not that I object to it, but I don't know for sure if it is legal." After a minute of reflection, the plea of defense counsel was granted.

. . .

It seemed to have been done so easily and peaceful like, but there was plenty of battle in the room. No sooner was the defense motion granted than the state's attorney got to his feet and addressed the bench. He suggested that there should be a time limit to this examination by Dr. McCollough, mental specialist of Jacksonville, and that the place of examination should be fixed.

"Oh, I can settle that right now," Judge Adams remarked, sitting informally half way around in the big swivel chair. "She's not going to be moved anywhere. She's going to stay right here in the custody and care of the Suwanee County Sheriff."

"As to the time limit of this examination," the state's attorney persisted.

His Honor agreed with this. He picked up a small notebook, but before he studied it, he fixed both prosecution and defense with his sharp eyes.

. . .

"I WANT DEFENSE and state to produce whatever witnesses they feel necessary, and any testimony that will throw light on the case, but no piddling along. We can't be just meeting here forever. I have tried many a homicide case, and I want this one to proceed like all the others. I can't see why this one should have any special interest. Let's have some definite arrangements."

After consulting the pages of the book, Judge Adams suggested Monday, Oct. 6, as the date when Dr. McCollough, the specialist in mental diseases, must appear before him to state his findings in the case of Ruby McCollum, where he may be examined by the defense counsel, and cross-examined by the state.

. . .

HE TALKED FOR THREE WHOLE DAYS

Though everybody concerned spoke in quiet, conversational tones, it was obvious that the state considered that defense was sparring for time, and seeking every opportunity to improve the case for Ruby McCollum, and the prosecutor was there to block his every move.

This was obvious when Judge Adams said that he was ready to hear any witnesses to Ruby McCollum's mental state, "not specialists. Dr. McCollough is a front rank man, but he was before me on another case and talked for three whole days. Any witnesses that either side cares to present?"

The short, plump state's attorney got to his feet.

. . .

"WE OFFER NO witnesses at this time because we are convinced that the defendant is perfectly sane. We are sure that she knows right from wrong." Then he sat down.

The Negro spectators, all seated in the gallery, leaned forward and all looked in the same direction like cows in a pasture. Their eyes were fixed hopefully on P. Guy Crews. The defense attorney sat quietly in his seat beside the defendant.

"That lawyer is no good," a man in the second row murmured in deep disappointment. "Why don't he let her talk?"

. . .

"THAT'S WHAT I say," a woman's voice grumbled. "He ain't let her say a word since the mess started. Shucks! You can learn more from the newspapers than you can from her."

. . .

THAT MAN IS A GETTING FOOL

"Shhh," your correspondent whispered over her shoulder. "This is not the kind of session for her to do any talking. Then, too, at regular sessions, at times a lawyer will not allow his client to take the stand to keep the prosecution from getting at them."

"Yeah, he had better watch his step," a tall, middle-aged man beside me agreed, and pointed at the state's attorney. "That man is a 'getting' fool. Never mind about that bald head and him being short and squaddly. He can go for broke. I done heard him at it."

. . .

THIS OPENING move in a tense and terrible struggle for Ruby McCollum's life was about over. Opposing counsel went up to the bench for a brief, whispered conference with the court. Then one could see the people in the room at leisure. Down stairs, where the white people sat, the room was nearly full. Four matrons sat together on the front row. They all leaned forward hungrily to catch every word and every move. The one on the center aisle had brought her son, about two years old.

. . .

MOST OF 'EM CAME TO HEAR RUBY TALK

The third one in the row had brought her knitting, and one was reminded of the females of the reign of terror of the French revolution, who brought their knitting and worked away as they cheered the sharp, heavy blade of the guillotine as the bloody heads fell into the basket.

Most of the Negro spectators had come from a distance to hear what Ruby McCollum was going to tell.

Six had come from as far as Fort Lauderdale. Jacksonville was well represented.

Local Negroes were conspicuously absent, with one or two exceptions.

. . .

THE CAST OF MAIN characters was interesting. The state's attorney in a rumpled blue suit, with a bald spot on his head so round and exact that it might have been the tonsure of a monk, sat very quietly in his chair.

But when he moved or spoke, he emitted alertness so intense that it was electrical.

. . .

CREWS FIGHTING FOR HIS CLIENT

P. Guy Crews, defense lawyer, tall, heavy of body, with his eyes sloping down so sharply at the outer corners that one got the impression that a slight blow, and his eyeballs would run down into his collar. His general expression was the kindly mournful one of a bloodhound. Soft-spoken and obviously careful not to antagonize the court, he was, nevertheless, mentally on his toes every moment. He was in there fighting for his client.

. . .

HIS HONOR, Judge Adams, was the most dynamic personality in the room. Slender, with his thick graying hair parted on the left and swept back in the manner of the Southern gentleman of generations past, high cheek bones and keen, bright eyes.

His face is reminiscent of three men whose pictures are familiar in history books . . . Patrick Henry, Abraham Lincoln and Jefferson Davis. He chewed his tobacco all through the hearing, kept notes in a ledger-like book, but heard everything that went on. When called on to make a decision, he thought a moment and came out positive and clear. His sentences were grammatically correct, but richly idiomatic. No effort was made to lard them with heavy legal terminology. He talked the common man's law. He allowed his human side to show on occasion.

. . .

FOR EXAMPLE, when defense counsel mentioned the use of two specialists in mental diseases, Judge Adams asked quickly, "Is it necessary to have two? Both of the men you mention are front rank men, but I had to listen to three days of testimony from Dr. McCollough in another case. I don't want to do anything to cut anybody off from offering testimony, but I want reasonable progress. Non-expert testimony is wanted today. I'd like to hear it." Who among us has not been bored with tedious, lengthy, technical testimony?

Court was dismissed and some went out hurriedly while others lingered in disappointment of sensational revelations.

The woman who knitted was the last to leave the room.

OCTOBER 11, 1952

Victim of Fate!

SUWANEE COUNTY COURTHOUSE, LIVE OAK, Fla.—Mrs. Ruby McCollum, central figure in the tragic drama of life and death which envelops this little community, is a strange figure. I watched her for an hour and a half this morning and she did not utter a single word, except in a short whispered conference with her lawyer.

She sat almost motionless in the large arm chair with her right elbow resting on the arm and her head inclined to the right in her hand.

Her plumpish body was clothed in a pale yellow cotton dress and low-heeled yellowish shoes.

. . .

CIRCUMSTANCES HAD not permitted her to have her hair done, but she had done the best she could to keep it in order by a hairnet. The presentation of her profile fixed the high cheekbones of her face in the mind. Otherwise, the face is short . . . and feline.

SILENT!—MOTIONLESS

Silent . . . all but motionless . . . the undiscerning could gain the impression that this Ruby McCollum was actually indifferent to her fate.

Her powers of restraint were evident when, near the end of the hearing, Ruby McCollum heaved a deep sigh and cast her eyes upward imploringly.

In this one gesture, one glimpsed the pattern of her whole life.

The quiet, restrained matron whom Live Oak had known prior to Aug. 3 . . . and the intensely emotional woman who had gone to Dr. Adams' office on that Sunday morning with a gun, and had shot him down with one bullet, then stepped over his body and fired others into his body in her rage.

. . .

THEN, TOO, there in the courtroom, she had held the pose that she had decided upon during the entire time, and it was evidently tiring by being unchanged. So when His Honor announced that court was dismissed, Ruby McCollum stretched out the right hand that had wielded the gun so effectively, and flexed all of the fingers of her hand.

THE HUMAN FACE HAS ITS PHASES

The dark brown color of her skin was accentuated where the dark skin of the back of her hand showed sharp contrast with the pale color of her palm. She looked at it, then rose and went away quietly with the sheriff.

Never once had she looked at the Negroes in the gallery! Never once had she looked at the assembly of curious whites behind her!

The human face has its phases like the moon. There is the high flame of a revealed soul on occasion, at other times a waning . . . and even complete absence . . . of light.

. . .

FOR THE space of the sigh and the uplifted, imploring eyes, it was full moon in Ruby McCollum's face. Then a sudden return to the absence of light.

It is possible that all during that grim session, she sat there feeling sorry for herself . . . a victim inadequate to her fate, utterly stripped, useless and despised.

But if Ruby McCollum, indicted for first degree murder, has delved into human nature, she must realize how wrong she is. Great misfortune, like poverty, is possessed of a miraculous substance by which others are made cheaply rich and fortunate. It is all in a sense of values. It is there in a point of view.

. . .

THE HUMAN soul is the creature of that mother of monsters and angels . . . the mind! As she sat there with her thoughts wandering around among the underpinnings of the world, with voiceless echoes shrieking around her, the spectators in the courtroom, and thousands who had heard of her plight . . . forgetting their own worries . . . felt lucky for not being in her shoes.

. . .

HAS ASKED FRIENDS TO PRAY FOR HER

It is said that Ruby McCollum has asked for the prayers of those who wish her well. She sighed and lifted her eyes like she was hoping.

. . .

HOPE IS like the hen making a nest on the Ark when there appeared small chance for a future. So far, she has said little about her reason for taking the life of Dr. Adams.

Perhaps she has reason to hope. It is possible, even probable, that we shall get some glimmer of this from her sanity hearing this week!

OCTOBER II, 1952

Ruby Sane!

LIVE OAK, Fla.—Mrs. Ruby McCollum has been adjudged SANE by the local courts . . . and she will face trial for the Aug. 3 killing of Dr. C. LeRoy Adams, prominent white physician and political leader.

The trial date has been set for Tuesday morning, Nov. 18.

As I sat upstairs in the section of the courtroom "reserved for Negroes" last Monday morning, I viewed with interest the central figure in this dramatic case . . . Ruby McCollum.

Mrs. McCollum is not a beautiful woman. She can't even be called pretty. Well-dressed and groomed, she is the kind who would best be described as attractive.

As I listened to the testimony being given which ended in the new trial date, I thought of the rumors which are still sweeping this town. I wondered what had made Ruby McCollum kill!

Had it come as a result of an argument with Dr. Adams over an alleged bill . . . as first stated . . . or is there any truth to the whispered rumors which have made Ruby's home life an "open secret."

Is there any truth to the rumor that what has happened in her life had to come from a strong inner drive to dominate . . . to possess?

. . .

OBVIOUSLY, Mrs. McCollum is a most self-contained woman. Never much of a mixer. She had her social life, it is true. But no one places her as ever about town . . . just talking with any and everybody.

"We saw her passing along in her car," is the way the average citizen of Live Oak refers to her.

"Her husband, Sam McCollum, was all over town and we all knew him well. A nice, friendly kind of a man! But we never saw them together. They never went 'nowhere' like a couple.

. . .

"BUT IT WAS STRICTLY understood that she was real jealous-hearted over Sam, though. She didn't make it a practice to go out and start no fights in the street, or nothing like that, but folks said she really made it hot for him when he got home."

Again and again, gossip has put Ruby McCollum down as a very jealous woman under her own roof. She was possessive, and what was hers . . . or she considered hers . . . she must RULE! Could that be the explanation, in part, of what took place in the office of Dr. Adams on the Sunday morning of Aug. 3?

Is Mrs. McCollum being correctly pictured by neighbors and acquaintances as 'the woman scorned' . . . with the 'Hell Hath No Fury' implications?

No living person knows the answers to these questions but Ruby . . . and Ruby ain't talkin' yet.

. . .

ON SEPT. 29, two local white doctors took the stand. They stated that Mrs. McCollum had been their patient for the past year. On Oct. 6, testimony developed that she had also visited doctors in Jacksonville.

Though only the psychiatrist, Dr. William H. McCollough, put a name to it, it was possible to see that Ruby McCollum was a hypochondriac . . . feigning sickness for a purpose of her own. Why? Was she seeking pity? No one can answer that question.

. . .

A CHARMING, worldly, ladies' man once told your reporter:

"There is nothing so dead . . . as a dead love. And no woman can be so thoroughly hated as she who tries to cling when you have let her know that you no longer want her.

"The smart women of the world," he continued, "are those who let you go immediately and without cries of reproach. That kind has a fine chance to win you back, by their indifference.

"No," he concluded, "men have no pity for a woman they have ceased to love. They can be more cruel to her than anything on earth. That is why you read so often of men murdering wives and sweethearts. She has committed the unforgivable crime . . . trying to hold him after he is tired of her."

. . .

NOW THAT Mrs. McCollum has been declared "sane," just what course will the prosecution take when the trial begins?

Certainly, they have heard the rumors which anyone walking into the town can get! Will these stories be aired, in an attempt to develop a "romance" angle? Or will the prosecution stick to the story, first disclosed immediately following the murder of Dr. Adams, that he was killed because of a "bill"?

Will the prosecution attempt to prove the rumors going the rounds, that Ruby and her husband had not lived together "as man and wife" since the birth of the last child?

. . .

PEOPLE TALK freely to me . . . but will they say the same things if they are put on the witness stand?

Will Sam McCollum be depicted as the husband who precipitated no scandal, for the sake of his children . . . but had told friends shortly before his death of "strange dreams"?

Nobody knows but Ruby McCollum! . . . and she ain't talkin' yet!

OCTOBER 18, 1952

Ruby McCollum
Fights for Life

LIVE OAK, Fla.—The curtain rises this week on the main act of perhaps the most sensational trial ever held in the South. At that time in the County Courthouse of Suwanee County, in the town of Live Oak, Fla., Ruby McCollum goes on trial for her life for the slaying of Dr. C. LeRoy Adams, a prominent and wealthy white physician and politician.

Ruby McCollum has admitted the slaying of the doctor, but still her act and her defense are saturated in mysteries. Judge Hal Adams, the presiding judge, has announced his determination to see that her trial is fair and just. To that end, he has steadfastly and firmly denied all members of the press access to the defendant. He says that the case is not going to be tried in the newspapers before a jury can be selected.

. . .

BUT THIS ORDER can have no power over the imagination of the people in and around Live Oak. Ruby McCollum has been enshrined in legend and folk-lore even before her trial begins. Something on the order of Jesse James, the very boldness of her act and the mystery surrounding it, have made of her a heroine to some.

People have had visions about her probable fate, and others have dreamed dreams. "She's going to come clear. I was spoken to in a vision, and the voice told me so. God moves in mysterious ways, His wonders to perform."

A woman who said that she was a "praying and pentecostal" soul asserted, and many others agreed with her. "It did not look possible by the white man's law, but that was the way it was going to be. Look how Paul and Silas had been in jail. The dungeon shook and their chains fell off, didn't they?"[1]

Another school of prophets swear that Ruby McCollum will surely be executed. Several claim to have already seen her spirit in one form or another wandering around free from her imprisoned body. She had appeared in the bedroom of a couple one night in the form of a cat-like animal, and was crying piteously.

To another, she appeared as a woman with the head of an eagle with a flaming sword in her hand in flowing robes, circling their house for three nights running and crying out in defiance. That meant that she was innocent and was going to her death unjustly somehow or another.

NOVEMBER 22, 1952

Bare Plot Against Ruby

LIVE OAK, Fla.—The trial of Mrs. Ruby McCollum, indicted for first degree murder in the shooting to death of Dr. C. LeRoy Adams, prominent white physician and political figure of West Florida, was dramatically halted here on the first day last week.

Judge Hal W. Adams, eminent jurist, in whose courtroom the trial was to be held, revealed to a startled, jam-packed audience that Atty. Guy Crews, Mrs. McCollum's lawyer, had been disqualified for working against the possibility of a fair trial for his client.

. . .

AS A stunned silence swept the room, Judge Adams said:

"This case cannot proceed further at this time. There has been an unexpected happening.

"Attorney for the defense is precluded from acting further for the defense. The court is astonished, in view of the wide-spread reports and publicity of this case, from Maine to California, and every part of the United States, that a plot such as I have every reason to believe exists, should have been attempted. The date for this plot is set at more than two weeks BEFORE Oct. 6." (It was on this that the sanity hearing was fixed.)

Judge Adams then proceeded to excoriate Attorney Crews for his bad faith . . . for his attempt to obstruct justice . . . and against a fair and impartial trial for Mrs. McCollum.

The dramatic announcement, which was made late in the afternoon, came as a complete surprise to spectators and reporters.

. . .

ALL MORNING long, examination of the 112-man panel of prospective jurors had proceeded. From this list, forty-six had qualified. Then came the noon recess, with court scheduled to reconvene at 2 P. M.

It was at 2 o'clock that the first break came . . . when Judge Adams announced a further recess because:

"Information has reached me that there is a document in existence, which, if true . . . may bring about a new development in the case."

At 3 o'clock, Judge Adams announced that the document had been discovered in Tallahassee . . . that a state trooper had been dispatched for it!

At 4:15 came the announcement, as Judge Adams once again entered the courtroom with a sheet of paper in his hands. The tightly packed courtroom fell into a tight silence. Tenseness fell like a shroud. Then, stern of mien, the white-haired jurist, standing in the lectern looked long and earnestly into the face of Mrs. McCollum before his gaze shifted to the face of Crews.

. . .

AFTER MAKING his dramatic announcement, Judge Adams praised two Associated Press reporters . . . Malcolm Johnson and Bernard Parker . . . "for their forethought and alertness in discovering the plot and promptly notifying the court of its existence."[1]

Continuing, apparently under great strain, the jurist said further:

"I have further and other expressions that I would like to make, and I am gutting my teeth to keep from making them."

Then came the statement which revealed Crews' disqualification:

"Defense counsel is precluded from speaking to me as an attorney . . . but he can speak to me as a man.

"Ruby McCollum must have a fair and impartial trial. She must have new counsel to represent her. I will see to it that she gets it. In justice to the defendant, it is not fair to attempt to proceed until she has sufficient time to be fairly represented. The state attorney concurs with me in this."

Judge Adams originally had planned to delay the trial for ten days, but when the state attorney objected, he set the new trial date for this Monday, Nov. 24.

NOVEMBER 29, 1952

Trial Highlights

Court session was set for 10 o'clock. At 9, though, clusters of people were outside on sidewalk and on edge of lawn.

. . .

Inside . . . courtroom jammed with spectators and prospective jurors. Two-thirds of seats in galleries, reserved for Negroes, occupied. Below . . . where whites sat, eyes of morbidly curious glued to door through which Ruby must come.

. . .

DIXIE ODDITY!—White enameled bucket of ice water on corner of rail, with white dipper rimmed with black. It serves as drinking fountain, and ANY and ALL who get thirsty, drink from same dipper.

. . .

Nine: fifty-eight—Ruby led in by state trooper. She walks briskly to seat without looking to right or left . . . up or down. Hair unpressed, bunched under hair-net. Wears expensive blue wool dress . . . long bright green coat of camel's hair . . . modish black suede low-heeled pumps on her small feet.

. . .

Only sign of Ruby's nervousness . . . pre-occupation with hands and rhythmic swinging of feet. Noticeably thinner and paler than at sanity hearing.

. . .

TEN o'CLOCK—Judge Adams enters . . . room rises respectfully.

"It has long been custom of this court to open a case of this kind with prayer," he says. Small man in tan coat prayed briefly.

NOVEMBER 29, 1952

Justice and Fair Play
Aim of Judge Adams as
Ruby Goes on Trial

LIVE OAK, Fla.—Ruby's first trial day is ended . . . but a strange melody lingers on!

And while Ruby is billed as the main actor in the drama being enacted here, it was this same Judge Adams who proved the principal actor, with a chorus of 112 prospective jurors.

The thing I'll always remember was Judge Adams' approach to the case. Once the panel had been placed under oath, he lost no opportunity to stress the serious and conscientious business of serving as a juror. Time and time again, he reminded the men of the compelling necessity of justice and fair play. Time after time he called upon them to "examine their consciences" with deep care.

"Remember that Dr. Adams was a white man," he said. "He was a descendant of an old, highly respected family in this area. He was a wealthy, cultivated man . . . Ruby McCollum (Mrs. omitted) the one who allegedly killed him, is a Negro woman. Can you honestly put aside all emotions and feelings and confine your verdict to what you hear from the witness chair and my charge to the jury?"

Twice, Judge Adams mentioned "reasonable doubt." He said that Ruby

McCollum was due the benefit of "reasonable doubt" . . . if any were to be found!

Were the prospective jurors willing to give it to her? Justice, worked out and handed down by long tradition, must prevail! If any man there could not cleanse his mind of pre-conceived ideas and be guided by the evidence . . . it would be a civic duty and an evidence of good citizenship to excuse himself right away.

Several excused themselves . . . and were complimented by the judge . . . as an evidence of honesty and good citizenship.

But the forty-six men from which a jury was to have been picked were dismissed following the afternoon's developments. Judge Adams will have to do it all over again!

NOVEMBER 29, 1952

McCollum-Adams
Trial Highlights

LIVE OAK, Fla.—Here are some McCollum-Adams trial highlights:

Defense Attorney Frank Cannon, Duval County Solicitor for the Criminal Court of Records . . . hired the evening before the trial started.

. . .

Mrs. Thelma Curry, receptionist for Dr. Adams, was obviously nervous and spoke almost in a whisper. Judge Adams, with one of his colorful colloquialisms, admonished her:

"Can't you talk no louder than that? Outside of here, we could hear you for half a mile . . . against the wind."

. . .

Vero Musgrave, used by the prosecution to place Ruby at the scene of the [crime], keeled over during vigorous cross-examination by Cannon . . . and was carried to the hospital on a stretcher. Musgrave is white.

Perfect order prevailed at all times during the trial . . . in court and in the city!

DECEMBER 27, 1952

Ruby Bares Her Love Life

LIVE OAK, Fla.—Before a courtroom packed with hushed spectators, Mrs. Ruby McCollum told of her love life with Dr. C. LeRoy Adams, white physician and politician . . . the man she killed "during a struggle over a pistol he had produced."

The jury heard her story and then found her guilty of first degree murder. The jurors refused to ask that she be shown any mercy.

In a thin, low voice, she told her story with frequent interruptions of objections by the state. Several times she was admonished to speak louder by the court so that the jury might hear her distinctly. The story did not come out in narrative form as Defense Counsel Frank Cannon, solicitor of the Duval County (Jacksonville) Criminal Court of Record, had planned.

The court sustained an objection by the prosecution that it would be too rambling, and confined it to interrogation by defense counsel and answers by Ruby McCollum.

. . .

RUBY STATED that she had known Dr. Adams for seven years. He had attended her at the birth of her third child, Sonya McCollum, but the alleged intimate relations between them did not begin until 1948 and continued until his death Aug. 3, 1952, she testified.

CANNON: Beginning with the birth of Sonya, did you have sexual relations with Dr. C. LeRoy Adams?

(State, through Assistant Prosecution Counsel Edwards declares the question is improper.)

CANNON (Heatedly): Since the state objects to all testimony, I propose to show by this witness (Ruby McCollum) that Dr. C. LeRoy Adams did begin to have sexual relations with this woman at the time of the birth of Sonya and continued down to the time of his death, and nobody knows this any better than the State of Florida!

(State objects, but is overruled by Judge Adams.)

RUBY: No, not right then.

CANNON: Later on, did you?

RUBY: Yes, I did.

(State objects through Attorney Edwards on the ground that such testimony would tend to confuse the jury and prejudice the jury; further, that it was no defense to the charges.)

CANNON: The defense contends that it is a mitigation and most material.

(Objection overruled.)

CANNON: Now tell the jury when and where.

RUBY: About three months after Sonya was born I continued sick. Dr. Adams was waiting on me. I asked if something could be done for me to get me well. Dr. Adams told me, "I can do something for you if you will do something for yourself. I will get you well provided you do as I say." I pretended not to understand and he told me, "You are not green. You understand what I mean." No, I'm not green and I guess I do understand, but I can't hardly believe my ears.

CANNON: Tell us if you ever had intercourse with Dr. Adams, and when and where?

RUBY: At my home in 1948 for the first time.

CANNON: Use the exact words that the Doctor said. Did he solicit you? You solicit him? Or did he take it away from you? I'll put it all three ways.

(State objects that testimony of witness does not bear on case.)

CANNON: You have to have something to make a case for the jury. It is silly to just say she shot him.

(Objections overruled. Ruby can answer.)

RUBY: Dr. Adams came to my house that afternoon on a professional call. He told me, "I'll be back in the morning after I finish my work and show you what I mean." He came the next morning and took me upstairs and laid me across the bed. Afterwards he told me, "I'll be back. You must call me sometimes." The second time he came I was not at home, but saw him later. After that intercourse he told me, *(implication that he was annoyed by her absence and thought she was avoiding him)*, "I want you to understand right now that I don't want no foolishness. And further, rest assured that nobody is going to bother you, and nobody is going to bother me. Nobody is going to be any trouble to neither one of us." I couldn't tell him no. He came to my house in 1950 one day and I told him "I am almost afraid." "Afraid of what?" he asked.

RUBY: You know. We haven't been using a diaphragm.

DR. ADAMS: To hell with that. I don't bother with things like that. In October I told him that I was pregnant. "I know it," Dr. Adams said.

In reply to a question by Attorney Cannon, Ruby stated that she had intercourse with Dr. Adams very frequently. Sometimes at her home and sometimes at his office. They had a scheme worked out to hide their relationship from others.

RUBY: Dr. Adams told me, "When I call you, you say, 'yes, I need you. Somebody is sick.' If there is somebody around to hinder what we want to do, say it has been a mistake" (wrong number). If my husband picked up the phone, Dr. Adams would hang up.

CANNON: Did you become pregnant after beginning relations with Dr. Adams?

RUBY: Yes, I became pregnant in October, 1950.

CANNON: Did Dr. Adams make any provisions for you in your condition?

RUBY: Yes, Dr. Adams told me to go to Dr. Saunders at Valdosta, Ga., at the Griffin Hospital. I stayed there three days then returned home. Dr. Adams came out next morning and continued to treat me and the baby for twelve days, till we were all right.

CANNON: Was this Dr. Saunders a friend of his?

RUBY: He told me that Dr. Saunders was his first cousin.

CANNON: Did you ask Dr. Adams for a birth certificate for the child?

RUBY: Yes.

(Strenuous objection by state sustained.)

CANNON (to Court): On what ground is objection sustained so that counsel may be guided? Then let the jury retire so that I can say what I want to. Judge Adams ordered the jury to retire and Ruby was then allowed to answer.

RUBY: One month after the birth of my last child I told him (Dr. Adams) that I had not received the birth certificate. Had he had it made out? He said yes, that he had. It was made out in his name (as the father) and the child's name was made out as Loretta Adams, but he was not going to give it to me unless I did as he said.

(State objects, no defense, immaterial.)

CANNON: Testimony of this nature would weigh heavily with me if I was on the jury. It is testimony of the highest authority, and impossible to be manufactured. The state's attempt to bar all testimony by this witness is unreasonable, and unfair. The attempt on the part of the state to bar all testimony from the ears of the jury is unpardonable. The accused should be protected by law.

(State objects and is sustained.)

CANNON: Is all testimony by Ruby McCollum concerning the birth certificate going to be barred by the court?

JUDGE ADAMS: It is.

CANNON: No disrespect intended to the court, but I am going to ask it all over again to protect the rights of the defendant.

(Jury recalled to the courtroom.)

CANNON: The baby, this youngest child of Ruby McCollum's, is here in court at this moment. It is in the balcony, and I desire to present it to its mother for identification and as evidence.

(Assistant State's Attorney Edwards is on his feet immediately and objecting strenuously. The objection is sustained angrily and Ruby is warned about answering before counsel has time to object.)

CANNON: Were you pregnant on Aug. 3, 1952, and by whom?

RUBY: By Dr. Adams.

CANNON: Did you have any conversation with Dr. Adams concerning this second pregnancy?

(Objection sustained.)

CANNON: How far had your pregnancy developed on Aug. 3, 1952?

RUBY: Two months.

CANNON: Had you been going to see him (professionally) a long time before?

RUBY: No, but he treated members of my family.

CANNON: During the month of May, 1952, did Dr. Adams strike you?

(Objection. Ruby not permitted to answer.)

CANNON: He struck you in March? May? July? February, January of 1952?

(Objections. Ruby not permitted to answer.)

CANNON: Did you have intercourse with Dr. Adams in his office in 1952?

RUBY: Yes sir, I did.

CANNON: Tell the jury the circumstances.

(Objection sustained.)

CANNON: Remember what day Aug. 3, 1952 was?

There came a long and tense silence. It smothered the room like a London fog. Ruby dwindled visibly in her seat. Her small hands trembled, her features assembled a piteous look. Finally she managed to speak.

RUBY: Yes, I remember that day. It was a Sunday.

CANNON: Did you go to Dr. Adams' office?

RUBY: Yes.

CANNON: What time?

RUBY: I, I don't know the exact time. I believe it was around 10 o'clock. I had Sonya and Loretta, my two youngest children with me. I had to take them with me because there was no one at the house to look after them.

CANNON: Had you continued sexual intercourse with Dr. Adams until Aug. 3 of this year?

RUBY: Yes.

CANNON: About how often?

RUBY: Many times. I don't know how many times at home and at the office.

(Continued Next Week)

JANUARY 3, 1953

Ruby's Story:
Doctor's Threats, Tussle
over Gun Led to Slaying!

LIVE OAK, Fla.—The courtroom remained hushed and tense as Mrs. Ruby McCollum told the jury why she killed Dr. C. LeRoy Adams, wealthy physician and politician.

She was constantly cross-examined by her attorney, Frank Cannon, solicitor of Duval County (Jacksonville).

> ATTY. CANNON: Prior to Aug. 3 did Dr. Adams try to induce you to leave your husband and move out on his farm?
>
> *(Objection sustained.)*
>
> CANNON: Did Dr. Adams ever ask you to leave your husband?
>
> *(Objection sustained.)*
>
> CANNON: On Aug. 3 were you indebted to Dr. Adams?
>
> RUBY: Yes, for $6. I know about that . . .
>
> *(Objection sustained.)*
>
> CANNON: (With a look of pain and frustration on his face) Tell us what you did when you left home.

RUBY: (After long pause) With my two youngest children, Sonya and Loretta, with me, I drove down to Dr. Adams' office. I saw that the office was closed and some people waiting. I stayed in the car with the children for awhile. Then I went into the office and sat and talked awhile. There were some white patients ahead of me. I sat and talked until Dr. Adams came out (into colored waiting room). He told me to come on in. I turned to the people in the room and asked, "You people ahead of me?" but Doctor said it was all right for me to come on in. I had a pain in my right shoulder, and he gave me a shot of penicillin in my right arm. I owed him for three shots ($9) and started off. Then I thought about the bill. Then I said, "The bill came out to the house to Sam."

(This exchange of words could be interpreted as a veiled lover's quarrel about something else. It is common talk that Dr. Adams never dunned anybody for payment. The McCollums had a reputation for always paying their bills. They had been using Dr. Adams as family physician for six years. They had received numerous bills from his office on the first of the month, and had never failed to pay promptly. The only logical conclusion is that they were talking in code for the benefit of the three colored women in the waiting room. It must not be forgotten that Ruby was two months pregnant by Dr. Adams for the second time nor that Dr. Adams was pressing her to leave her husband and move out to his farm. Also it has been hinted that the envelope in which the bill was mailed to Sam McCollum contained something beside that bill. It was intended for Sam's eyes, but Ruby intercepted it by opening the letter herself on Saturday night. Friday was Aug. 1. Hence, Ruby hastened down to the office on Sunday morning as soon as Sam went to church. Do not forget that Thelma Curry kept the records of the Negro patients and mailed out statements. She was a witness for the state and on the stand said she did not know the amount of the bill, nor even who mailed them out.

It is very strange, as Defense Lawyer Cannon remarked, that the bill which the state claimed precipitated the killing was not offered in evidence.

Nor was the over-the-shoulder, large handbag which it was claimed contained the pistol when Ruby went to the office. Cannon made it plain that he thought that peace officers stole that bag and its contents, including eighteen $100 bills that Sunday morning after Ruby was rushed away from her home, leaving only her small children there.)

UNDER cross-examination by Attorney Cannon, Mrs. McCollum said she entered Dr. Adams' office on this fatal morning and reminded the doctor that he sent the bill "out to the house to Sam" (her husband).

She quoted Dr. Adams as saying: "Yes, and I'm going to get mine if I have to go to the county judge."

Ruby said she asked him the amount of the bill. She quoted the doctor as saying "$116."

Ruby then said she paid Dr. Adams $9 for the three shots and an additional $100 and asked him for a receipt. She said he wrote the receipt and handed it to her. She said the doctor detained her after she put the receipt into her pocketbook and headed for the door. She told the court that the doctor led her back of a screen hiding the operating table and demanded that she get on the table.

She testified that she told the doctor her shoulder was hurting, but that the doctor accused her of being cold.

"I'll do what I can, but how about some other time?" Ruby said she told Dr. Adams.

. . .

IN A hushed courtroom, Mrs. McCollum said Dr. Adams told her: "No! I want you up there today! You get up there." Ruby said the doctor then slapped her, and that she again told him that she didn't "feel up to it," but that she would try to do what he wanted her to do.

She said Dr. Adams struck her again. It was then, Mrs. McCollum said, that I said, "Never again" and began to fight in earnest. She said the doctor "picked up the gun from somewhere back there and poked it in my stomach, demanding that I get on the table."

Mrs. McCollum said it was at this time that the two of them tussled over the gun. (Dr. Adams was shot four times.) Mrs. McCollum said she blacked out after this and did not "come to" until she reached her home. She had changed her dress and prepared the baby's formula when police officers arrived.

> RUBY: Mr. Gray (deputy sheriff) went upstairs behind me and told me, "I know your husband and have often done things for him."

> After Ruby had revealed where she had disposed of the gun, she was placed under arrest and taken away. What became of the pocketbook that she had carried that day and its important contents, she has no idea.

Afternoon session, Friday, Dec. 19, 1952.
Cross-examination of Ruby McCollum by state.

> O. O. EDWARDS: (Assistant State Attorney) What time did you say you left home?

> RUBY: I can't say exactly. 10–10:30 I guess.

> CANNON: (Defense) You have testified that you were pregnant on Aug. 3. Are you pregnant now?

> *(Objection sustained.)*

> CANNON: Did you have a miscarriage?

> *(Objection sustained.)*

> STATE: (Edwards) On Aug. 23, while you were at Raiford, did you write a letter to Mr. Crews (P. Guy Crews, formerly Ruby's lawyer) which came into the hands of R. H. Cox, director of the Women's Department at the State Prison Farm, denying sex relations with Dr. Adams and further stating that he was not the father of your child and that you had never had sex relations with any other man than your husband?

RUBY: I did write something like that.

CANNON: (On his feet with hot challenge in his eyes) Tell the jury now WHY you wrote that letter.

(Objection sustained.)

CANNON: (Seeking a way past the barrage of objections) Did anyone do anything to you to force you to write that letter?

RUBY: I was afraid. A nurse gave me 2ccs of some drug and it put me to sneezing and feeling very ill. I wrote it because I was afraid.

CANNON: Before you wrote that letter, were there officers continually asking you about your pregnancy?

RUBY: I don't know whether he was an officer or not, but he came in and held my arm while the nurse gave me the shots.

CANNON: What were you afraid of?

RUBY: Being in such a place, then too, what they would do to me. I was told that I would get a bigger dose next time if I didn't get rid . . .

(Objection sustained.)

This matter dealt with her miscarriage while at Raiford.

3:25 p.m.: Atty. Odis Henderson, for the defense, begins to address the jury. He developed the theory of the defense that the killing of Dr. Adams could not be sustained as premeditated murder. The state had not proved its case as such. The physical facts, the wounds, indicated that they had happened during a scuffle. He made a vigorous attack upon the four Negro witnesses for the state, charging that what they testified had been put into their mouths by State Attorney Black. Henderson accused them with scorn of "dealing with cesspools of truth." "A deaf woman (Carrie Dailey) testified to what she heard, and a blind woman (Della Arnold) swore to what she saw." Young Henderson, after stating that this was his very first murder trial, ended his speech before the jury touchingly by saying, "Do unto her as you would have done unto you under similar circumstances."

. . .

Assistant State Atty. O. O. Edwards made a more emotional appeal to the jury. He played on emotional traditions. The jury was urged not to believe that Ruby McCollum had been the mistress of Dr. Adams. The thought was repugnant. "She killed him to keep from paying the $116 medical bill. Shot this prominent citizen in the back, gentlemen, a man whom we all knew and loved so much and was so needed by white and colored, then went home, changed her dress and fixed her baby's formula."

He demanded the death penalty without mercy.

Saturday Morning,
Dec. 20, 1952

A. K. Black, State Attorney for Suwannee County, began to address the jury as soon as court opened. Black developed no further theory of the killing. He based his case on the medical bill, and Ruby's resentment at being asked to pay it.

"You all know that a good way to get them (Negroes) mad is to try to collect. You've seen 'em mull" (sullen). Black cautioned the jury not to be deceived by Ruby's behavior on the stand. He warned that she was highly intelligent, wealthy, but full of sly cunning. She had been planning for her day in court from the time that she went to Dr. Adams' office on the morning of Aug. 3, 1952. He began his address by paying compliments to the defense attorneys, and ended by demanding the death penalty for Ruby McCollum.

. . .

YOU COULD see the spectators coming alive the moment that Frank Cannon, Florida's legal Jack Barrymore, got to his feet.[1] The man's dynamic personality exploded for a 100 feet around him in all directions. He could easily pass for the brother of Senator Spessard L. Holland.[2] They look that much alike, but possibly Frank Cannon is a trifle more striking. Cannon is destined to go a long way in this world.

First he acknowledged the compliments of State Attorney Black, then added dryly, "Even if he did almost get me run out of town the day this case opened."

Attorney Cannon referred to Black's attempt to bar him from the McCollum case.

Defense Attorney Cannon continued to develop the theory that the killing was not one of premeditated murder, but of sudden passion. That the contention of the state was ridiculous.

"Loretta Adams (Ruby's last child), the fruit of the body of Dr. C. LeRoy Adams, was right here in this courtroom, but prevented of being exhibited to the jury. The state has sought to keep all evidence from reaching your ears. It is a matter of conscience. God forbid that I ever sit as a prosecutor and withhold evidence from a jury!" Attorney Cannon said.

"The case of Dr. Adams and Ruby McCollum is not rare. A lot of white men, as you call it, get messed up with n——r women. You men of the jury know men who are doing the same thing," Cannon said.

Insisting again that the state had not proved premeditation, and that it had excluded all evidence that might be helpful to the defendant, Atty. Frank Cannon asked for a second-degree verdict, and took his seat.

Judge Adams charged the jury, and the twelve jurors retired for consideration at 11:24 a.m. The jury filed through the courtroom at noon on the way to dinner. At 2:21 the knock came on the door and the jury brought in its verdict.

"We, the jury, find the defendant, Ruby McCollum, guilty of murder in the first degree. So say we all."

Defense Attorneys Cannon and Henderson indicated immediately that they would file an appeal within the fifteen days allowed by the court. Attorney Cannon went into conference with the male members of the family without leaving the courtroom.

JANUARY 10, 1953

Ruby's Troubles Mount: Named in $100,000 Lawsuit!*

LIVE OAK, Fla.—The attorney whose prosecution placed Mrs. Ruby McCollum in the shadow of the death chair . . . A. K. Black . . . has resigned as state attorney for Suwanee County, Fla.

He quit his post to represent Mrs. Florrie Adams (white), widow of the slain Dr. C. LeRoy Adams, who has filed a $100,000 suit against Mrs. McCollum and the estate of her dead husband, Sam McCollum.

Mrs. McCollum shot and killed Dr. Adams last August and was found guilty, without mercy, at her trial last month. Mr. McCollum died of a heart attack shortly after the doctor had been killed.

During her trial, Mrs. McCollum testified that she and Dr. Adams had been "lovers" and that he was the father of her last child.

. . .

UNCONFIRMED REPORTS intimate that Mrs. McCollum is worth close to $200,000, while McCollum's estate is in excess of this amount.

And as Ruby's troubles mounted, these other developments were evident:

---

*The same story appears on the same date in an alternate edition of the paper as "Ruby, Facing Chair, Now Sued by Medic's Widow for $100,000."

1. Releford McGriff, first Negro attorney to enter the case, joins former county solicitor Frank T. Cannon to appeal for a new trial based on different points. Judge Hal Adams is set to hear the motion this week.

2. The inheritance of the three McCollum children is threatened by the suit of Mrs. Adams.[1]

3. Danger of little relief from the mandatory death sentence faced by Mrs. McCollum is seen . . . and a judgment against her monies would tie up future defense funds.

4. If Mrs. Adams' suit is successful, a large block of stock owned by McCollum in the Central Life Insurance Company of Tampa would fall into the hands of white ownership.

Mrs. Adams is asking the $100,000 damages on the grounds that she has been deprived of the "comfort, association and protection of said husband, and also has been deprived of the support of said husband."

. . .

AS EVENTS continue to unfold in the gripping tragedy, Mrs. McCollum found her troubles mounting. A faint ray of hope still gleamed through all the darkness as her attorneys prepared to argue a motion for a new trial on fifty-eight different points before Judge Hal Adams.

Two top attorneys are making the appeal . . . Atty. Frank T. Cannon, former county solicitor, and Releford McGriff, noted Florida Negro attorney . . . the first Negro attorney to enter the picture.

The suit filed by Mrs. Adams came as a bombshell as interested spectators were anxiously awaiting the judge's decision on a new trial. Sentence has yet to be passed on Mrs. McCollum, found guilty of first degree murder of Dr. Adams by an all-white jury.[2]

. . .

Mrs. McCollum has testified that she had shot Dr. Adams during a scuffle over a pistol in his office in August. She told the Court during her tense trial that she had been intimate with the highly respected white physician over a period of years and that he was the father of her 15-month-old daughter.

Her husband, Sam McCollum, wealthy businessman, died shortly after the slaying, leaving a large estate in both real and personal property. The joint holdings of the McCollums has been estimated by close associates as being close to $500,000.

The suit for $100,000 has been filed against both Mrs. McCollum and the estate, which under normal circumstance would go to the three McCollum children if the mother were to die in the electric chair. A successful suit on the part of Mrs. Adams would deprive the youngsters of this inheritance.

. . .

INCLUDED IN THE McCollum estate are large blocks of stock in the Central Life Insurance Company of Tampa, one of the largest Negro-owned insurance companies in the Deep South. If Mrs. Adams' suit is successful, it would place her in the position to become the first white director of a Negro-owned life insurance company. McCollum was a director of the company at his death.

Matthew L. Jackson and Roland J. Yates were appointed administrators of the estate of Sam McCollum. There has been wide-spread conjecture on the attitude of the civil court toward the suit.

JANUARY 17, 1953

The Life Story of Mrs. Ruby J. McCollum!

". . . And on the signing of the order by the chief executive of the State of Florida, you shall be removed to the State Prison at Raiford, and there electric currents shall be caused to pass through your body sufficient to cause your death, and to continue until you are dead . . . and may God have mercy upon your soul!"

It was the climax to the life of a Florida woman whose story has rocked an entire nation. These were the fateful words pronounced by Judge Hal W. Adams of the Circuit Court of Suwannee County, Fla., as the shadows of dusk fell on Jan. 17, 1953.

The words were pronounced in a choked voice and they were addressed to Ruby McCollum as she stood before the bar to hear her sentence. Her left elbow rested casually on the back of the chair that had been occupied during all of the hearings and the trial by the state's attorney who had prosecuted her and won a first degree murder conviction against her.

What thoughts passed through her mind as she fleetingly reviewed her entire lifetime in one brief moment, only Ruby McCollum knew. Outwardly calm and self-possessed, Ruby McCollum returned to her seat at the counsel table.

The discerning spectator could not avoid the conclusion that here was no ordinary mortal, no cringing victim inadequate to its fate. There sat an extraordinary personality. Here was a woman—a Negro woman with the courage to dare every fate, to boldly attack every tradition of her surroundings and even the age-old laws of every land.

Her closest relatives and the people who said they knew her as a friend or associate express profound and painful surprise. Mrs. McCollum—born Ruby Jackson—had nothing of daring, nothing out of the ordinary in her. Always quiet with little to say and utterly absorbed in ordinary domestic affairs. What had changed the Ruby they knew so strangely?

. . .

THE TRUTH IS that nobody, not even the closest blood relations, ever really knows anybody else. The greatest human travail has been in the attempt at self-revelation, but never, since the world began, has any one individual completely succeeded. There is an old saying to the effect that: "He is a man, so nobody knows him but God."

As to what has come about in the life of Ruby Jackson McCollum, perhaps it was inevitable from the hour of her birth, as humble and unnoticed as it was. An Arab proverb says: "The fate of every man hath he bound about his neck."[1] We are born with our own climate which shall create the outward circumstances which people carelessly say determined our course.

Fatalists contend that it was there in the seed already at birth. So, the ordinary-seeming little boy born on the island of Corsica becomes the Emperor of France and the most renowned military figure the world has ever known, but he died on St. Helena still imperfectly known.[2]

So, thus, we go to search out the beginnings, the early years of Ruby Jackson McCollum who has been sentenced to die for the fatal shooting of Dr. C. LeRoy Adams, nominee for the Florida State Senate, the most successful doctor of medicine in Live Oak, Fla., and easily the most popular man in the entire area.

. . .

HE WAS A broad-shouldered six-footer with magnificence of body and an irresistible smile, a handsome white man whom women of both races seemed unable to resist. He was the charmer who fell before the blazing gun of Ruby McCollum, the slightly built, soft-spoken Negro woman, on a Sunday morning in August, 1952.

Certainly, there was nothing in the early life of Ruby Jackson, later to become Ruby McCollum, that made relatives and neighbors see anything that might hint of what was to later come to pass. That, probably, was because the average person takes it for granted that silence means a lack of thought and internal activity.

It is rarely considered that thoughts too profound for words might be surging through an individual like a great underground river, unless and until it bursts forth like an awesome Niagara or Victoria Nyansa.[3] That was undoubtedly the case with Ruby McCollum.

. . .

RUBY IS THE second child of William and Gertrude Jackson. Clemon Jackson, now a barber in Ocala, Fla., was the first-born. Ruby followed him into the world two years later. William Jackson, who died in 1949, was a successful farmer with his farm hard beside Fessenden Academy at Martin, Fla., about eight miles from Ocala.

The Jacksons were never rated as wealthy, but were classed as good-livers by all standards. Cattle was the chief product of their farm, but William Jackson saw to it that there was always an abundance of foodstuffs planted so that his family of seven would have plenty to eat.

There were four girls and three boys in the clan. There was always plenty of milk and butter and country-cured meats; they raised sugar cane, and so had plenty of syrup to go with hot cakes on winter mornings.

The house was small for such a large family. It had four rooms with two set aside for bedrooms. Typical of the Southern country home, there was a front porch with a swing under the big oak tree that stood beside the front porch. A smaller china-berry tree was on the other side.

. . .

BOTH PARENTS of Ruby Jackson were deeply religious. Her father was a steward in the African Methodist Church, while her mother was a Sunday School superintendent and later a stewardess. Their home was one of those where family prayer and group singing of hymns and Sunday School songs were habitual.

Naturally, in such a home, nobody danced, and playing cards was not even allowed on the premises; not even dominoes or checkers were allowed. That would seem to this stern and devout couple too much like harboring the very works of the devil.

Such was the atmosphere surrounding the very early life of the child Ruby Jackson, later to become Ruby McCollum.

FEBRUARY 28, 1953

The Life Story of Mrs. Ruby J. McCollum!

WHAT AMUSEMENTS did Ruby Jackson and her brothers and sisters find to offset the fact that they weren't allowed to dance or play cards or such other amusements? How did they provide an outlet for their pent-up emotions and desires, filled as they were with the vigor of virile youth?

They often played those old games which came from England centuries ago: "Heavy, heavy, hangs over they head"; or "Old horse, I ride him," and the like. These staid games have entertained millions of rural families in the South of the past. They were the kind of games played in Ruby's home at night when the family gathered after supper was over and the dishes washed. It was "sport" around the big kerosene lamp in the living room.

. . .

ON AFTERNOONS when there was nothing for the children to do on the farm, they gathered around and played the usual games loved by small children, "London bridge is falling down," and the like.

One thing was mandatory, however, for the Jackson children: they HAD to study the Sunday School lesson and they had to know it in advance. That was a "must." Nonetheless, in their way, they had fun.

Around Live Oak today they will tell you that Ruby McCollum seldom visited people. Perhaps this is because she was not allowed to do much visiting during her childhood. Her parents did not approve of such, and her mother, Mrs. Gertrude Jackson, was the disciplinarian of the family. She whipped soundly when she thought it was called for.

Clemon Jackson, the oldest son, said most of their whippings came when they stole off to visit and play with other children.

"Mama promised us a whipping if we did that," he said, "and we sure got it, too. Oh, yeah, Ruby got her share. When Mama said a thing she meant it. She was not the kind that had to wait for Papa to come home to get order. She got order right now when she spoke or we caught it."

. . .

RUBY JACKSON, on the whole, was an obedient child, but when the kids did make up their minds to steal off and play at someone else's house, Ruby took her licking with fewer tears and less noise than any of the others.

She had had the fun she had determined upon, and she took her whippings in payment without the usual screams and without seeking to justify herself.

She made an outcry, however, when her mother—as harried and hardworking mothers sometimes do—erred in fixing the blame for some misdemeanor. When Ruby was not guilty, or felt justified, she defended herself with a few well-chosen words. It was an insight then into the type of woman she was to later develop into.

Ruby gave no trouble either about attending church and Sunday School, nor Fessenden Academy. She was quick in study and quiet in classes. Something of a "lone wolf," she made few intimate friends. But the few friends she did make she held close! Sarah Davis, who later married her brother Clemon, and Tommie Lee Ward were her close and constant chums.

. . .

BUT AS QUIET as Ruby was, so far as conversation was concerned, she was quite much of a "tom-boy." She shinnied up trees and "skinned the cat" with the best that any of the boys could do. She gave a good account of herself in a fight with them, and played marbles with them ardently. She was very enthusiastic when it came to yard-ball.

When Ruby reached the age of ten, she took to reading a lot, and the things Ruby Jackson read were always romantic love stories. As time went on she spent more and more time that way and less on the romping games with the other children. It was a sign of "things to come"—even at that early age.

William Jackson seldom meted out physical punishment to his children, but there was one occasion on which he took his three oldest ones into camp.

They were Clemon, Ruby and Hattie. Father Jackson had some fine watermelons which he had ordered not to be touched.

But the three youngsters looked upon them and yielded to temptation, taking a melon under the shade tree where they gorged themselves until it was all gone. Father Jackson came upon the rinds before the children had prepared a good excuse. Ruby explained it by saying that Maybelle, their dog, had eaten the melon. But this wasn't good enough for their father and he gave the three children a sound thrashing.

. . .

WITHOUT MUCH schooling themselves, both Mr. and Mrs. Jackson wanted their children to be well-schooled. The Rev. Leonard Francis Morse, now dean of Lee Theological Seminary at Edward Waters Junior College, in Jacksonville, Fla., taught the Jackson children at Fessenden Academy.[4]

The Jacksons did not handle too much cash money, but they found a way to pay for the tuition of their children. Since they had more cows than anything else, they swapped beef for tuition. Before Florida's "no fence" law went into effect, the family ran from fifty to sixty head of cattle at a time. When the law was passed, they found they did not have enough land for pasturage, so the herds were drastically cut—by necessity.[5]

But Ruby Jackson went on to finish high school and then her normal course at Fessenden in 1929 when she was nearly nineteen.

Three traits of personality emerged strongly in Ruby Jackson's early life.

The most striking of these manifestations is that she revealed little about herself. Even when she made words with her mouth, she really said nothing about her inside feelings. She was extremely self-contained.

The second thing was that she was retentive in material things. Never did she make any great clamor for clothes and the things most girls crave, but what she did get, she took very good care of. The little pocket change which came to her, she could hold onto much longer than any of her brothers and sisters.

The third strong characteristic was a maternal instinct which was out of the ordinary.

(Continued next week)

MARCH 7, 1953

The Life Story of Mrs. Ruby J. McCollum!

The maternal instincts which swept over Ruby Jackson when she was a child were to play the key role in her later life. Dean L. F. Morse of Fessenden Academy says that by the time Ruby passed puberty "she was a very pretty girl, but seemed unaware of it."[6]

Boys and even grown men began to give her looks, but she went on being quiet and reserved, a good student in school and busy about the home and church. She didn't seem to realize that she was possessed of a power that made itself felt upon the men and boys who saw her. How fatal it was to be, time alone was to tell, as the world now knows.

Ruby's mother allowed her to start receiving company when she was 18, and she seemed satisfied to wait until then with utmost calm. While she was 17, a local youth made an open bid for her heart and hand and boldly made an earnest proposal of marriage . . . only to be turned down.

Her [r]easons? They were a sure indication of what was to come in her heart and what was in her mind: the young man just didn't attract her. He lacked the different strengths she admired in a man. Utterly female, she wanted to be conquered and have a great store of strength to lean upon.

He lacked the "get-up" that she hoped for, and, she felt, a lack of mental vigor, too. "I could not see myself loving a man who could see ten things and not even understand one. I wanted a man who could see one thing and understand ten, a mate would could [sic] cope with life and give me protection."

During this period, she was seldom allowed to go to Ocala, the nearest sizable town, but she found activities around Martin where she lived. She became extremely active in church work and was a delegate to the Sunday school conventions and the district conferences.

. . .

HER CLOSE friends were the heart of all the activities in Martin. Ruby organized a singing-band made up of her brother, Clemon; Sarah Davis, Tommie Lee Ward, Helen Smith and herself. Ruby sang lead and the group became widely known in the area, singing for churches and other such gatherings.

Upon her graduation from Fessenden Academy, Ruby taught school for a year at New Chapel, a small community not too far from Ocala. She must not have cared much for this experience because there is no comment about it, nor is there any comment about why she did not continue after that one year. She just taught school for a year, period.

But it was about this time that the new shape of things to come began to make itself felt within her. Internally, she began to sense a lack. There was no one around whom she could drape her intense feelings, her great capacity for love.

. . .

SHE HAD made a tremendous discovery! She found that she had a singular power over men: It was no trouble at all to bend them to her will. She was one of those females who appear now and then in human history. Something drew men to her and bound them!

Ruby had confidence in her powers, but it was a disappointment in a way: she moved men, but—so far—no man had ever moved her! It was in this period around eighteen that she began to have recurrent dreams. There were four to the series. Of these, only one was clear so that she could remember the details upon awakening.

In this dream she found herself in a strange community and entering a large, beautifully furnished home. She was not only expected there, she was welcome. A muted, throbbing rhythm said over and over: "Come to me! Come to me!" Somehow it seemed to be her home. Love and satisfaction radiated the place.

This dream troubled Ruby. Walking around the little four-room house that was her home in Martin she could not imagine why she should dream of so much comfort and luxury being hers. At that time, Ruby Jackson had never heard of the sub-conscious. It never occurred to her that she might have wishes that had never emerged into the conscious.

. . .

SHE SAW [the dreams] as a prophetic sign, though how such a thing was to come about, she had no idea. The years were to prove her right! Years later, in Live Oak, she was to recognize the house the moment she saw it. It had been built by a Negro bolita banker, whom her future husband—Sam McCollum—was to defeat and come into possession of the nine-room house!

At eighteen Ruby was full of internal conflict. As yet she had been attracted to no man. She had been trained to despise and fight against physical pleasures and desires as sinful things inspired by the devil. She had had that background at home.

So, naturally, she was distressed to find that so many men leered at her. Young and old men of her own race just could not seem to pass her unnoticed; white men winked their eyes at her and followed her, or secretly nodded at her to please follow them. The only part she liked was the secret knowledge that she had power over men.

. . .

NOW, WITH her shapely and well-developed body blooming, she felt herself a woman. She had laughed and worked and suffered to a certain extent. There had been a hurt in her life which she had revealed to no one. Her tears had been in utter secrecy. Now she began to feel an emptiness in her existence.

She felt like a blossom on the bare limb of a pear tree in the spring . . . opening her gifts to the world, but where was the bee for her blossom?[7] Yes, numerous men had gazed upon her with open desire, but so far their looks had raised no mingling-blood call in her.

She wanted beauty and poetry mingled in her life, something to make her everyday side-meat taste more like ham. Sometimes, deep, deep in the mood of her strange yearning she would picture herself reclining on soft grass in a beautiful rose-scented setting on a white moonlit night.

Ruby Jackson was now ready for life and love!

(Continued next week.)

MARCH 14, 1953

The Life Story of
Mrs. Ruby J. McCollum!

The sun had gone home, leaving its footprints in the sky. The drifting mists gathered in the west to arm with thunders and march forth against the world. Lightning flashed against the horizon and the thunder rolled into crescendoes. Ruby Jackson stood at the front gate of the Jackson home in Martin, Fla., seemingly unconscious of the approaching summer storm.

She stood there questioning fate. For some time now she had been living at her own front gate, ready for departure. Internally, she had outgrown the confines of Martin, Fla.

The horizon of the world was her hatband. Ruby longed for fulfillment of her natural desires, and so she was restless beneath her always outward calm. Neither relatives nor friends suspected the intense fires that raged within her.

Ruby now was twenty years old and, yes, she wanted a mate. She was of a good respectable family in Martin, unsoiled by the lap and wash of slander. She was considered physically attractive, and what there was to choose from in her community she could have had.

. . .

BUT BY NATURE, Ruby did not walk in footprints. Secretly she saw no reason why her life must follow the pattern of her surroundings. Her family and friends did not know the real Ruby and she was conscious of it. "Often," Ruby said, "you can make people follow you, but almost never can you make them understand."

Yes, like all girls of her age, she had flirted briefly here and there. She found something dead about the young men she had known so inside her she drew away as mortals do from a corpse. She was looking for LIFE.

Now she had met Sam McCollum, a young man a little older than herself. The McCollums were prosperous farmers over at St. Peters, a small

community near Ocala, Fla. That had been nearly a year ago and Ruby was
still thinking Sam over.

. . .

SHE WAS attracted to him, but she debated whether or not he had what
she wanted. She wanted many things that her life and surroundings so far
had not afforded her. At times she felt that Sam had in him that which
would bring fulfillment of her dreams and then, again, she wondered.

Was Sam McCollum masterful enough? That was what she debated
within herself as she stood at her father's gate that day at sundown. Internally, she was ready to set out on her journey to the big horizon. Was Sam
the vehicle to take here where she wanted to go?

Sam attracted and charmed her more than any man she had met so far.
He had both mental and physical vigor. Secretly, he stirred her tremendously.
He was full of things. Sam made a little summertime out of a seemingly
nothing and they both lived off of it for the hours that they were together.

. . .

SILENTLY unsatisfied by her narrow surroundings, she had been fum-
bling around the door-knob of life and Sam McCollum had opened that
door! If only she could be sure of his capacities, she would love him for it.
But so strong were her desires that she felt that she was not yet ready to
commit herself. Better wait and see.

She had met Sam McCollum at her church. He had come to attend a
special program that Sunday afternoon nearly a year ago. Ruby had a leading part in the program in addition to her group singing. Sam saw her and
he liked what he saw.

It took nearly two more years for Ruby to finally make up her mind to
marry Sam McCollum. In that time she discovered that Sam had what she
wanted. He was witty and gay, and beneath his casual exterior Ruby found
that he had drive and ambition in him. He had a way of commenting and
saying things that were always entertaining. And his small community did
not satisfy him, either.

. . .

THOUGH THE McCollums had a going farm, Sam took to picking oranges—quick and generous pay—and construction work. He often went away from home on jobs like that and came back with a pocket full of money and stories of what he had heard and seen in the larger and outer world where he had been.

Outside of her own requirements in her future husband, Ruby had another obstacle to overcome. The McCollums were something less than enthusiastic about her. After she began to go steady with Sam, his brother, Buck, and his father came over to Martin frequently, but the rest of his family held aloof.

But, even Buck warned Sam that Ruby was inclined to be too possessive and domineering. They accused her of seeking to cut him off from his family. She must be the "be-all" and ruler of his mind and Buck saw his brother crumbling before the determined Ruby Jackson, for all her quiet ways.

"I told Sam years ago that woman was going to kill him." Buck raged when he heard of his brother's death. "He had got so under her influence that he wouldn't listen to me." (Sam McCollum died of a heart attack after his wife killed Dr. Adams.)

. . .

RUBY JACKSON felt certain of two things when she became Mrs. Sam McCollum at a quiet home wedding at Martin in 1931: she was sure that she had a go-getter, a winner in economic ways and a vigorous mind, and she felt certain that the opposing McCollums could be no trouble to her with Sam.

Sam had a construction job in New York and he and his bride went North immediately. In the years to follow, they went many places together. In those years Ruby was almost completely happy. Her world had expanded marvelously and, by comparison, she handled plenty of money.

From the very beginning, Sam brought home his money and handed it over to Ruby and she managed things. There was only one tiny dissatisfaction in Ruby's love of her husband . . . Sam did not rule her enough.

. . .

THE GREAT tragedy that engulfed them in 1952 might have been avoided had Sam only understood Ruby better! From the beginning of their life together, the tiny seed of despisement had already been planted.

Ruby proved a good and industrious wife. She was a wonderful cook, sewed well and kept a clean house. She was a very devoted mother. Without too much taste in clothes, she was neat and attractive in her clothes.

But there, perhaps, she was wiser than most people thought. When a female body is too gaudily dressed, it is possible for the male mind to lose the connection.

Ruby, brought up in a very religious home, knew even before she married Sam that, though he always worked, he gambled on the side and thus increased his income. Her femaleness is such that she accepted all parts of her man. She did not gamble herself, but she was with him in spirit.

. . .

IF SHE CUT him off to an extent from his own blood relatives, she also cut herself off from her own, in her loyalty to her man. Their families knew nothing of the relations between Sam and Ruby.

Her attitude was such that even her stern, religious parents came to look upon Sam McCollum as the perfect husband and son-in-law. The young couple led their own life as they moved about from job to job, now Florida, now North again and back.

What with his work and successful gambling, they were [h]oarding up money without as yet making any flash. Both were proud and happy about their son, Sam Jr., who came to them more than a year after they were married. He was born into a loving, affectionate, charmed family circle.

(Continued next week.)

MARCH 21, 1953

The Life Story of
Mrs. Ruby J. McCollum!

The terrible years of World War II were about to burst upon the world when Sam and Ruby McCollum settled down in Live Oak, Fla. The coming of the McCollums was so quiet and inconspicuous that even Negro Live Oak was scarcely conscious that they were there. Privately, Ruby and Sam felt that they were now ready for bigger things.

Looking over their new town one day, Ruby suddenly pointed and then said to her husband: "Look, Sam! There is my house!"

Sam McCollum could neither understand her claim nor her excitement until she explained about the dream she had long years ago.

"Know whose house that is?" Sam asked. "A fellow named Hopps built it and lives in it. A bolita banker and the biggest colored man in town, I hear."

"But it is mine," Ruby told him with conviction, "I am going to live in that house."

Sam studied the large nine-room house on Ninth Street and chuckled.

"Pretty high kick for a low cow, Ruby. That calls for a lot more cash money than we got. But, then maybe it's a sign that I'm going to beat Hopps out at his own game. I've been looking around and listening a lot. Looks like Hopps has just been lucky and had nothing to buck but dumb guys. Maybe . . ."

"Oh, you can take him, Sam. I've noticed in many ways how smart you are. Go ahead and put him out of our house."

. . .

AND THE time came around . . . not so long after . . . when it was so. Sam McCollum went about things in a different way. He hung about places where he could learn things about Hopps' methods of operation. He made connections among the Suwannee County authorities. He sat up nights and mapped out plans.

The boastful, flashy bolita king, Hopps, had no fear of the quiet, incon-spicuous Sam McCollum. He looked upon him as a country tin-horn and scorned him.

Nevertheless, the time came shortly when Sam McCollum took over Hopps' bolita empire, and in later years expanded it to dimensions which Hopps had never dreamed of.

. . .

PEOPLE IN Live Oak still tell of that night when Sam and Hopps sat down at a gambling table and fought with their skills for what was left. When the air cleared, Sam was master of the big house on Ninth Street, and Hopps—clean as a fish—walked out of town with his coat over his arm.

Ruby McCollum, reared in the little four-room house at rural Martin, eight miles outside of Ocala, walked into the big house that a gambler had built, and which a gambler was again to own and occupy . . . and Ruby found the end to her dream!

It is obvious that long years of "protection" had blinded Ruby McCol-lum to the gravity of her situation on the morning of Aug. 3, 1952, when she shot Dr. C. LeRoy Adams, prominent white doctor, to death.

This conclusion is further verified when the sheriff, his deputy, and a state highway patrolman arrived at her home to make the arrest.

"But I can't go now," Ruby told them, "there is nobody here to look after the children."

. . .

SAM AND RUBY prospered. The weekly "take" grew to enormous pro-portions for a couple who had had so little cash in their early years. Ruby spent thousands in refurnishing and renovating the big house as time went on. However, they made no great splash in the manner done by most people who gain money in such a fashion.

They were well dressed for Live Oak, but there was no flash, and no glitter of diamonds, no multiplicity of showy cars, no drinking and loud social affairs.

But some changes had been made, nevertheless. The weekly "take" from bolita was always carried directly to the McCollums' home by an employee of Sam's and turned over to Ruby. She took care of all the tabulations and the tickets. It was she who wrote "give" on the tickets to be paid off, and "get" on the tickets still to be paid for on a play. She had charge of large amounts of cash.

. . .

THE McCOLLUMS bought some real estate, bought stock in the Central Life Insurance Company and made other investments, but Ruby held tightly on to the cash. This habit of hers led to some of the hatred shown her by some townspeople.

"She was just too stingy for words. Close-fisted and stingy as all get-out. Wouldn't even give God an honest prayer without snatching back amen," some people said of her.

For his high school education, Sam Jr. was sent to Fessenden where Ruby herself had gone, but otherwise no great outlay was made upon the boy. Ruby was laying it away to benefit her children in later years, when they would need it more.

. . .

BUT THINGS in the home were changing. Ruby was to find out that her husband was having brief—but numerous—outside affairs. She was not the kind to make any visible display to the world, but she was deeply hurt.

Ruby accused Sam and warned him that money and power were going to his head. Sam's retort was that she had the big house and had charge of all his money; she had servants and anything she wanted, so why should she care?

But, though she never discussed it nor admitted it to her relatives or closest friends, Ruby was hurt. She knew that people in Live Oak noticed that never were she and Sam seen together. They even belonged to different churches; she was a Methodist and he a Baptist, though he seldom went to church. Ruby was ever at home, a tireless wife and mother, while Sam was around town night and day.

. . .

SAM WAS either blind to character traits and natural reactions, or he just didn't care any longer how Ruby felt. From their past life, he could easily have been too sure of her. It is easily possible that he had analyzed her as caring more for material things than for him as a man.

All right, she had the money packed away secretly in the house and allowed him little for spending money. She had what she craved, so he would go out and do as he pleased. Then they both would be satisfied—or, so it would seem.

Live Oak says as a matter, of course, that Sam McCollum had "gangs of women," though he put very little on them. That was because Ruby had a tight grip on the purse-strings. Almost without exception, Sam's temporary "lights-of-love" were very young girls.

. . .

HE TOOK the pick of the new crop of young "friers" every year as they emerged and dropped them when the next crop of adolescents arrived. But not one of the number ever had anything to brag about after falling under the spell of McCollum's great wealth, nothing, that is, but hope.

And, women were offended at Sam McCollum for another reason: he was always bragging about the upright character of his wife. He gloated over other men:

"I wouldn't even put up with the kind of wife you got. My wife is always at home no matter when I get there. She's home and acting like a wife ought to act. Clean in her living and looking after her husband and children. The kind of woman you got—"

. . .

"MRS. McCOLLUM must have something on Sam," some women gritted their teeth in envy and malice, "the way he acts scared of her and lets her hold on to all that money and allowance him out what she wants him to have, which sure ain't much, God knows."

So life went on with the McCollums fourteen years after they had married and through the birth of their third child and second daughter. They showed a solid front to the outside world. Neither of them would break the circle of loyalty, that is, by slighting remarks about each other. Both of them had their pride.

(Continued next week.)

MARCH 28, 1953

The Life Story of Mrs. Ruby J. McCollum!

YEARS before the birth of Sonya, the second daughter of Ruby and Sam McCollum, something had happened which destroyed the very texture of their union, though Sam, perhaps, never realized how serious it was.

Naturally, Sam now had many connections and even friends among people who were connected with gambling in one way or another among both races. Sam decided to entertain a group of them in his home. His wife, Ruby, was then five months in pregnancy with their second child who was their first daughter. Young Sam Junior was then seven years old.

The noisy group gathered in the McCollum home which Ruby had recently refurnished. This was the acid test! And in it both Sam and Ruby failed each other!

The quiet, reserved Ruby did not care for these people. She did not even want them in her home to begin with. Sam, more self-confident than formerly, and possibly flushed with his new importance, forced his will upon his wife, knowing that it would not please her, but feeling that as a very capable bread-winner and master of his home, she should place his success before her feelings.

Ruby said that in the course of the riotous evening too many of the guests drank too much liquor—incidentally, she never touched a drop in her life—and two or three of the overloaded guests callously vomited over her floor and furnishings.

. . .

SHE TOOK Sam aside and protested, demanding that he get that bunch out of the house. Sam did no such thing. Behind the closed door of their bedroom, Sam is said to have whipped her soundly and forced her to clean up the mess the guests had made.

"For that I never forgave Sam McCollum," Ruby once said. "Though his hard blows hurt me outwardly, luckily I was not injured internally and I had no miscarriage. He might have injured me and my unborn child, but for the time he seemed not to care.

"Spiritually," she continued, "the incident gave me a terrible fright. Always after that I felt insecure with Sam. I no longer felt myself to be the woman of his heart. A useful tool as a wife and mother, but no longer the mistress of his inner heart. It was a terrible shock."

An image—something sacred and precious—had fallen off the shelf in Ruby's heart. After a few days they carried on as usual, but now they were different in Ruby's concept.

. . .

SAM WAS a good provider and Ruby was a good wife and mother and that is just the way things rested for six years.

Then Ruby found she was pregnant for the third time and she did not know whether she was glad or sorry. She loved children and motherhood. It was five years since she had had an infant to demand what she loved to give in that way, but try as she would, she could not recapture the old feeling of intimacy towards Sam.

She was distressed that this was so. Her hope was that the birth of the child would bring it all back again. So, she looked forward to the birth of Sonya, her second daughter and third child.

Sam was very pleased about the prospect of another child now. Perhaps he felt the slackening off of the old ties and hoped the same thing that Ruby did. However, he stayed away from home just as much as ever, which was no help at all.

. . .

WHEN HE did come home he talked about a new white friend he had just made. Dr. C. LeRoy Adams was Sam's new white friend. Dr. Adams had moved to Live Oak in recent months and set up his practice of medicine.

Oh—from all he could learn from both white and colored, he was a wonderful doctor, and Sam knew for himself that Dr. Adams was a fine friend to have. That is one of the anomalies of the Southland, this friendship which exists between individual whites and individual Negroes.

And, besides, this Dr. Adams was accepting Negro patients, something Live Oak's two other white doctors did not do. This Adams was very nice and friendly to Negroes. Sam made up his mind that he was not only going to bring Dr. Adams to the house to meet Ruby, he was going to place her under his care during the birth of their child.

The hand of Fate had come into their lives.

. . .

SOMETHING electric passed between Ruby McCollum and Dr. Adams the very first time they met! A few minutes after the introduction in her living-room, Ruby happened to lift her eyes to find the big, handsome white man studying her with a look that she knew only too well. It was so intense it flustered her.

Warm inside, to such an extent that her whole body seemed to be on fire, she excused herself with the excuse that she did not feel too well and went upstairs, leaving her husband and Dr. Adams laughing and talking below.

Perhaps a quarter of an hour later, she heard them go outside to the driveway. She peeped out from behind the drapes and saw Dr. Adams stealing glances up at the second floor while he complimented Sam on his home and its surroundings.

. . .

THE BABY was delivered and did well, but, somehow, Ruby found her own recovery slow and fitful. One day she felt fine, then suddenly she felt as if she were about to die. She could not understand it. She had had little trouble after her first two babies. It was hard to understand, this new condition.

She took the medicine faithfully that Dr. Adams prescribed, but, strangely, sometimes immediately afterwards she had that desperately sick, sinking feeling.

Months went by this way, and Ruby began to feel very worried. Suppose she never got well and left her three children motherless? It was a terrible thought. She just **must** get well for their sake! And in her anxiety she had a revelation.

. . .

SOMETHING WAS laid bare that she had been hiding from herself for years. Her solicitations were for her children—not for her husband. She was not in love with Sam McCollum any longer!

condition, Ruby summoned Dr. Adams to her home and told him of her peculiar sensations and her worry about her health.[8]

He sat opposite her and took a long time to answer.

With his big hands clasped and his elbows resting on his knees, Dr. Adams looked down at the floor for a while, then looked her squarely in the face and told her:

"I can get you well if you will do as I tell you."

. . .

SAM McCOLLUM's neglect of her for the past few years had caused her to lose confidence in her power to sway men. She had come to feel that she no longer mattered to anybody that way.

And now, this big, handsome, aggressive man restoring it to her again.

She did not consider whether she wanted Dr. Adams or not. He was handing back to her a precious treasure that she had considered lost forever.

She felt warm and grateful toward him. And this man was no trash out of the streets of Live Oak. This was an important, outstanding man, and—in addition—physically equipped to be desirable to many women.

(Continued Next Week)

APRIL 4, 1953

The Life Story of Mrs. Ruby J. McCollum!

LIVE OAK, Fla.—Looking at her three children after Dr. Adams left her home, Mrs. Ruby McCollum was overcome with shame. She would do penance . . . some very special act to punish herself. She'd pray to God to forgive her.

Ruby went into her kitchen with the intention of preparing a very special meal to delight her husband and children. The telephone shrilled out . . . and as she picked it up, an unkind voice came to her through the receiver:

"Ruby McCollum, you think you're some big hen's biddy, don't you? Humpf! Riding 'round in your big shiny car, and holding on to Sam McCollum's money tighter'n' Dick's hat-band. You may grab all his money, but I'm here to tell you that you ain't got Sam! Right now he's laying up with somebody young enough for your daughter, and he don't . . ."

Ruby flung the receiver back on the hook and turned away from the phone. She had no idea who her caller was, and never was she to know, but the malicious voice had done something to and for her. She felt less guilty right away. It called back her feeling of neglect over the last few years. She didn't feel blameless, but she did feel a certain justification.

Let Sam stay home more and she would never do it again.

Then the phone rang again. This time it was Dr. Adams.

The next afternoon was a riot of passion. She was wanted as she had always yearned to be wanted.

. . .

THAT NIGHT in bed alone, Ruby thought things over. She saw the net closing around her. Did she want to escape? Well, yes, and, then again, no.

But how, without bringing on the very disgrace to her children that she wished to avoid?

How had the former Ruby Jackson gotten into such a fix? My God, her mother and father, brothers and sisters would die of shame if they dreamed of it. And Sam, the community of Live Oak would explode like gasoline. Please God, help her! For, she didn't see how she could help herself.

Ruby felt uncomfortable at times about the community; Dr. Adams was at her home so often. But who would dare to accuse her to Sam McCollum? Not a soul. But did Sam himself suspect . . . or know? From the way he acted she could never be sure.

. . .

THEN CAME the day when she knew she was pregnant, and she knew that it could not possibly be the child of her husband. Very nervous, and very diffident, she went to the office of Dr. Adams for an examination. He made it very briefly.

The seven months that followed were alternately Heaven and Hell to Ruby McCollum.

But alone, Ruby was worried. What would Sam say or do? What about her own family? What about the Negroes of Live Oak? She was heavy in spirit as the weeks dragged on nearer and nearer to her period of confinement.

(Continued next week.)

The Life Story of
Mrs. Ruby J. McCollum!

LIVE OAK, Fla.—Our scene changes rapidly, and, now Ruby Jackson McCollum is the tragic star of a world-shaking drama. In a blindingly brutal light, America has discovered that the separation of the races is a fiction. The pretensions have met the realities of that courtroom at Live Oak, Fla. Many things had happened between that pregnancy and this day.

Now, a Southern white lawyer, born within twelve miles of the Suwannee County Courthouse, fights brilliantly and ruthlessly to save the life of a Negro woman who has shot to death a prominent and popular white man.

This white attorney is flanked by and ably assisted by a Negro lawyer not appointed by any special group.

In close and amiable association, Col. Frank T. Cannon and Rutgers-educated Releford McGriff, the latter a Negro, fight shoulder-to-shoulder for the life of Ruby McCollum. The battle rages up and down and across all the lines once held too sacred to even put into words. For the first time, too, Suwannee County, Fla., had Negroes on a jury panel. Yes, it is a world-shaking drama.

But time is long by the courthouse clock. Ruby McCollum is a little woman by some standards. She sits in the large armchair with her small feet crossed and often swinging clear of the floor. For hours she sits almost motionless as the battle between defending and prosecuting counsel rages on and on.

She loses the thread of the conflict for her life and goes back into her drama. Ruby McCollum has lived so many lives and died so many deaths since that unforgotten afternoon nearly seven years [ago].

"O memories that bless and burn!"⁹

. . .

AS RUBY sits there with her head resting on her right hand, she lives those years again and tries vainly to shut out the truth from her heart.

Ruby McCollum, sitting in the courtroom in peril of her very life, had memories that blessed and burned. Perhaps she would die a repugnant death, but . . . she had lived!

And, materially, Ruby McCollum had lived. She had seen the day laborer she had married twenty years before rise to a place of wealth and a certain kind of power. She had sought for a go-getter when she chose Sam McCollum, and she had not been disappointed in that way.

"I picked you from the very top," she would say. "I was married to the top Negro of Suwannee County, and Dr. Adams was the top white man. When I tie up with a man, I have influence with him. Men love me when they get to know me."

. . .

AND SAM McCOLLUM, too, was bound. Live Oak knew all about his philandering, but it was axiomatic that Ruby ruled the roost. Even in the very act of his misdeeds, he bragged about his wife—to the intense irritation of numerous females.

He could not even bear to let her go when she bore a child which he knew could not possibly be his own. Nor did Sam McCollum denounce his wife about town. He quietly moved into a separate bedroom and, after a while, began to complain of pains about his heart.

Live Oak might speculate about the light-skin baby that Ruby rode about town in the flashy car that Sam had bought her, but Sam gave nobody the satisfaction of hearing him talk about his wife.

. . .

GRADUALLY people found out that Sam's doctor had warned him that he had heart disease. He must slack off and take things easy. But this he could not do. He had fought and gotten for his Ruby what she wanted, wealth and the power that goes with it.

He now held important real estate; he held a sizeable block of stock in a thriving business, the Central Life Insurance Company; and he had nearly a hundred thousand dollars in cash hidden away in his house.

But Ruby Jackson, the Ruby whom he had had to court for so long before she consented to marry him, was lost to him except in name only. So he was not so lively about town, not so full of wisecracks as he once had been. In quiet moments, the sides of his face looked limp and sagging, like wet-wash hung out to dry from his ears . . . "oh, barren gain . . . !"[10]

. . .

RUBY IS snatched back from her musings on her female mastery of men by the bitter, snarling voice of Edwards, the assistant state's attorney. He gives her a vengeful look and points a scornful finger at her as she sits between her attorneys:

"This Ruby McCollum is just as sane as I am. She knew what she was doing when she went to Dr. Adams' office that Sunday morning to kill him. That, gentlemen of the jury, is a sly, calculating killer. Every movement she made that morning was a preparation for this day.

"Look at her sitting there. You heard from her from the witness stand, so meek and mumbling with a little voice that the court had to order her to speak so that you could hear her. She was not like that when she pumped four bullets into the back of Dr. Adams.

"I ask you gentlemen to bring in a first degree verdict against this brutal slayer and without a recommendation of mercy . . . this savage . . . as she sneaked back and forth from her car to the office door to see if anybody recognized her, to await her opportunity to get into the treating-room with that gun which she had carried there in that big, tan hand-bag slung across her shoulder . . ."

. . .

THE BITTER tirade, the scalding denunciation of her went on and on, and Ruby sank back once more into her reveries.

Her children. Ruby had always loved children. Her ultra femininity gave her a very strong maternal instinct. Now her four children, three of them of tender years, were snatched away from her. She could not see them often, let alone give them her tender care. She had thoroughly dominated the lives of two able men, she had her memories that blessed, but she also had those which burned. A very barren gain. The folds of fate were closing in.

APRIL 18, 1953

The Life Story of Mrs. Ruby J. McCollum!

LIVE OAK, Fla.—Mrs. Ruby McCollum does not feel that she is entirely at fault for what has happened.

"I was between two guns that morning (the morning of the killing) . . . the gun of my husband and the gun of Dr. Adams."

She makes no explanation further. But, she does mention the terrible scene when Sam McCollum, her husband, found that she was again pregnant.

. . .

THEN THERE was that communication which Ruby cannot bring herself to talk about freely. The prosecution contends that it was a monthly bill for services rendered by Dr. Adams to the McCollum family, but Ruby's reaction to it clearly indicates that there was something more in that envelope than merely the doctor's monthly statement.

Incidentally, that piece of paper has disappeared, although it could have been important evidence in the state's case.

One thing she did reveal, however, was that Dr. Adams was insisting that she move out to his farm.

. . .

OTHER HORRIBLE memories tore at her as she sat there and heard the violent denunciation of the prosecution. The fear-filled flight from Live Oak to the State Prison at Raiford that Sunday morning.

Then the treatment that she had received at Raiford.

"Oh, just the thought of being in such a place!"

Did she have a miscarriage in prison?

"The state knows all about that," she murmured, and she could not say anything more. Her breath came in soft gasps. She thought of all the brutalities of mind and body that had been visited upon her when, as much as she regretted the deaths of Dr. Adams, and her husband, she had only acted instinctively, to—in a way—protect her own life.

"I was between two guns."

Now both of the men were dead, and she was here fighting for her life . . . "Oh barren gain . . . and bitter loss!"[11]

. . .

ADMITTING THAT she had ceased to love Sam McCollum, Ruby still deeply regrets his death, resulting from shock.

"Dr. Adams stole my love away from Sam. It would not have happened if Sam had not killed off my love years before I ever met the doctor, though. I was a good wife to Sam and a very loving mother to our children as everybody in Live Oak knows."

Ruby now is deeply pained over the suffering she has caused her mother and family. At times it almost drives her out of her mind, so that for hours everything seems to be in a mist.

"I do not grasp that it is myself at times. I seem to be walking in somebody else's dream, a bad dream, like a heavy fog around me. Sometimes I feel that I will wake up and find that it is nothing but a nightmare that I've had. I have died many deaths since that Sunday morning, but sooner or later I find that I am still alive and the bars are still around me."

Ruby feels—and, perhaps, justly so—that she has been the victim of a trap. Soon after she began seeing Dr. Adams, he began [to] prescribe medicines for her that had a queer effect upon her. She was sometimes very exalted and sometimes unnaturally depressed, but never her old, usual self.

Sam McCollum had sensed the same thing about his wife, and perhaps that was why he was so patient with her. He went to several intimate friends and told them that he was worried about Ruby. She had become so strange.

(Continued Next Week)

APRIL 25, 1953

The Life Story of Mrs. Ruby J. McCollum!

LIVE OAK, Fla.—In the search for her old self and her efforts to free herself from the daze in which she lived, it is a matter of record that Ruby McCollum went from doctor to doctor all over her section of Florida. Therefore, it is highly probable that Mrs. McCollum is a victim, instead of the cold, ruthless killer that the state claims her to be.

"I do not know why I killed Dr. Adams. I remember very little about what happened," she says.

Ruby says that Dr. Adams feared assassination from some other source for several weeks prior to Aug. 3, 1952. If he named the source of his fears to her, she is still keeping it a secret.

. . .

SHE IS TELLING the truth, because a hardware dealer states that Dr. Adams came to him to buy a gun about a week before he was killed. However, the hardware dealer disclosed, he had none in stock, but ordered one for Dr. Adams. It had not arrived at the time of his death.

"I was not his enemy," Ruby says significantly, "I loved him."

Rumor is rife in Live Oak that Ruby McCollum was the agent of some as yet secret person or persons in the slaying of Dr. Adams, and many express the hope that the name or names of whoever they may be will be bared soon.

. . .

NOBODY BELIEVES that she was induced to kill him for pay, but the general feeling is that her mind was so badly torn and twisted over a period of time that she would feel from what she was led to believe that to commit the act would be justified.

And that is where that missing communication plays such an important role . . . what has become of it? . . . Ruby McCollum, herself, does not know. It disappeared along with the handbag of cash and other important papers in it.

The secret of the slaying of Dr. C. LeRoy Adams may lie in that piece of paper! . . . what became of it? . . . and why was it destroyed?

Ruby says that all she knows is that she seemed to come alive again as she drove into her driveway on that fatal Sunday morning with her two youngest children—and that gun—in the car with her.

. . .

VAGUE, TROUBLED dreams came to her. The report of a gun . . . had she killed anybody? The thought was repugnant to her, and she hurled the gun into the bamboo hedge and went inside her home to take care of her children.

If her statement that she could not go with the arresting officers then because there was no one to take care of the children sounds stupid, it is because she had not as yet realized that she had killed a man.

It took the terrifying drive to Raiford, the brutal blows of a State Trooper and repeated accusations of others to make her realize, or believe, that it was so. Ruby still wonders at times if it was really her own hand that fired the bullets into Dr. Adams. Did she really do it, or, was there somebody else there who did it, and then imposed upon her the sense of guilt during her dazed condition? She earnestly wishes she could be sure.

. . .

WHILE THE appeal from her conviction was being argued before Judge Hal Adams, Ruby, herself, argued the whole thing through in her own mind. But, she went further back than the actual killing of Dr. Adams.

She says that she reviewed her entire life from conscious beginning to the hour of her conviction on Dec. 20, 1952. She had been trying to find the answer to her life ever since she had been brought back from Raiford to the county jail in Live Oak.

She still did not have the answer when, in the deepening dusk of the evening when Judge Adams denied the appeal of her lawyers, she was told to stand and receive the sentence of death in the electric chair.

. . .

HIS VOICE husky with emotion, Judge Adams began hurriedly to pronounce the sentence of death upon Ruby McCollum, then catching himself, he interrupted to ask the formal question:

"Have you anything to say why death should not be imposed upon you?"

"I, I, I don't know whether I was right, or wrong . . ."

The prisoner at the bar, Ruby McCollum, said this with the hesitation of uncertainty, then stood calmly while the dread sentence was imposed upon her. Then she went calmly back to her seat, intertwined the fingers of her hands and looked up at the ceiling.

. . .

SHE HAD been—and still is—questioning the meanings of her life and what has happened to her . . . and why . . . and the answer has not been found.

She does not know what turn, what attitude brought her to where she is—she who was ever quiet and gentle. She shares with the public which knew her in surprise at the killing of Dr. Adams.

As she said, she does not know whether she did right or wrong, and neither does anyone else, including the sovereign State of Florida, in spite of the verdict. Nobody will ever know until something more is revealed of the mystery that surrounds the death of Dr. Adams.

Ruby, bewildered and suffering in her cell at Live Oak, assured us that she strives at last to learn to kiss the cross. To submit to unfathomable Fate with courage and dignity.

. . .

WHATEVER HER final destiny may be, she has not come to the bar craven and whimpering. She has been sturdy and strong . . . a woman who ruled the lives and fate of two strong men, one white, the other colored. She had dared defy the proud tradition of the Old South openly and she awaits her fate with courage and dignity.

. . .

(Continued next week.)[12]

MAY 2, 1953

My Impressions
of the Trial

My comprehensive impression of the trial of Mrs. Ruby McCollum for the murder of Dr. C. LeRoy Adams was one of a smothering blanket of silence.[1] I gained other vivid and momentary impacts, but they were subsidiary and grew out of the first. It was as if one listened to a debate in which everything which might lead to and justify the resolution had been waived. Under varied faces, one was confronted with the personality of silence. Some conformed by a murmuration of evasions, some by a frontal attack that this was something which it would not be decent to allow the outside world to know about, and others by wary wordlessness.

It amounted to a mass delusion of mass illusion. A point of approach to the motive for the slaying of the popular medico and politician had been agreed upon, and however bizarre and unlikely it might appear to the outside public, it was going to be maintained and fought for. Anything which might tend to destroy this illusion must be done away with. Presto! It just did not exist.

I found myself moving about in this foggy atmosphere even before the first sanity hearing. Fearing to be quoted, many in the Negro neighborhood of Live Oak avoided even the suspicion of having told anything to a newspaper representative by fleeing my very presence. Others loudly denounced Ruby McCollum for hav[ing] slain Dr. Adams to make sure that if "the White folks" [hear] anything that they said about the case at all, it would be pleasing to them.

"Ruby (before August 3, they would not have dared to speak of her by her first name) had done killed the good-heartedest and the best White man in Suwannee County, if not the whole State of Florida. They won't be doing her right unlessen they give her the chair."

"Ruby McCollum knowed better than to go messing around with that White man in the first place. She didn't have no business with him. She knowed so well that she was a nigger. How come she couldn't stay among her own race?"

"I hope and pray that there ain't no salvation for Ruby at all. Killing up that nice Dr. Adams. You could always go to him when you was in a tight for money and he sure would help you out. And if you didn't have the change to pay him when he waited on you, he would scold at you and say, 'Did I ask you for any money? I'm trying to get you well.' And Ruby had to go and kill a nice man like that."

"No, Doc Adams never dunned nobody for money at all. You could pay him when you was able to do so. Never heard of him bearing down on nobody because they owed him money."

"But Sam and Ruby McCollum wasn't no bad pay neither. That's how come I can't see to my rest how come they got to fussing over her doctor's bill and ended up in this killing-scrape. Everybody knows that them McCollums paid what they owed on the dot. They had it to pay with and they sure Lord paid. Everybody in Live Oak and Suwannee County will give 'em that. They never owed nobody."

Hints like that now and then gave away the code. Let you understand that they were play-acting in their savage denunciation of Ruby McCollum. The sprig of hyssop was in their hands and they were sprinkling the blood of the rascal lamb around their doorway so that the angel of death would pass over them.[2] This was West Florida, and somehow the Negro population of Live Oak felt apprehensive.

And inside the Court-house on this December day of the trial, a man and a woman came up the stairs to the galleries reserved for Negro spectators and took a seat next to me. The woman murmured for my benefit, "Don't be surprised what might come off here." They looked down the row of seats and the man said, "I hope no fool don't go and block up that aisle."

"But you wouldn't run off and leave me, honey, if something was to take place, would you?"

"Not if you can keep up, baby."

The local White people felt no such timidity about physical violence. But fear was there. It stemmed from what the outside world would say about the trial, and hence the banning of the Press.

"Ruby has killed a White man, and not only that, the most prominent White man in Suwannee County. She ought to die for that, and she's going to die. You can't let women go around killing up every man they take a notion to. We don't want these newspaper men coming in here and printing lies about us. Ruby was just full of meanness, shooting Dr. Adams to death rather than to pay her doctor's bill."

"We don't have to believe a word that a woman says who will murder a good man because he sent her a bill for waiting on her. That baby of Ruby's is not Dr. Adams' child. Oh, it might look sort of bright-colored like some say it does, but it could be Sam McCollum's like them other three. Sometimes babies are throw-backs, and take after their fore-parents, you know."

"Oh, we all recognize that Doc Adams got around right smart among the ladies, that woman in Lake City, this one and that one, was separated from his wife for years until he put in to run for the State Senate, but he never had anything to do with Ruby like that. She's lying. She just hated to pay him that money she owed the man. A brutal murder without any excuse for it at all."

It was like a chant. The medical bill as a motive for the slaying was ever insisted upon and stressed. It was freely admitted by all that the McCollums had always been good pay. Paid what they owed with promptness every time. Yet, there was this quick and stubborn insistence that the medical bill, and that alone, could have been the cause of the murder. It was obviously a posture, but a posture posed in granite. There existed no other, let alone extenuating circumstances. This was the story, however bizarre it might appear to outsiders, and the community was sticking to it. The Press was discouraged from "confusing" the minds of the State and the nation by poking around Live Oak and coming up with some other loose and foolish notion.

I sought and had more than one interview with Judge Hal Adams. My impression of the Judge who was to preside over this trial [was] as a man possessed of many substances marketable in the human bazaar. I found no fault with the broad black Stetson, the black string tie of a past century, his chewing tobacco, his witty turn of mind, nor yet his mouth full of colorful Southern idioms. To me, these were externals, and need not of necessity indicate his type of mind. How he conducted the trial was my yard-stick. We got along very well indeed.

But naturally I was disappointed when he told me that he could not allow me to interview Ruby McCollum. I had been sent to Live Oak by the *Pittsburgh Courier* first and foremost to get an interview with the defendant in this case, and second, to report the case as it developed. But Judge Adams was not short and harsh with me in his refusal. He said that the very nature of the case made it advisable to deny access to Mrs. McCollum to the Press. He did not want the case tried in the newspapers before it was presented in court. Justified or not, I gave house-room to the impression that if/when the Court permitted Mrs. McCollum to be interviewed, that I was as likely as anyone else to get the scoop. Then seeking to co-ordinate my favorable impression of Judge Adams with the impression I had gained of an existing dogma of Ruby McCollum's motive for slaying Dr. Adams so widespread in the locality, I pondered as to whether the Judge had lent his prestige to its acceptance. But if so, was it native to his spirit, or was he a captive of geographical emotions and tradition?

There was abundant precedent for this mental query, the case of General Robert E. Lee being the most notable. This truly great military genius and greater of soul than as a soldier, hated the institution of slavery, and had freed his own slaves. He believed devoutly in the Union and regarded Secession as the ultimate death of the nation. Yet, when the struggle came, his decision was that he would go along with Virginia and his neighbors.

From the instant minute that this matter occured to me, Judge Hal Adams became the foremost figure of interest to me in this trial. The slaying had been admitted by the defendant, and the degree only of her guilt was left to be decided by the trial. To me, the real drama was inherent in the reactions to the evidence presented, or allowed to be admitted by Judge Adams. There is an old southern saying to the effect that a man ain't got no business pulling on britches unless or until he's got guts enough to hold 'em up. Another saying that considers the strength and calibre of a man says, "The time done come when big britches got to fit Little Willie." Could big britches fit Hal Adams, or to put it another way, could Hal Adams fit himself into big britches? [handwritten addition: Fate had issued the challenge to him.] My mind was entirely open. I would let the Judge furnish the answer.

The confirmation for me that I was right in the impression that the Negroes of the locality were using protective coloration and concealing their own thought about the case lay in the fact that the closer the association had been with the McCollums, the more violent were their denunciations of Ruby. One example was furnished by Hall, the Negro undertaker. Rumor informed me that Sam McCollum's money had set Hall up in business and though it was thoroughly established that Hall had also been on very intimate and confidential terms with the late Dr. Adams,—often driving the Doctor at night to his amorous rendezvous,—he could scarcely have felt as violent towards his benefactors, the McCollums, as he pretended. It was only natural for Hall to realize that he might be suspected of sympathy with Ruby because of past benefits, and thus he was giving himself a cat-bath to forestall possible retaliation from White Live Oak. That is, washing himself off with his tongue.

I had the same impression concerning the janitor of the County Courthouse. It was said that this Negro had been the McCollum yardman for twelve years. He had, during those years, enjoyed numerous favors from the McCollums. Yet, nobody could match his violent denunciation of Ruby nor his expressed hopes and prayers for her speedy execution. He was waiting for Judge Adams every morning when he arrived at the Court-house to carry in his brief-case or any other luggage that the Judge might have, and being most obsequious every moment, to the extent that the Judge publicly thanked him at the end of the trial. No one could blame him for looking after the comforts of the Judge, and certainly I do not, for Judge Adams appears to me to be a most kindly and considerate person, and easy to like. The motives of the janitor are under scrutiny here, and especially so when it is considered that this man had been an inmate of Raiford himself, so it is alleged, and it certainly would do him no harm to be on the good side of a County Judge when the police rapped on his door the next time. Rightly or wrongly, the janitor believed the Judge wanted such an expression from the Colored folks, and he was moved to furnish it with great enthusiasm. Nor do I censure the man for his behavior. I merely examine it as part of a pattern which I found. A fleeing away from fear.

But in all fairness to this humble man, human nature cannot be ignored. The McCollums were wealthy and otherwise stood high in the community. These local Colored people were for the most part, little people, the kind of people, irrespective of race, who have only the earth as their memorial. With death, they go back to the ground to rejoin the countless millions of other nameless creatures who are remembered only by the things which grow in soil. There is ever a residue of resentment against the successful of the world. It has nothing at all to do with race. Thus from the beginning of time the most popular story is one in which the poor triumph over the more fortunate. Heaven, where the humble will be equal or superior to the powerful on earth; Dives and Lazarus; the Cinderella story; the peasant boy who wins the princess and with her, one half of the kingdom; Dick Whittington, Lord Mayor of London.[3] So there was a certain amount of revengeful satisfaction in seeing Ruby, the high-cockalorum brought low. No longer could she stir their envy.

And there was bound to be a certain amount of grim satisfaction on the part of many Negro men when it was found that Ruby had gotten into "bad trouble" by giving herself to a White man. This complaint has come down from slavery days, the advantage that White men have over Negro men with women. In a cafe with the jook-box playing at the top of its voice, the case was mentioned, and a tall man pleasuring himself with breaded pork chops and side dishes of collard greens and fried sweet potatoes snarled,

"I ain't got no sympathy for Ruby at all. All before, she wouldn't even wipe her feet on nobody like us. If she just had to have herself a outside man, she could of got any kind she wanted right inside her own race. That's one thing about our race, we're just like a flower-garden, you can get any color from coal-scuttle blonde to pink-toed white." (This is a very common boast made always when Negroes rail against miscegenation, not seeming to realize that by it, they are endorsing the very thing which is being denounced, for obviously if there had been no miscegenation, there could be nothing but black individuals among us.) "Naw, she had to go and have that White man, and when she knows so well how these White men don't allow us no chance at all with *their* women. Colored women ought to be proud to stick to their own men and leave these White men alone."

"And more 'specially when they ought to know that White men ain't no trouble at all. They can't do nothing in the bed but praise the Lord. Nothing to 'em at all."

A loud guffaw of gloating laughter from all the men at this. The thought sort of evened things up.

"We gets all we want for nothing, but they got to pay for it, and they had better, 'cause they sure can't bring nobody to testify and neither sinners to repentance. Shucks!"

Even louder laughter. But a tipsy woman (she would not have dared to speak had she been sober, or of a more discriminating type.)

"Talk that you know and testify to that you done seen. Some of these here White men got lightning in their pants. In the 'Be' class, be there when you start and be there a long time after you done fell out."

In the shocked and enraged silence that followed this, the slattern must have jolted to a soberer state, for she added sheepishly, "That is, it could be. I wouldn't know my own self."

"You better say 'Joe' 'cause you don't know," one man growled, giving expression to the emotions of his male hearers. "Tain't a thing to the bear but his curly hair. Don't you stand there and tell that lie that a White man can do with a woman what a Colored man can do. Oh, I reckon some of these trashy Nigger women who is after their money will lay up under 'em, like Ruby McCollum done, and moan and squall and cry to make 'em think they're raising hell, but that's only because of the spending-change the White man puts out. But everybody knows that a White man ain't no trouble, not a damn bit, and any old nappy head that tells me they is in my hearing, is going to get a righteous head-stomping."

No female picked up the challenge, naturally, so the men took the floor and the occasion to give the matter a vimful "reading." They poured out all the stored resentment of the centuries between 1619, when the first batch of Negro slaves were landed in the English colonies, to September 1952. This age-old hobby-horse was flogged along from Ginny-Gall to Diddy-Wah-Diddy. (Mythical places of Negro folklore reputedly a long way off. Like Zar, which is on the other side of Far.)

In a way, but a limited way, these men had a point. However, by the measuring stick of history, which is the recording of human behavior, their contention had no standing for the simple reason that force was lacking to back it up. From the cave-man to the instant hour, usage has hallowed the rule that to the victor belongs the spoils, and the primest spoils are women. Perhaps when the millennium has arrived, and the lamb is sharing an apartment with the lion, some changes will be made.[4] In fact, we will know that that blessed age has arrived when the males of every species relinquish the pick of the females to the weaker individuals. But it has not yet arrived, and the victor, whether the war be of a military or of an economic nature, takes his pick of the females, or to be more accurate, the females pick *him*.

The men felt that this angle figured in the case, and it did to an extent, but nobody at all even hinted that there was any exchange of monies between Ruby McCollum and Dr. Adams. It was not that kind of an affair. But anyway, the scab of the age-old sore was scraped off and it oozed blood afresh.

"I'm glad that Ruby killed him. He ought to of been satisfied with his own women. But he had to keep on messing around until he got killed. Good enough for him."

"Yep, Ruby shot him lightly and he died politely. If Ruby comes clear of this, I aim to buy her a brand new gun."

Then somebody became conscious of my presence in the place and the talk was turned off like a faucet. A voice demanded stewed new peanuts. This was a dish that I had never heard of, but found that it was well known fare in the area, fresh-dug peanuts shelled and cooked with side-meat or ham-hock like any other pea. The owner of the cafe snorted that he didn't bother with no trashy grub like that.

Then a woman moaned, "Poor Ruby is going to get the 'lectric chair. I seen it in a dream last night. I seen her with her head all shaved."

"Nobody didn't tell me, but I heard that Ruby meant to kill Sam before she died."

"But she never kilt him. Appears to me like Sam just naturally scared hisself to death. You know how scared Sam was of White folks. He figured they might lynch him for what Ruby had done."

"Well, anyhow it was told to me that that wasn't none of Sam's gun she killed Dr. Adams with at all, 'cause Sam had done carried his gun out of the house more'n a week before the killing."

"Oh, you can hear anything except sinners praying for repentance. Let's squat that rabbit and jump another one."

"I help you to say! Some folks around here must of been raised on mocking-bird eggs." (There is a folk-belief that if a person eats a mocking-bird egg, they can no longer keep a secret.)

Vainly did I seek the witnesses for the Defense. There were none. Even the closest friends of the McCollums must have been persuaded that the risk was too great. The names of eight Negroes were added to the jury panel, seven of whom disqualified themselves as being friends of Dr. Adams. The eighth somehow seemed to have been overlooked until the jury box was filled.

All this indicates the climate of the Ruby McCollum trial outside of the Court-house, and had bearing upon the trying of the case, or so it seemed to me.

From my seat in the balcony on the east side of the building, I could get a good view of the court-room. It was clean and comfortable enough as court-rooms go. Provisions had been made for custom and comfort. Tobacco-chewing and snuff-dipping are common enough in the area not to be apologized for, so spittoons were handy in the official section. The Judge and jury had been taken care of that way.

The substantial building had evidently been constructed before drinking fountains were common in public places, so the janitor presently came up the center aisle toting a bucket of ice water and balanced it on the corner of the railing enclosing the court officers. One glance told me that a man had bought the outfit, for the bucket was white enamel with a red rim around the top, and the dipper was white enamel, but with a dark blue handle. A woman would have seen to it that they matched. The janitor passed the bucket and dipper around to the jury when he thought that they wanted a drink, and also to court officers, but White spectators could go up and get a drink if they wanted it.

I was in my seat as soon as the door was opened to spectators, a good hour before court set at ten o'clock. The main floor of the room began to fill immediately. The first to enter was four matrons. The[y] took seat[s] on the very front row. One brought a son about two years old. Another had her knitting along. A few rows back, a young woman had an infant child in her arms. It got to fretting and crying during a tense period in the trial and many people frowned at her, wondering why she did not take that fretting child out of there so they could hear better. When Judge Adams looked pointedly in her direction, she reluctantly rose and went out.

But long before court set, eyes kept going to the door behind the railing through which both Judge Adams and Ruby McCollum must eventually enter. Bring her in and let them have their pleasure of her. It was plain that none doubted the conclusion of the trial. The uprising was going to be put down. The emotion rose like a fume from the lower floor.

As the hour drew near, Judge Adams passed through the room from his office at the front of the building and exchanged pleasantries with friends in the audience and behind the rail. I could see that his presence was tantalizing, for everybody, including myself, was anxious for the business to begin. Visiting lawyers, including ex-Governor Caldwell, occupied seats on the right of the platform inside the railing and passed off the time with chat.[5]

A side drama relieved the tedium somewhat. A. K. Black, state's attorney for Suwannee County, was already there with the assistant state's attorney, O. O. Edwards. Naturally they were attracting much interest on the occasion. Black is a short, plump man in a rumpled blue suit and with a bald spot on top of his head so perfectly round that it might be the tonsure of a monk. H[e] is not impressive in appearance, but a Negro seated behind me murmured, "Don't let that sleepy look fool you. He's just playing possum. He's going to get Ruby unlessen her lawyer watches out right smart. That Black is a 'getting fool.' I done seen him at it."

O. O. Edwards did little greeting and mixing around. He appeared to be preoccupied, not to say dedicated, and his look was grim.

Into this scene and setting Frank T. Cannon made his entrance, and I mean entrance. You could hear people commoting around him as he came through the door. Cannon, Solicitor for the Court of Record of Duval County (Jacksonville), was Ruby McCollum's new lawyer, replacing the recently disbarred P. Guy Crews. There were three more additions to this item. First, Frank Cannon had been born and raised in the vicinity of Live Oak. Therefore, he was being honored as a home boy who had matched his wits with the lawyers in the big city and made his mark. Then the looks of the man. Cannon is possessed of a challenging head of thick, wavy hair, an extra-handsome profile, becoming sun-tan, tall graceful body clothed in raiment of good taste and good quality. Then the man has what is known in the theatre as stage presence. Something exudes from him and grabs hold of the audience. So smiling voices called out to Cannon, and hands were extended. He was a one-man procession down the aisle. He had stolen the scene from the prosecutor.

Perhaps in this commotion, I was the only one who looked to see what effect this had on his legal opposition. Edwards registered nothing, but I caught a brief look on the face of Black, that if that look had been a bullet, it would have worked Cannon three times before it killed him and five times afterwards. Nor, was I mistaken in my interpretation of it, for after the trial began, it came out that Black had not only attempted to have Cannon barred from the case, but Cannon publicly accused him of also trying to have him run out of town. A free translation of the reaction of Black to Cannon's sudden entrance into the case read that Cannon, being who he was, had probably had things pretty much his way around Live Oak before he moved to Jacksonville to seize his share of success and glory there, and now, when this sensational case gave Black his chance at the center of the stage with national publicity, here comes that handsome, wavy-haired fool to hog the spotlight. Black had behaved little, but human. That is my impression of the incident.

Court set at last, and Ruby McCollum was led in by a State Trooper. Then the place came really alive. I could tell that she had been given the opportunity to groom herself with more care this time. Her hair was pressed and hung in a long bob to her shoulders, but confined loosely by a net. It was parted on the left side neatly, and being December, she had on a coat, a camel's hair coat bright green in color over a pale yellow wool dress. Her small feet were encased in low-heeled black pumps. She is small, and looked almost childish in her seat.

She had walked in briskly with an expressionless face. She looked neither to right nor left as she moved to the Defense table and took her seat beside Cannon.

In the course of the two sanity hearings, the balked first trial, and the real trial, I have striven to enclose my impression of Ruby McCollum in a sentence, but failed. A[s] to her physical appearance, I would say that she is attractive, but not beautiful. A sort of chestnut brown in color, with the breadth of face I think of as feline, though there was nothing sly or calculating in it that I could discern. Even under her terrible strain, she appeared to be possessed of dignity. As before, she seemed to set herself as she took her seat in a resolved position. Her right elbow rested on the arm of the chair, and her head resting on her hand, [s]lightly inclined to the right. I did have the impression that every muscle of her body was consciously set and locked in place as she sat lest they betray her disturbed condition. The only signs of nervous-strain that she exhibited was an occasional [s]win[g]ing of her crossed feet. Again, she would extend her right hand at full length and examine it in minute detail. Flex and extend the fingers and look at them very studiously; turn it and examine palm and back as if it were something new and interesting to her. Since it was the hand which had wielded the gun on Dr. Adams, I could fancy that she might be regarding it as having a separate existence, a life and will of its own and had acted without her knowledge or consent.

There was one poignant, one heart-rending scene while Ruby was on the stand. She maintained her shut-in, expressionless mask through the questioning by State and her own counsel until one could say that Ruby McCollum was a woman without nerves until that time, as Cannon led her through her story to the moment of the actual slaying. Ruby did not break down and weep piteously; she did not scream out in agony of memory with her voice, but there was an abrupt halt in her testimony and something rushed forth from the deeps of her tortured soul and inhabited her face for a space. The quintessence of human agony was there. I witnessed momentarily the anguish of the hours, perhaps the days and weeks which preceded the slaying on August third, the indescribable emotions of the resolve to slay, to b[l]ot out from the world that which she had come to know and which tightened her hand upon the gun, and the memories of it all which lived down deep in barred cavities in the cellar of her soul with an eternal life which she could bestow but could not take away. What I beheld in the eyes of Ruby McCollum in that instant when she balked into silence, when the agony of her memories robbed her of the power of speech for a time, may God be kind and never permit me to behold again. In a flash, I comprehended the spacial infinity of the human mind, that mother of monsters and angels, and the ineffable glory and unspeakable horror of its creations.

That illuminated minute was the hub and life of the trial for me. It gave sense and meaning to all that had gone before and everything that could possibly follow. Now, I could discern that what was going on in the courtroom was nothing more than a mask, and that the real action existed on the other side of silence. The defendant had freely admitted the slaying of Dr. Adams. She was in the hands of the law, and therefore, there was no reason for the legal machinery of Suwannee County and the State of Florida to be operating here on the case except to fix the degree of the guilt of Ruby McCollum.

Yet, thirty-eight times Frank T. Cannon had attempted to create opportunity for the defendant to tell her story and thus throw light on her motives, and thirty-eight times the State had objected, and thirty-eight times Judge Adams had sustained these objections. My faith was in Judge Adams, and after that terrible minute, I looked at him confidently and said, almost audibly, no, Judge Adams will never allow this. He will never consent for any human being to be sent to their death without permitting the jury to hear their side of the story. He won't! Judge Adams won't! It was only when he exhibited anger and threatened Cannon with contempt of Court if he persisted, that I wilted back, first in my soul and then in my seat. My disillusionment was terrible. My faith had been so strong. Cannon's words, uttered with tragic resignation, "May God forgive you for robbing a human being of life in such a fashion. I would not want it on my conscience."

But race had nothing to do with me. Had Judge Adams been as black as Marcus Garvey, and Ruby McCollum as fair as the lily-maid of Astolat, still I would have felt the same.[6] And my discomfort increased when I then recalled that A. K. Black and Dr. C. LeRoy Adams were on the most intimate terms, and so it was hardly possible that he did not know all that Ruby had to tell already. He made no denial when Cannon charged directly, "And nobody knows this any better than the State." (Black)

This impression was deepened when Thelma Curry, the late Dr. Adams' Colored receptionist, was called to the stand as a prosecution witness. Things went on very well until she tried to tell about a quarrel she overheard between Dr. Adams and Ruby a few days before the slaying. Immediately she was recalled from the stand by Edwards, and Adams added his bit by telling her, "Get down and go back where you came from!" Nor had she been allowed to tell what she knew about the letter.

That letter! It is still my impression and ever will be that the motive for the killing was in that letter. The State based its whole case on that letter, maintaining that it was nothing more than a monthly medical bill for $116 and that Ruby shot Dr. Adams four times rather than pay it, still this letter was never introduced in evidence by the State. This added to my mounting feeling of horror. I could not avoid the conclusion that the omission had purpose, and that purpose was sinister. Therefore, I lacked the feel of solemnity I should have felt when Ruby stood and was sentenced to death by Judge Hal Adams. And her statement when she was asked if she had anything to say, and she said [simply], "I do not know whether I did right or not when I killed Dr. Adams," held a mountain of meaning for me.

I sum up my impressions of the trial of Ruby McCollum for the murder of Dr. C. LeRoy Adams in the clearest way I know how, now. It was as if I walked in a dream, somebody else's dream, of fog and mist with occasional heavy drops of cold rain. Somewhere beyond me, and hidden from me by the enshrouding mists, the tremendous action of this drama of Ruby McCollum and Dr. Adams, and the actors in the trial went on. It was behind a sort of curtain, on the other side of silence.[7]

Chronological List
of Essays

"Bits of Our Harlem," 1922

"The Hue and Cry About Howard University," 1925

"The Emperor Effaces Himself," ca. 1925

"The Ten Commandments of Charm," 1926

"Noses," 1926

"How It Feels to Be Colored Me," 1928

"Characteristics of Negro Expression," 1934

"Conversions and Visions," 1934

"Shouting," 1934

"Spirituals and Neo-Spirituals," 1934

"Race Cannot Become Great Until It Recognizes Its Talent," 1934

"You Don't Know Us Negroes," ca. 1934

"Fannie Hurst," 1937

"Art and Such," ca. 1938

"Stories of Conflict," 1938

"The Chick with One Hen," ca. 1938

"Now Take Noses," 1939

"Ritualistic Expression from the Lips of the Communicants of the
    Seventh Day Church of God," 1940

"Lawrence of the River," 1942

"High John de Conquer," 1943

"The 'Pet Negro' System," 1943

"Negroes Without Self Pity," 1943

"The Last Slave Ship," 1944

"My Most Humiliating Jim Crow Experience," 1944

"The Rise of the Begging Joints," 1945

"Crazy for This Democracy," 1945

"Jazz Regarded as Social Achievement," 1946

"Review of *Voodoo in New Orleans* by Robert Tallant," 1947

"The Lost Keys of Glory," ca. 1947

"What White Publishers Won't Print," 1950

"I Saw Negro Votes Peddled," 1950

"Mourner's Bench," 1951

"A Negro Voter Sizes Up Taft," 1951

Ruby McCollum Series "Zora's Revealing Story of Ruby's 1st Day in Court!," October 11, 1952; "Victim of Fate!," October 11, 1952; "Ruby Sane!," October 18, 1952; "Ruby McCollum Fights for Life," November 22, 1952; "Bare Plot Against Ruby," November 29, 1952; "Trial Highlights," November 29, 1952; "Justice and Fair Play Aim of Judge Adams as Ruby Goes on Trial," November 29, 1952; "McCollum-Adams Trial Highlights," December 27, 1952; "Ruby Bares Her Love Life," January 3, 1953; "Ruby's Story: Doctor's Threats, Tussle over Gun Led to Slaying!," January 10, 1953; "Ruby's Troubles Mount: Named in $100,000 Lawsuit! " January 17, 1953; "The Life Story of Mrs. Ruby J. McCollum," February 28, March 7, March 14, March 21, March 28, April 4, April 11, April 18, April 25, May 2, 1953

"My Impressions of the Trial," ca. 1953

"The South Was Had," ca. 1954–59

"Court Order Can't Make Races Mix," 1955

"Which Way the NAACP?," ca. 1957

"Take for Instance Spessard Holland," ca. 1958

Acknowledgments
from the Editors

Any volume such as this is indebted to the contributions of many. We appreciate the Zora Neale Hurston Trust entrusting us with this important project, and we stand on the shoulders of Hurston scholars who came before us, particularly Robert Hemenway, Hurston's first biographer. We appreciate our team at Aevitas Creative for advocating for us and this project. Our editor at Amistad, Tracy Sherrod, lent her keen eye to shape the final vision for the volume.

Archivists and librarians around the nation were generous and helpful as we sought to identify and collect published and unpublished writings by Hurston and worked to identify the many historical figures mentioned in them. For guiding us through collections, helping locate missing items and otherwise unidentifiable people, responding quickly to emails (even in the midst of a global pandemic, no less), arranging scans and processing payment, we are so very appreciative. Those who provided assistance include: Tara C. Craig, Head of Public Services Rare Book & Manuscript Library, Columbia University; Loretta Deaver, Reference Librarian, Manuscript Division, Library of Congress; Joellen ElBashir, Curator, Manuscript Division, Moorland-Spingarn Research Center, Howard University; Simon Elliott, Public Services, UCLA Library, Department of Special Collections; Kathy Lafferty, Reference Librarian, Spencer Research Library, University of Kansas; Tomaro Taylor, Head of Special Collections,

University of South Florida Libraries; Florence M. Turcotte, Literary Manuscripts Archivist, Department of Special and Area Studies Collections, George A. Smathers Libraries, University of Florida; Vianca Victor, Rare Book & Manuscript Library Public Service Specialist, Columbia University; Sonja N. Woods, Archivist, University Archives, Moorland-Spingarn Research Center, Howard University; Lewis Wyman, Reference Librarian, Manuscript Division, Library of Congress; and the staff of The University of Alabama Libraries Annex, including Kevin Ray, Institutional Records Analyst, Hoole Special Collections; and Jon Ezell, Coordinator of Creative Media & Instructional Design Librarian.

Transcribing, editing, and annotating this volume was made possible by three creative arts and humanities research grants at Texas Woman's University. Tracy Lindsay, Director of Operations, and Rocío Gutiérrez, Senior Grants Analyst, in the Office of Research and Sponsored Programs provided tremendous support. The funding supported smart and dedicated graduate research assistants in the department of language, culture, and gender studies. Each contributed to the project in important ways. Allyson Hibdon, MA in English, provided early support for the project, helping with organization and scanning that got the project up and running. Daniel Stefanelli, MA in English, transcribed, collated, and annotated texts and served as a valuable resource for thinking through the theoretical challenges posed by the materials. Jodi Meyer, MA in English, saw the project to completion—transcribing, collating, and annotating, as well as assisting with the apparatus. Elizabeth D. Headrick, MLS, in her role as a graduate research assistant in the Blagg-Huey Library's digital services department, put her skills to good use with OCR software to help improve the quality of images we relied upon. We are so grateful for their assistance. Professor West dedicates her work here to Leland Evan Fager II, her late brother.

Harvard University provided tremendous support as well. Julie Wolf, our stellar copy editor, helped us to blend our voices in the introduction. Dr. Abby Wolf, Executive Director, and Dr. Sheldon Cheek, Curator, Image of the Black Archive & Library, both at the Hutchins Center for African & African American Research, proved enormously helpful as we worked to identify intellectual property rights for the essays published here. It was a significant undertaking. Their time and insights are much appreciated. Professor Gates also extends a special thanks to his wife, the brilliant historian Dr. Marial Iglesias Utset and to Dr. Kevin Burke, Director of Research at the Hutchins Center.

Acknowledgments from the Zora Neale Hurston Trust

The trustees for the Zora Neale Hurston Trust gratefully acknowledge the contributions made by so many people who conducted extensive research to recover and compile relevant essays written by Zora Neale Hurston in order that they could be shared with Zora's multitudes of fans and the reading public in the outstanding work, *"You Don't Know Us Negroes" and Other Essays.*

We are deeply indebted to Dr. Henry Louis Gates Jr. for lending his considerable editorial skills and literary talent to this project. It is because of his appreciation of Zora as an author, ethnologist, filmmaker, and critic that he has so ably undertaken the task of introducing these essays in a context that helps us to understand the historical period in which they were written as well as Zora's perception of the realities of the time. We are also deeply indebted to Dr. Genevieve West, a professor of English who brought her talents in English and languages to bear on this project. Her research and presentation of Zora's essays capture Zora as a prescient observer of universal thinking and behavior that exists throughout time.

We want to acknowledge the many others who worked to bring these essays to light for today's readers. We especially want to thank Tracy Sherrod, Amistad's editorial director, and her staff, whose dedication to the success of *"You Don't Know Us Negroes" and Other Essays* is evident throughout this work. In addition, we want to acknowledge Tara Parsons, Amistad's associate publisher, and her staff for their contributions to the success of this outstanding compilation of essays. We also want to acknowledge Judith Curr, HarperOne Group president, for her continuing advocacy of Zora's works. We especially want to acknowledge our agent, Joy Harris, and those in her agency who work tirelessly each day to assure the success of Hurston projects. We thank you for bringing *"You Don't Know Us Negroes" and Other Essays* into the sunshine.

We extend our heartfelt appreciation to Zora's family of authors, journalists, teachers, academicians, scholars, literary reviewers, and her many legions of fans and readers whose continuing appreciation of Zora's works has brought her name to its well-deserved place in the literary canon of the twentieth and twenty-first centuries.

Credits

"A Negro Voter Sizes Up Taft." *Saturday Evening Post*, December 8, 1951, 29, 150–52.

"Art and Such," typescript (9 typed pages). Box 12, Item 6. Zora Neale Hurston Papers. Special Area Collection. George A. Smathers Libraries, University of Florida.

"Bare Plot Against Ruby." *Pittsburgh Courier* (Pittsburgh, PA), November 29, 1952, 1.

"Bits of Our Harlem." *Negro World* (New York City, NY), April 15, 1922, 6.

"Characteristics of Negro Expression." In *Negro Anthology*, edited by Nancy Cunard, 39–46. London: Wishart, 1934.

"Conversions and Visions." In *Negro Anthology*, edited by Nancy Cunard, 47–49. London: Wishart, 1934.

"Court Order Can't Make Races Mix." *Orlando Sentinel* (Orlando, FL), August 11, 1955, 10.

"Crazy for This Democracy." *Negro Digest*, December 1945, 46–48.

"Fannie Hurst." *The Saturday Review*, October 9, 1937, 15–16.

"High John de Conquer." *The American Mercury*, October 1943, 450–58.

"How It Feels to Be Colored Me." *The World Tomorrow*, May 1928, 215–16.

"I Saw Negro Votes Peddled." *The American Legion Magazine*, November 1950, 12–13, 54–57, 59–60.

"Jazz Regarded as Social Achievement." *New York Herald Tribune Weekly Book Review* (New York City, NY), December 22, 1946, 8.

"Justice and Fair Play Aim of Judge Adams as Ruby Goes on Trial." *Pittsburgh Courier* (Pittsburgh, PA), November 29, 1952, 5.

"Lawrence of the River." *Saturday Evening Post*, September 5, 1942, 18, 55–57.

"McCollum-Adams Trial Highlights." *Pittsburgh Courier* (Pittsburgh, PA), December 27, 1952, 4.

"Mourner's Bench, Communist Line: Why the Negro Won't Buy Communism." *The American Legion Magazine*, June 1951, 14–15, 55–60.

"My Impressions of the Trial," typescript. Box 1, Folder 2. Documents Relating to the Trial of Ruby McCollum for the Murder of Dr. LeRoy Adams, Live Oak, Florida. Special and Area Studies Collections. George A. Smathers Libraries, University of Florida.

"My Most Humiliating Jim Crow Experience." *Negro Digest*, June 1944, 25–26.

"Negroes Without Self-Pity." *The American Mercury*, November 1943, 601–3.

"Noses." *The X-Ray*, 1926, 22–23.

"Now Take Noses." In *Cordially Yours: A Collection of Original Short Stories and Essays by America's Leading Authors*, edited by Thomas Page Smith, 25–27. Philadelphia: Boston Herald Book Fair Committee, 1939.

"Race Cannot Become Great Until It Recognizes Its Talent." *The Washington Tribune* (Washington, DC), December 29, 1934.

"Review of *Voodoo in New Orleans* by Robert Tallant." *Journal of American Folklore* 60, no. 238 (October–December 1947): 436–38.

"Ritualistic Expression from the Lips of the Communicants of the Seventh Day Church of God," typescript. Margaret Mead Papers and South Pacific Ethnographic Archives, 1838–1996, Container C5. Manuscript Division, Library of Congress.

"Ruby Bares Her Love Life." *Pittsburgh Courier* (Pittsburgh, PA), January 3, 1953, 1, 4.

"Ruby McCollum Fights for Life." *Pittsburgh Courier* (Pittsburgh, PA), November 22, 1952, 1, 4.

"Ruby Sane!" *Pittsburgh Courier* (Pittsburgh, PA), October 18, 1952, 1, 4.

"Ruby's Story: Doctor's Threats, Tussle over Gun Led to Slaying!" *Pittsburgh Courier* (Pittsburgh, PA), January 10, 1953, 1, 4.

"Ruby's Troubles Mount: Named in $100,000 Lawsuit!" *Pittsburgh Courier* (Pittsburgh, PA), January 17, 1953, 1, 5.

"Shouting." In *Negro Anthology*, edited by Nancy Cunard, 49–50. London: Wishart, 1934.

"Spirituals and Neo-Spirituals." In *Negro Anthology*, edited by Nancy Cunard, 359–61. London: Wishart, 1934.

"Stories of Conflict." *The Saturday Review*, April 2, 1938, 32.

"Take for Instance Spessard Holland," typescript. Zora Neale Hurston Collection, Special and Area Studies Collections. George A. Smathers Libraries, Gainesville, Florida.

"The Chick with One Hen," typescript carbon, corrected. Zora Neale Hurston Collection. JWJ MSS 9 Series II, Writings, Box 1, Folder 8a. Yale Collection of American Literature. Beinecke Rare Book and Manuscript Library, New Haven, Connecticut.

"The Emperor Effaces Himself," typescript carbon. Zora Neale Hurston Collection. JWJ MSS 9 Series II, Writings, Box 1, Folder 15. Yale Collection of American Literature. Beinecke Rare Book and Manuscript Library, New Haven, Connecticut.

"The Hue and Cry About Howard University." *The Messenger*, September 1925, 315–19, 338.

"The Last Slave Ship." *The American Mercury*, March 1944, 351–58.

"The Life Story of Mrs. Ruby J. McCollum!" *Pittsburgh Courier* (Pittsburgh, PA), February 28, 1953, 3.

"The Life Story of Mrs. Ruby J. McCollum! Part 2." *Pittsburgh Courier* (Pittsburgh, PA), March 7, 1953, 2.

"The Life Story of Mrs. Ruby J. McCollum! Part 3." *Pittsburgh Courier* (Pittsburgh, PA), March 14, 1953, 2.

"The Life Story of Mrs. Ruby J. McCollum! Part 4." *Pittsburgh Courier* (Pittsburgh, PA), March 21, 1953, 3.

"The Life Story of Mrs. Ruby J. McCollum! Part 5." *Pittsburgh Courier* (Pittsburgh, PA), March 28, 1953, 2.

"The Life Story of Mrs. Ruby J. McCollum! Part 6." *Pittsburgh Courier* (Pittsburgh, PA), April 4, 1953, 3.

"The Life Story of Mrs. Ruby J. McCollum! Part 7." *Pittsburgh Courier* (Pittsburgh, PA), April 11, 1953, 3.

"The Life Story of Mrs. Ruby J. McCollum! Part 8." *Pittsburgh Courier* (Pittsburgh, PA), April 18, 1953, 2.

"The Life Story of Mrs. Ruby J. McCollum! Part 9." *Pittsburgh Courier* (Pittsburgh, PA), April 25, 1953, 3.

"The Life Story of Mrs. Ruby J. McCollum! Part 10." *Pittsburgh Courier* (Pittsburgh, PA), May 2, 1953, 2.

"The Lost Keys of Glory," typescript. Helen Worden Erskine papers. MS#0400, Box 2. Rare Book and Manuscript Library. Columbia University Library, New York City, New York.

"The 'Pet Negro' System." *The American Mercury*, May 1943, 593–600.

"The Rise of the Begging Joints." *The American Mercury*, March 1945, 288–93.

"The South Was Had," typescript. Zora Neale Hurston Collection. Special and Area Studies Collections. George A. Smathers Libraries, Gainesville, Florida.

"The Ten Commandments of Charm." *The X-Ray*, 1926, 22."Trial Highlights." *Pittsburgh Courier* (Pittsburgh, PA), November 29, 1952, 5.

"Trial Highlights." *Pittsburgh Courier* (Pittsburgh, PA), November 29, 1952, 5.

"Victim of Fate!" *Pittsburgh Courier* (Pittsburgh, PA), October 11, 1952, 4.

"What White Publishers Won't Print." *Negro Digest*, April 1950, 85–89.

"Which Way the NAACP?," typescript. PP 487, Box 7, Folder 22. Kenneth Spencer Research Library, University of Kansas.

"You Don't Know Us Negroes," typescript. Lawrence Spivak Papers. MSS40964, Container 37. Collections in the Archive of Folk Culture. Manuscript Division, Library of Congress, Washington, DC.

"Zora's Revealing Story of Ruby's First Day in Court!" *Pittsburgh Courier* (Pittsburgh, PA), October 11, 1952, 1, 4.

Notes

Introduction

1. Frantz Fanon, *Black Skin, White Mask*, trans. Richard Philcox (New York: Grove Press, 2007), 90.
2. Du Bois W. E. Burghardt. *The Souls of Black Folk*. Second Edition. Chicago: A. C. McClurg & Co., 1903. p. VIII.
3. Zora Neale Hurston, "Full of Mud, Sweat and Blood," review of *God Shakes Creation* in *New York Herald Tribune Books*, November 3, 1935, 8.
4. Zora Neale Hurston, "Mother Catherine," in *Negro Anthology*, ed. Nancy Cunard (London: Wishart & Co, 1934), 54–57.
5. E. Franklin Frazier, *Black Bourgeoise* (Glencoe, IL: The Free Press, 1957).
6. James Weldon Johnson (1871–1938) was an influential writer, editor, activist, and leader of the National Association for the Advancement of Colored People (NAACP). He is best remembered for his novel *The Autobiography of an Ex-Colored Man* (1912), a collection of his poems *God's Trombones* (1927), his anthology *The Book of American Negro Poetry* (1922), and his study of Harlem *Black Manhattan* (1930).
7. William Bradford Huie, *Ruby McCollum: Woman in the Suwannee Jail* (New York: EP Hutton & Co, 1956), 27.
8. Zora Neale Hurston, *Zora Neale Hurston: A Life in Letters*, ed. Carla Kaplan (New York: Doubleday, 2002), 705.
9. Hurston, *A Life in Letters*, 705.
10. Huie, *Ruby McCollum*, 185–88.

Bits of Our Harlem

1. Henry Wadsworth Longfellow (1807–1882) was a celebrated white nineteenth-century American poet.
2. Lenox Avenue, which is also known as Malcolm X Boulevard, runs north and south from Central Park through the center of Harlem to the Harlem River on the island of Manhattan.
3. 181st Street runs east and west, but it does not intersect with Lenox, raising the possibility that the street number is incorrect.

High John de Conquer

1. Hurston published this essay in October 1943 and alludes to the challenges of being a nation at war.

2. "Old Cuffy" is a historical, generic name for an enslaved person.

3. The Middle Passage is the journey of slave ships from West Africa to the West Indies.

4. Albatrosses were associated with good luck on sea voyages, though the superstition was popularized by Samuel Taylor Coleridge in *Rime of the Ancient Mariner* (1798).

5. Hurston alludes here to Mark 8:36 in the Bible.

6. Hurston likely alludes to 1 Corinthians 15:38–40, which distinguishes between consuming human flesh and that of other creatures.

7. John Henry is an African American folk hero who dies racing a steam-powered steel-driving machine building a railroad. He appears in many folk songs and stories. An audio recording of "John Henry" in the Library of Congress, collected by John A. Lomax in 1939, can be found at www.loc.gov/item/lomaxbib000326/?.

8. John the Conqueror root is a popular ingredient in hoodoo, or root working, which Hurston studied. Though the root is said to be native to the Southern US, its precise historical origin is unknown, and most roots sold under the name today are unlikely the original plant. See Carolyn Morrow Long, "John the Conqueror: From Root-Charm to Commercial Product," *Pharmacy in History* 39, no. 2 (1997): 47–53.

9. "The Tar-Baby" is an Uncle Remus story first recorded in writing by Joel Chandler Harris, although the story originated in oral tradition. In the folktale Brer Fox creates a doll made of sticky tar to entrap Brer Rabbit. See Joel Chandler Harris, *The Tar-Baby and Other Rhymes of Uncle Remus* (Ontario: Copp, Clark, 1905), 3–16.

10. An earlier version of this story appears in Hurston's short fiction. See "Possum or Pig" in *Hitting a Straight Lick with a Crooked Stick*, ed. Genevieve West (New York: HarperCollins, 2020), 141–42.

11. Jason and the Argonauts, from the ancient Greek epic *Argonautica*, embark on a long journey to steal the Golden Fleece and use it to claim kingship over Iolcus.

The Last Slave Ship

1. In May 2019, the wreck of the *Clotilda* was confirmed to be found near Twelve Mile Island along the Mobile River. See Allison Keyes, "The 'Clotilda,' the Last Known Slave Ship to Arrive in the U.S., Is Found," *Smithsonian Magazine*, May 22, 2019, smithsonianmag.com/smithsonian-institution/clotilda-last-known-slave-ship-arrive -us-found-180972177.

2. In *Barracoon*, Hurston spells Cudgo's name as "Cudjo." See Zora Neale Hurston, *Barracoon: The Story of the Last "Black Cargo,"* ed. Deborah G. Plant (New York: Amistad, 2019).

3. Kossula was not the last survivor of the *Clotilda*, as Hurston claimed here. Her letter to Langston Hughes dated July 10, 1928, indicates that she knew of another, who is likely the one referred to in recent scholarship. See Hannah Durkin, "Uncovering the Hidden Lives of Last *Clotilda* Survivor Matilda McCrear and Her Family," *Slavery & Abolition* 41, no. 3 (2020), doi.org/10.1080/0144039X.2020.1741833. For Hurston's letter,

see *Zora Neale Hurston: A Life in Letters*, ed. Carla Kaplan (New York: Doubleday, 2002), 123. For additional information on the other survivor, see Hannah Durkin, "Finding Last Middle Passage Survivor Sally 'Redoshi' Smith on the Page and Screen," *Slavery & Abolition* 40, no. 4 (2019), doi.org/10.1080/0144039X.2019.1596397.

4. Dahomey, located in present-day Benin in West Africa, was ruled by King Ghezo from 1818–1858.

5. Abomey is a city in the south of Benin.

6. Hurston's original use of the alternate spelling of *barracoun*, rather than *barracoon*, has been retained throughout this essay.

7. A short piece matching Hurston's quotation appears in the *Alabama Beacon* on November 19, 1858.

8. Timothy Meaher (1812–1892) was the Maine businessman who hired Captain William Foster to undertake a slaving voyage aboard the *Clotilda* after the US had outlawed the transatlantic importation of enslaved people. For more biographical information on these men and the other Meaher brothers, see Sylviane A. Diouf, *Dreams of Africa in Alabama: The Slave Ship* Clotilda *and the Story of the Last Africans Brought to America* (Oxford: Oxford Univ. Press, 2007), ACLS Humanities E-book.

9. A narrative written by Captain William Foster in 1860, titled "Last Slaver from US to Africa. AD 1860" and held in the Clotilda Collection of the Mobile Public Library Digital Archives, shares Foster's experience of visiting Dahomey. This piece includes reference to the snake collection, though there is no mention of the human-skin drum. Hurston may have accessed another unknown source for this detail. See also Diouf, "An Essay on Sources," in *Dreams of Africa in Alabama*, 245–49.

10. Roger Taney (1777–1864) was the Supreme Court justice who penned the infamous final majority opinion ruling against enslaved man Dred Scott's case for freedom in 1857. Taney argued that Dred Scott was not a US citizen and therefore had no right to sue in federal court.

11. John Dabney (ca. 1809–1881?) was a Virginian plantation owner who helped smuggle the Africans further inland. See Diouf, *Dreams of Africa in Alabama*.

12. A biography of James Dennison (1840–1915), written by his granddaughter, Mable Dennison, and titled *Biographical Memoirs of James Dennison*, is available online via the Mobile Public Library Digital Archives in the Clotilda Collection. See Mable Dennison, *Biographical Memoirs of James Dennison*, Clotilda Collection, Mobile Public Library Digital Archives, http://digital.mobilepubliclibrary.org/items/show/1806.

13. The repetition of the name Tim in this sentence is an obvious error in the original text, but it is impossible to know whether Tim or Tom took the larger number of enslaved pairs as property.

Characteristics of Negro Expression

1. *Paradise Lost* (1667) is an epic poem published by John Milton that recounts the Fall of Man from Genesis, the first book of the Bible. Thomas Carlyle's novel *Sartor Resartus* (1836) comments on difficulties with the philosophic system.

2. Louis XIV (1638–1715) was king of France from 1643–1715.

3. Hurston's note: From pang.

4. The Occident is the Western world, especially Europe and America.

5. Victrola is a brand of record player; a "console victrola" is a furniture-sized machine, as opposed to a small tabletop version.

6. The Treaty of Versailles, signed in 1919, ended World War I for the Allies and Germany. Waterman is a luxury fountain pen company founded in 1884.

7. Langston Hughes (1901–1967) was a Harlem Renaissance poet, novelist, and activist. The poem Hurston quotes is titled "Evil Woman," collected in Hughes's second volume of poetry, *Fine Clothes to the Jew* (1927). The lines of Hughes's poem break differently than they appear here, but Hurston's rendering of the poem has been retained. See Langston Hughes, *The Collected Works of Langston Hughes 1921–1940*, ed. Arnold Rampersad (Columbia: Univ. of Missouri Press, 2001), 99.

8. Hurston refers to the "St. Louis Blues" (1914), a song composed by W. C. Handy that deviates from common twelve-bar blues.

9. Bill "Bo-Jangles" Robinson (1878–1949) was a dancer, singer, and actor. Earl "Snakehips" Tucker (1906–1937) was a dancer most famous for his signature dance, the "snakehips."

10. John D. Rockefeller (1839–1937) and Henry Ford (1863–1947) were very wealthy business magnates and industrialists.

11. In John 18:10 of the Bible, Peter draws a sword and cuts off the ear of one of the men who came to arrest Jesus.

12. Each of the African American folk heroes Hurston mentions here may have been based on a real person. John Henry, the "steel-driving man," races a steam-powered steel-driving machine on a railroad. "Stacker Lee," also known as "Stagger Lee," was a real person, Lee Shelton (1865–1912), whose murder of another man inspired a song of the same name recorded by many artists. See Cecil Brown, *Stagolee Shot Billy* (Cambridge: Harvard Univ. Press, 2003). "Smokey Joe" is most likely "Smokey" Joe Williams (1886–1951), a baseball pitcher for the Negro leagues whose career spanned almost three decades. Finally, it is unclear whether "Bad Lazarus" corresponds to "Old Bad Lazarus" as recorded by the Works Progress Administration (WPA) in 1939, on deposit in the Library of Congress, or the work song "Po' Lazarus" about a man shot unjustly by a sheriff, recorded by Alan Lomax in 1959. The version Lomax collected later appeared in the movie *O Brother, Where Art Thou?* (2000). It appears at the Library of Congress at www.loc.gov/folklife/lomax/lomaxiconicsonglist.html.

13. Rudolph Valentino (1895–1926), an Italian immigrant, was an actor who starred in the silent film *The Sheik* (1921). Hurston refers to the slicked-back hair Valentino was known for outside the film, which closely resembles the "conk" hairstyle popular with Black men beginning around the time *The Sheik* was made.

14. Paul Whiteman (1890–1967) was a white bandleader popular in the 1920s and 1930s, sometimes called the "King of Jazz," although the title has attracted controversy.

15. Hurston's reference to a "sixth race" is connected to theosophy, an esoteric movement popular during the Harlem Renaissance that theorized five "root races" of the world, most of which are extinct, and predicted the rise of a "sixth root race" in the future. The songs Hurston refers to are "Shake That Thing," copyrighted in 1926 by Papa Charlie Jackson, and "It's Tight Like That," recorded in 1928 by Tampa Red and Georgia Tom.

16. Roland Hayes (1887–1977) was an African American singer and composer who met with significant commercial success, touring abroad from the 1920s–1940s.

17. Hurston references the first point of Woodrow Wilson's (1856–1924) "Fourteen Points Speech" (1918), "open covenants of peace, openly arrived at."

18. Hiding a light under a bushel alludes to Matthew 5:15 in the Bible.

19. *Dixie to Broadway*, a musical revue, ran for seventy-seven performances in the Broadhurst Theatre on Broadway from 1924–1925.

20. "Octoroon" literally refers to those who are one-eighth Black and seven-eighths white, but Hurston's use of the term here points out a colorism or shadeism in mainstream appropriations of Jook culture.

21. Hurston's note: Elegant (?).

22. Hurston's use of "black" here refers specifically to colorism or shadeism through which Black women with dark complexions face discrimination from those who privilege a lighter complexion.

23. In construction, a mud-sill is a supporting sill that rests directly on the ground; here, Hurston uses its metaphorical meaning, a person of low social standing.

24. Hurston's note: Yaller (yellow), light mulatto.

25. A lumber camp would have a commissary for purchasing everyday supplies. In this context, the woman with a dark complexion is purchasing items on credit on his account, despite his denigrating women who look like her.

26. *Sex* (1926) was a risqué Broadway play written and starred in by Mae West.

27. Hurston likely refers to "Let Your Linen Hang Low" (1937), recorded by Rosetta Howard with the Harlem Hamfats.

28. George Gershwin (1898–1937) was a popular white American composer and pianist.

29. Ann Pennington (1893–1971) was a white Broadway actress, singer, and dancer.

30. The Fisk Jubilee Singers are an a cappella group first formed in 1871 from students at Fisk University, a historically Black university in Nashville, Tennessee. The group traveled the world, and continues to do so, singing the spirituals first sung by enslaved Africans and African Americans.

31. Butterbeans and Susie was a comedy/musical duo of husband and wife Jodie Edwards (1893–1967) and Susie Edwards (1894–1963), active in the first half of the twentieth century.

32. Hurston refers to minstrelsy, a racist genre of entertainment in which white actors put on blackface—sometimes using burnt cork as a pigment—and performed sketches as Black caricatures.

Conversions and Visions

1. Emilie Townes explains the phrase "rimbones of nothing" as "that space in which creation itself enters our lives in ways too deep for words and only sounds and images roil our souls, challenges our vision" in "Walking on the Rim Bones of Nothingness: Scholarship and Activism," *Journal of the American Academy of Religion* 77, no. 1 (March 2009): 1–15.

2. Hurston's note: Brick dust is used in New Orleans to seam steps. It leaves them reddish.

3. Catamounts are a species of American big cat, also known as cougars, pumas, and mountain lions.

4. This is likely an allusion to the Bible, John 12:24–25.

Spirituals and Neo-Spirituals

1. The song title as it appears here also appears in the Library of Congress at www .loc.gov/item/afc9999005.1832. To demonstrate Hurston's point, a variant title appears as "The Dying-Bed Maker," and a similar version of the song, "Jesus is a Dying Bed Maker," also appears to have circulated.

2. Harry T. Burleigh (1866–1949) was a composer and singer known for classical arrangements of spirituals. J. Rosamond Johnson (1873–1954) was a composer and pianist known for composing the Black National Anthem, "Lift Ev'ry Voice and Sing" (ca. 1900). Lawrence Brown (1907–1988) was a jazz trombonist. R. Nathaniel Dett (1882–1943) was a conductor and musician. Hall Johnson (1888–1970) was a composer known for arranging spirituals. "Work" may refer either to John Work Jr. (1871–1925), a professor and director of the Fisk Jubilee Singers at Fisk University, or to his son John Wesley Work III (1901–1967), a composer, ethnomusicologist, and professor also employed by Fisk University.

3. Hurston's note: Evening dress.

4. Hurston's original note directs readers to "The Sermon as Heard by Zora Neale Hurston from C. C. Lovelace, at Eau Gallie, Florida, May 3, 1929," which she published in *Negro Anthology*, ed. Nancy Cunard (London: Wishart, 1934), 50–54.

5. "O That I Knew the Secret Place" (1721) is a hymn written by Englishman Isaac Watts (1674–1748).

Ritualistic Expression from the Lips of the Communicants of the Seventh Day Church of God

1. "Seventh Day" churches are a subset of Christianity that, among other differences from mainstream Protestantism, keep a holy day of rest, called the Sabbath, on Saturday rather than Sunday, which is the day most Christian denominations observe a day of rest.

2. A doxology is a short, traditional prayer of praise often employed at the end of a church service or song.

3. The parentheses throughout this essay are Hurston's.

4. R. A. R. Johnson (c. 1854–1940) is the founder of the Holy Church of the Living God, the Pillar and the Ground of the Truth, the House of Prayer for All People, which Hurston visited in the spring of 1940 to conduct research in collaboration with Jane Belo (1904–1968), an anthropological field-worker, photographer, and videographer who studied religion and trances, particularly in Bali; the church is still active today, mainly under the shorter name the House of God. Sanctification is a doctrine that differs between many denominations of Christianity, mainly in whether the sanctification is complete or partial in life, whether it occurs instantly or over time, and whether it is permanent or dependent upon continuously seeking God.

5. It is unclear whether this church was ever officially begun; a church under a similar name, the Evangelical Methodist Church, was founded in 1946, but it appears similar only in name to the one mentioned by Johnson.

6. Abyssinia is an exonym for the Ethiopian Empire, one of the first places to adopt Christianity as the official religion; the Empire fell in the 1970s after seven hundred years.

7. Adam and Eve were the first two people created by God in the Jewish, Christian, and Muslim faiths.

8. Enoch is one of the ten antediluvian patriarchs discussed in the book of Genesis; the verse Genesis 5:24 reads that "Enoch walked with God; and he was not, for God took him," (New American Standard Bible), which many have interpreted as God taking him to heaven without first dying.

9. Noah, Abraham, Isaac, and Jacob are all biblical patriarchs; Jesus Christ is held as the Messiah figure by Christianity, but not by Judaism or Islam.

10. The King's Highway is referenced in the biblical book of Numbers as the path the Hebrews followed after the exodus from Egypt.

11. Judgement Day is a prophesied event in Abrahamic religions, usually accompanied by the final salvation of believers and the end of the world.

12. The Commandment Keeper Church is also the subject of a documentary directed by Hurston and listed on the National Film Registry of the Library of Congress at https://www.loc.gov/static/programs/national-film-preservation-board/documents /commandment_keeper_church.pdf.

13. Washington likely preached at the Macedonia Baptist Church in South Carolina, a historical Black church that gained publicity in the 1990s after its arson at the hands of Ku Klux Klan members. The editors were unable to find records of a New Zion church in the state that mentions Washington as a former pastor, although there are several churches by the name that existed during his lifetime.

14. Consumption is a historical term for the disease now known as tuberculosis.

15. Deuteronomy 28 presents blessings for those who obey God and curses for those who disobey him.

16. Hurston's capitalization and use of the single quotation mark is irregular here. Her original reads: "'Yes Lord, Amen etc." It has been standardized to follow the conventions she establishes in the previous sentences.

17. The editors were unable to find records of a Baptist church called Pilgrim in Beaufort, South Carolina, during this time period.

18. This verse recounts the phenomenon of speaking in tongues (other languages) when filled with the Holy Ghost/Spirit, which is considered an aspect of God in most denominations of Christianity.

19. Daniel and John are both figures featured in their own books of the Bible; Daniel is an Old Testament text, while John is the author of the final Gospels and Revelation.

20. This "conversation" ends abruptly with significant white space remaining on the page. The next section begins on a new page, suggesting, perhaps, that Hurston intended to return to the "conversation" with Mary but never did so.

21. The "Children at the Red Sea" references Exodus 15, in which the Israelites dance and sing God's praises after the parting of the Red Sea by Moses to allow them final passage out of Egypt.

22. In Matthew 10 of the Bible, Jesus teaches his disciples to exorcise spirits and heal disease, then provides them with rules for how and when to apply these gifts.

23. Witch smellers were people who led rituals to expose evil witches within a community; the practice was most common in southern Africa, but legal and cultural shifts in the 1900s have lessened their prevalence.

24. The African Methodist Episcopal Church (AME) is an influential, historically

Black denomination of Christianity founded by Richard Allen (1760–1831), who was born enslaved. Allen was first ordained, though, in the Methodist Episcopal Church (MEC), a more mainstream (and white-dominated) denomination.

25. This hymn was originally published as a poem by Joseph M. Scriven (1819–1886).

You Don't Know Us Negroes

1. This essay was slated for publication in 1934 in *American Mercury*, but for reasons that remain unclear, the essay was never published.

2. The parable Hurston references tells of six blind men who encounter an elephant. Each mistakenly believes that the part he holds in his hands—the head or the tail, for instance—constitutes the entire elephant. It is a parable with Buddhist roots about perception and not seeing the entire picture. See John D. Ireland, *The Udana and the Itvuttaka: Two Classics from the Poli Canon* (Candy, Shri Lanka: The Buddhist Publication Society, 1997), 82–88.

3. For the story of Ishmael's conception see the book of Genesis, verses 16–17 in the Bible.

4. This phrase was popularized in the 1943 song "Is You Is or Is You Ain't My Baby." It also appears regularly in the work of the white writer Octavus Roy Cohen (1891–1959). See his *Florian Slappy Goes Abroad* (Boston: Little, Brown, 1928). For recent discussions of these stereotypes, see Bernard A. Drew, *Black Stereotypes in Popular Series Fiction, 1851–1955: Jim Crow Era Authors* (Jefferson, NC: McFarland, 2015).

5. Flournoy E. Miller (1885–1971) and Aubrey Lyles (1884–1932) were an African American vaudeville team who wrote and produced wildly popular Broadway musicals, including *Shuffle Along* (1921) and *Runnin' Wild* (1923).

6. See Hans Christian Andersen, "The Shadow," in *The Complete Stories* (New York: Anchor Books, 1974), 43–45.

7. Josh Billings is the pen and performance name of Henry Wheeler Shaw (1818–1885), a white American humorist.

8. David Belasco (1853–1931) was a white American producer and playwright. See James F. Wilson, *Bulldaggers, Pansies, and Chocolate Babies: Performance, Race, and Sexuality in the Harlem Renaissance* (Ann Arbor: Univ. of Michigan Press, 2011).

9. All of the authors Hurston mentions in this paragraph are white. DuBose Heyward (1885–1940) was the author of *Porgy* (New York: George H. Doran, 1925), which became the foundation for the Broadway hit *Porgy and Bess* (1935) and a film version. E. C. L. Adams wrote *Congaree Sketches* (Chapel Hill: Univ. of North Carolina Press, 1927). Julia Peterkin (1880–1961), an American novelist, won the Pulitzer Prize for *Scarlet Sister Mary* (New York: Bobbs-Merrill, 1928). Thomas Sigismund Stribling (1881–1965), a Tennessee novelist, won the Pulitzer Prize for *The Store* (Garden City, NY: Doubleday Doran, 1932). Paul Green (1894–1981) was a poet and playwright who also won the Pulitzer Prize for his play *In Abraham's Bosom*, which was first produced in 1926 (New York: McBride, 1927).

10. Roark Bradford (1896–1948) was a white American novelist best remembered for *Ol' Man Adam an' His Chillun* (1928), which became the foundation for the hit play *The Green Pastures* by Marc Connelly (1890–1980), who was also white. The play won the Pulitzer Prize in 1930.

11. Al Jolson (1886–1950) was a well-known white American performer and comedian who performed in blackface makeup.

12. Richard B. Harrison (1864–1935) was a prolific African American actor featured on the cover of *Time Magazine* on March 4, 1935, just days before his death.

13. In 1 Kings 17:2–5 of the Bible, Elijah remains in the wilderness at God's direction, drinking from a stream and fed by ravens who bring him food twice each day.

14. These are all biblical allusions. For the story of Jonah in the whale's belly, see Jonah 1–2. For the story of Isaiah lying on his left side, see Ezekiel 4:4–8. For the story of Jeremiah crying in the wilderness, see Jeremiah 31.

Fannie Hurst

1. Fannie Hurst (1885–1968) was a Jewish American author and activist known for focusing on gender issues. Hurston worked for Hurst as an amanuensis for a time after they met in 1925, though their friendship lasted longer than her formal employment.

2. The Queen of Sheba, first mentioned in 1 Kings 10:1 of the Bible, was said to have brought a caravan of spices, gold, and jewels to a talk with King Solomon.

3. *Humoresque: A Laugh on Life with a Tear Behind It* (1919); *Back Street* (1931); *Lummox* (1923); and *The Vertical City* (1922) were all popular works of fiction by Hurst.

4. Jacques S. Danielson (1875–1952) was a Russian immigrant pianist who married Hurst in 1915.

5. The Medicis were a powerful, wealthy family during the Italian Renaissance known for their extensive patronage of the arts.

6. Elisabeth Marbury (1856–1933) was an American literary and theatrical agent who worked with Hurston.

7. Vilhjalmur Stefansson (1879–1962) was a Canadian explorer and archaeologist.

Art and Such

1. The original text reads, "music tales and dances." The comma has been added to facilitate the reader's comprehension.

2. Reconstruction was the period from 1863–1877 after the Civil War during which the former Confederacy was rejoined with the Union, slavery was abolished, and attempts were made to establish civil rights for formerly enslaved people.

3. O. Richard Reid (1898–unknown) was a Black portrait painter. On Fannie Hurst (1885–1968) see notes 1 and 3 above. John Barrymore (1882–1942) was a white stage, film, and radio actor. H. L. Mencken (1880–1956) was a white author and scholar of American English.

4. Augusta Savage (1892–1962) was a Black sculptor and art teacher active during the Harlem Renaissance.

5. The parenthetical note here is Hurston's reminder to herself to add a list of Savages's works to the essay.

6. J. Rosamond Johnson (1873–1954) was a composer and pianist known for composing the Black National Anthem, "Lift Ev'ry Voice and Sing" (ca. 1900).

7. For information on James Weldon Johnson (1871–1938) see note 5 in the

Introduction. J. Rosamond Johnson was his brother. James wrote the poem his brother put to music for "Lift Ev'ry Voice and Sing."

8. Brooks Thompson (1860–unknown), an African American carpenter, is mentioned in a chapter on his daughter in Marsha Dean Phelts, *An American Beach for African Americans* (Gainesville: Univ. Press of Florida, 1997). Thompson's birth date was provided in an email to the editors.

9. Theodore Roosevelt (1858–1919) was president of the US from 1901–1909. His campaign song for the 1904 presidential election was written by Bob Cole (1868–1911), J. Rosamund Johnson, and James Weldon Johnson.

10. James Weldon Johnson, *God's Trombones* (New York: The Viking Press, 1927).

11. *The Autobiography of an Ex-Colored Man* (Sherman, French, 1912; repr., New York: Knopf, 1927) is a work of fiction. Johnson's *Black Manhattan* (New York: Knopf, 1930) is a work of history, while *Along this Way* (New York: The Viking Press, 1933) is an autobiography.

12. Hurston writes about herself in third person here.

13. *Mules and Men* (Philadelphia: J. B. Lippincott, 1935) is an ethnography of Hurston's time collecting folklore in the South. *Their Eyes Were Watching God* (Philadelphia: J. B. Lippincott, 1937) is considered Hurston's masterpiece. *Tell My Horse* (Philadelphia: J. B. Lippincott, 1938) documents Hurston's time studying folklore in Jamaica and Haiti.

Stories of Conflict

1. Richard Wright (1908–1960) was an African American writer of fiction and nonfiction whose work mainly focused on the lives of Blacks. When this review by Hurston appeared, Wright had recently panned Hurston's 1937 novel *Their Eyes Were Watching God* and accused of her minstrelsy.

2. Hurston likely refers to the Great Dismal Swamp, an area of marshland between Virginia and North Carolina that those who escaped from slavery lived in up to the Civil War. See Richard Grant, "Deep in the Swamps, Archeologists Are Finding How Fugitive Slaves Kept Their Freedom," *Smithsonian Magazine*, September 2016, www.smithsonianmag.com/history/deep-swamps-archaeologists-fugitive-slaves-kept-freedom-180960122.

The Chick with One Hen

1. Alain LeRoy Locke (1885–1954) was a professor of philosophy at Howard University, editor, and essayist. He was the campus advisor for *The Stylus*, the campus literary magazine that published Hurston's first short story. He introduced Hurston (as well as Langston Hughes and Claude McKay) to Charlotte Osgood Mason, who would financially back her work collecting folklore in the South in the late 1920s and early 1930s. Locke is best known as the editor of the volume that is often called "the bible of the Harlem Renaissance," *The New Negro: An Interpretation* (1925). For Locke's review of Hurston's novel, *Their Eyes Were Watching God*, see "Jingo, Counter-Jingo and Us," *Opportunity*, February 1939, 7–10.

2. In his review, Locke describes Hurston's characters in *Their Eyes Were Watching*

God as "pseudo-primitives" and dismisses the novel as art or literature, saying, "setting and surprising flashes of contemporary folklore are the main point." He takes Hurston to task, saying that "[h]er gift for poetic phrase, for rare dialect and folk humor keep her flashing on the surface of her community and her characters and from diving down deep either into the inner psychology of characterization or to the sharp analysis of the social background." He asks that Hurston "come to grips with motive fiction and social document fiction." "Having gotten rid of condescension, let us now get over over-simplification," he concludes. See his "Jingo, Counter-Jingo and Us," *Opportunity*, January 1938, 10.

3. Locke held bachelor's degrees from Harvard and Oxford Universities and a PhD from Harvard. He was also the first African American Rhodes scholar.

4. Sterling A. Brown (1901–1989) was an African American poet and essayist who taught at Howard University from 1932–1969. His first volume of poetry, *Southern Road* (1932), placed him alongside Langston Hughes as an important and innovative poet who found inspiration in working-class Black life.

5. Hurston writes about this controversy in her essay "The Hue and Cry About Howard University," which is included here.

6. *The Green Pastures* (1930) by Marc Connelly (1890–1980) won the Pulitzer Prize for drama. Connelly, who was white, based the play on stories by another white writer, Roark Bradford. His stories in *Ol' Man Adam an' His Chillun* (1928) depict rural Southern African Americans in simplified and stereotypical ways. *The Green Pastures* was subsequently adapted to considerable acclaim in 1936 in a film by the same name. See Marc Connelly, *The Green Pastures: A Fable Suggested by Roark Bradford's Southern Sketches, Ol' Man Adam an' His Chillun* (New York: Farrar and Rinehart, 1930); Roark Bradford, *Ol' Man Adam an' His Chillun* (New York: Harper, 1928).

Jazz Regarded as Social Achievement

1. Rudi Blesh (1899–1985) was a white jazz critic and educator.

2. Hurston explains in "Characteristics of Negro Expression," "Jook is the word for a Negro pleasure house. It may mean a bawdy house. It may mean the house set apart on public works where the men and women dance, drink and gamble. Often it is a combination of all these."

3. George Pullen Jackson (1874–1953) was a white musicologist and educator.

4. W. C. Handy (1873–1958), a Black composer and musician, is known for strongly influencing the development of the blues.

5. Ethel Waters (1896–1977), a Black singer and actress who got her start in the blues, went on to perform regularly on Broadway. Hurston wrote about her friendship with Waters in her 1942 autobiography, *Dust Tracks on a Road*. See Cheryl A. Wall, ed., *Zora Neale Hurston Folklore, Memoirs, and Other Writings* (New York: Library of America, 1995), 738–42.

6. Josh White (1914–1969) was a Black guitarist popular on race records performing the blues, among other genres.

7. Gabriel Brown (1910–1972) was a Black guitarist whom Hurston recorded performing "John Henry." The recording is available online at the Library of Congress.

See Gabriel Brown, "John Henry," recorded by Zora Neale Hurston, Alan Lomax, and Mary Elizabeth Barnicle, audio, www.loc.gov/item/ihas.200196392.

8. Huddie William Ledbetter (1888–1949), whose stage name was Lead Belly, was a Black musician and composer known for his use of the twelve-string guitar.

9. Alan Lomax (1915–2002) was a white ethnomusicologist who collected a significant body of traditional music now on deposit at the Library of Congress; Hurston collaborated with Lomax to record Gabriel Brown, noted above.

10. Louis Armstrong (1901–1971) was a Black trumpeter and singer whose musical career spanned five decades.

Review of *Voodoo in New Orleans* by Robert Tallant

1. Robert Tallant (1909–1957) was a white Louisiana writer of fiction and nonfiction.

2. Hoodoo and Voodoo are often used interchangeably, but Hurston distinguishes between them. Hoodoo refers to sympathetic root work, while Voodoo (as well as Vodou and other spellings) refers to a distinct syncretic religion.

3. William Seabrook (1884–1945) was a white writer known for his travel narratives, especially on the occult. See *The Magic Island* (New York: Harcourt, Brace, 1929) for his account of Vodou practices in Haiti.

4. Marie Leveau (1801–1881) was a free Creole woman often known as the Voodoo Queen of New Orleans.

5. The editors have been unable to definitively identify the person Hurston refers to here.

6. Hurston's *Mules and Men* (Philadelphia: J. B. Lippincott, 1935) is an autoethnographic account of her fieldwork in Florida and Louisiana, including her apprenticeship as a Voodoo priestess. Rockford Lewis (dates unknown) was a Voodoo practitioner active from the 1930s–1950s who served two years in a penitentiary for mail fraud, sending advertisements for his charms through the post office.

7. "Two-Headed Doctor" is a term for a skilled Voodoo practitioner. Billy Middleton explains the term this way: the name "two-headed doctors" indicates the "dualistic elements of their identity: their ability to do both good and harm . . . their existence on the dividing line between religion and folk magic, their dual Christian and African religious roots, and their ability to both heal and practice divinatory magic." See Billy Middleton, "Two-Headed Medicine: Hoodoo Workers, Conjure Doctors, and Zora Neale Hurston," *Southern Quarterly* 53, no. 3/4 (Spring/Summer 2016): 163.

8. High John de Conquer is a figure in African American folklore who also lends his name to a popular ingredient in hoodoo, or root working, which Hurston studied.

9. The Works Progress Administration Writers' Project was a Depression-era program begun in 1935 and funded by the federal government to provide jobs for out-of-work writers.

10. Hurston details the practices of the Sect Rouge, a group distinct from mainstream Haitian Voodoo, along with information on Legba, a loa or spirit in Haitian Voodoo, in her book *Tell My Horse* (Philadelphia: J. B. Lippincott, 1938).

11. Lyle Saxon (1891–1946) was a Louisiana writer heavily involved with the Works Progress Administration's Depression-era Federal Writers' Project.

What White Publishers Won't Print

1. *Scarlet Sister Mary*, a novel by white American author Julia Peterkin (New York: Bobbs-Merrill, 1928), won the 1929 Pulitzer Prize.

2. Charles Spurgeon Johnson (1893–1956) was a Black sociologist and founder of the magazine *Opportunity*. There he published Hurston's first story, "Drenched in Light," for a national audience. He served as the first African American president of Fisk University.

3. Carl Van Vechten (1880–1964) was a white American author, critic, and photographer who enjoyed a close friendship with Hurston. His best-known and most controversial novel is *Nigger Heaven* (New York: Alfred A. Knopf, 1926).

4. See *The Other Room* (New York: Crown, 1947) by Worth Tuttle Hedden (1896–1985).

5. Sinclair Lewis (1885–1951) was the first American to win the Nobel Prize in literature. See his novel *Main Street* (New York: Harcourt, Brace and Howe, 1920).

The Hue and Cry About Howard University

1. Howard University is a historically Black university in Washington, DC, founded in 1867; the same year, the Normal and Preparatory Department was founded to act as a prep school, which Hurston later attended.

2. May Miller (1899–1995) was an American poet and playwright active during the Harlem Renaissance.

3. Hannibal (247 BC–ca. 183–181 BC) was a Carthaginian general who led a contingent of elephants over the Alps to attack the Roman Empire in 218 BC. Napoleon Bonaparte (1769–1821) was a French military leader who claimed the title of Emperor of the French in 1804.

4. The Student Army Training Corps was a federal program designed to train university students for service in World War I. The editors have been unable to identify the Wienstein Hurston refers to here.

5. Charles H. Wesley (1891–1987) was an American historian, minister, and author of some two dozen books on African American history.

6. "There's A Long Long Trail A-Winding" and "K-K-K-Katy" were popular World War I–era songs. "Roll Jordan Roll" and "Ain't Gonna Study War No More" (sometimes called "Down by the Riverside") are both traditional spirituals.

7. "God of Our Fathers" was written in 1872 by Daniel C. Roberts (1841–1907), an Episcopalian priest, to commemorate the centennial of the US Constitution. "Immortal Love Forever Full" was written in 1866 by John Greenleaf Whittier (1807–1892), a Quaker poet and writer.

8. James Stanley Durkee (1866–1951), a minister, served as Howard president from 1918–1926. To date, he was the last white president of the university.

9. "The Heavens Declare the Glory of God" is Psalm 19:1 of the Bible.

10. Edward Hugh Sothern (1859–1933) was an American actor, considered one of the foremost Shakespearean actors of his time.

11. Lulu Vere Childers (1870–1946) was the director of music at Howard from 1905–1946, where she founded both the Conservatory of Music and School of Music. Roy W. Tibbs (1888–1944) was head of the Department of Piano and Organ and founder

of Howard's glee club. Charlotte Beatrice Lewis (dates unknown) was an accomplished pianist and assistant professor of piano and history of music at Howard.

12. John Marshall Miles (b. 1887) was a religious worker and educator.

13. Reed Smoot (1862–1941) was a United States senator from Utah from 1903–1933. He served as the chair of the Senate Finance Committee and on the Senate Appropriations Committee but was not the chair as Hurston suggests.

14. The Rand School of Social Science was founded in 1906 by members of the Socialist Party of America.

15. "The Hill" refers to Capitol Hill, a historic neighborhood and the location of the US Capitol, Supreme Court, and Library of Congress.

16. Thomas B. Dyett (ca. 1886–1971) went on to become a New York lawyer and civil servant. Norman L. McGhee Sr. (1897–1979) completed his bachelor's (1920) and law (1922) degrees at Howard and went on to become a successful businessman in Cleveland, Ohio.

17. Dr. Emmett J. Scott Sr. (1873–1957) was a journalist and close associate of Booker T. Washington at the Tuskegee Institute; he served as Howard's secretary-treasurer from 1919–1933. George W. Cook (1855–1931) was a Howard alum and served at various positions on Howard's faculty and administration for over fifty years. Edward L. Parks (1851–1930) served as a professor of economics, treasurer, and registrar during his time at Howard.

18. Hurston alludes to Tuskegee University, a historically Black university known at the time for providing students preparation for work in trades and applied professions.

19. Kelly Miller (1863–1939) was a writer, educator, and alumnus of Howard University and Johns Hopkins University who later held several positions at Howard from 1890 until his death.

20. Z. Alexander Looby (1899–1972) went on to become a lawyer, educator, and activist after earning his Bachelor of Law from Howard. George Winston Brown (b. 1898). Frederick Douglass Jordan (1901–1979) attended Howard as an undergraduate and would become a bishop in the African Methodist Episcopal Church.

21. William V. Tunnell (d. 1943) was a Howard alum and Episcopalian minister in addition to his work in Howard's history department.

22. An ancient tale tells of how Spartan boys were trained to hide their deceptions at all costs: a young boy once stole a fox, and, when confronted, hid the fox under his clothes rather than reveal it. The panicked fox slashed at the boy, killing him.

23. Calvin Coolidge (1872–1933) served as US president from 1923–1929.

24. Dwight Oliver Wendell Holmes (1877–1963) continued his career in college administration as president of Morgan State from 1937–1948. Frederick D. Wilkinson (b. 1890).

25. Lucy Diggs Slowe (1885–1937) was an English professor at Howard in addition to her role as dean of women.

26. Norman L. McGhee (1897–1979) was a lawyer and businessman educated at Howard. See note 16 above.

27. Oliver Otis Howard (1830–1909) was a Union general, namesake and former president of Howard University.

28. The Curry College (formerly School) is a private institution in a suburb of Boston founded in 1879.

29. Samuel Silas Curry (1847–1921) was a speech teacher and namesake of the Curry School referenced above.

30. Carl J. Murphy (1889–1967) was a Howard alum and former German professor there until he assumed operations of Baltimore's *Afro-American* newspaper following his father's death.

31. "Facts Howard University Washington, D. C. 1918–1926," 1926, Howard University Digital Archive, accessed May 17, 2021, https://dh.howard.edu/cgi/view content.cgi?article=1010&context=hu_pub.

32. Emory B. Smith (1886–1950) was a pastor and briefly a judge in addition to his work at Howard.

33. Dudley Weldon Woodard (1881–1965) was a mathematician and Howard alum who established Howard's math graduate program. Saint Elmo Brady (1884–1966) was a chemist who taught at Tuskegee, Howard, and Fisk during his career. Frank Coleman (1890–1967) was a physicist and veteran of World War I who served as head of Howard's physics department.

34. Robert Josselyn Leonard (1885–1929) was an educator and school administrator who studied vocational training.

35. Alain LeRoy Locke (1885–1954). Alonzo H. Brown (dates unknown) taught at both Howard and Fisk Universities during his career in math education. Metz T. P. Lochard (1896–1984) was an English professor at both Howard and Fisk in addition to his work in journalism. For additional information on Locke, please see note 1 in "The Chick with One Hen." Orlando C. Thornton (b. 1895).

36. The absence of first names has made it difficult to determine precisely who Hurston references here. We believe these to be Alfred Cromwell Priestly, Pezavia Eugene or Harry C. Hardwick, and James Garfield or Suzanne Y. Goins.

The Emperor Effaces Himself

1. Marcus Garvey (1887–1940), a Jamaican activist and Pan-Africanist, founded the Universal Negro Improvement Association (UNIA) in Harlem, which remains the largest grassroots Black Nationalist organization in American history. The organization promoted Black empowerment and created the newspaper *Negro World*. Middle-class Blacks often looked down upon the UNIA's elaborate costumes and parades. Garvey was convicted of mail fraud in 1925 and deported in 1927.

2. The flag of the UNIA depicts three wide red, black, and green horizontal bars. The flag is often called the Pan-African flag.

3. Hurston likely alludes to William J. Simmons (1880–1945), founder of the second Ku Klux Klan in 1915.

4. Lloyd George (1863–1945) was a Welsh attorney and politician who served as prime minister of the United Kingdom from 1916–1922.

5. William Edward Burghardt Du Bois (1868–1963) was an African American sociologist, historian, and activist, a founding member of the NAACP, founding editor of the organization's journal, *The Crisis*, from 1910–1934, and a critic of Garvey.

6. James Weldon Johnson (1871–1938) was an African American writer and civil rights activist. He was the first Black executive secretary of the NAACP, published poetry and novels, and taught at Fisk University. William Pickens (1881–1954), an African American who graduated from Talladega College and Yale University, was a linguist, educator, and autobiographer who later worked for the NAACP.

7. E. L. Gaines (dates unknown) was a Californian and war veteran who resigned

from his role as the UNIA's military commander in 1924 over a financial dispute with Garvey. Gaines had served in his role for four years and never been paid. The story made headlines. See "Garvey's Military Leader Quits Post," *New York Age*, August 2, 1924, 1–2.

8. William H. Ferris (1874–1941), a graduate of Harvard University, minister, civil rights activist, and educator, held the multiple roles outlined by Hurston in the UNIA.

9. The UNIA created the Black Star Line and purchased four battered ships on which goods and Black Americans could return to Africa. Garvey was convicted of using the images of one ship to sell stock in another ship—one he had not yet purchased but was preparing to purchase. The brochures containing the image of the ship were sent through the US Mail, which led to Garvey's conviction for mail fraud.

10. Booker T. Washington (1856–1915) was an African American educator, orator, and founder of the famed Tuskegee Institute.

11. Garvey was prosecuted by William Heyward (1877–1944), US attorney for the Southern District of New York, and Maxwell S. Mattuck (1893–1957), assistant US attorney. Heyward led the famed Black 369th Army Infantry Regiment, frequently called the Harlem Hellfighters, into battle during World War I. Mattuck was the lead prosecutor at Garvey's trial, so he is likely the person alluded to here.

12. John Sidney de Bourg (ca. 1852–unknown) was a Grenadian-born activist deported by the British from Trinidad at the age of sixty-seven for his role in the Workingmen's Association. De Bourg attended Garvey's 1920 International Convention of the Negro Peoples of the World where he was elected to office and then went to work for the UNIA. When Garvey went on trial, however, de Bourg testified against the UNIA leader and was later awarded nearly ten thousand dollars in back pay owed by the UNIA. See Tony Martin, *The Pan-African Connection from Slavery to Garvey and Beyond* (Dover, MA: The Majority Press, 1984), 68–69.

13. Judge Julian William Mack (1866–1943) served on the US Court of Appeals, 7th Circuit from 1911–1929, when he was reassigned to the 6th Circuit. For a biography of Mack, see Harry Barnard, *The Forging of an American Jew: The Life and Times of Judge Julian W. Mack* (New York: Herzl Press, 1974).

The Ten Commandments of Charm

1. Venus is the Roman goddess of love, beauty, and sex.

Noses

1. Like Coty, Rigaud and Hudnut were cosmetic companies. Both companies' names are misspelled in the original text as *Regnaud* and *Hudaut's*, respectively, and thus are corrected here.

How It Feels to Be Colored Me

1. *Hegira* in this context refers to Hurston's departure from Eatonville.

2. Hurston attended Barnard College, where she was the only Black student enrolled.

3. Peggy Hopkins Joyce (1893–1957) was an American actress, model, and dancer well-known during the Jazz Age.

Race Cannot Become Great Until It Recognizes Its Talent

1. Puck is a figure from British mythology. He appears in Shakespeare's *A Midsummer Night's Dream*. Caliban, half human and half monster, appears in *The Tempest*. Brer Rabbit is an important trickster figure in African American folklore.

2. Claude Neal (ca. 1911–1934) was an African American farmhand who was lynched before a crowd of spectators outside the Jackson County Courthouse in the Florida panhandle on October 26, 1934.

My Most Humiliating Jim Crow Experience

1. Charlotte Osgood Mason (1854–1946) was a white American socialite and patron of Harlem Renaissance artists, including Hurston, Langston Hughes, and Claude McKay. She financed much of Hurston's anthropological fieldwork in the Deep South and the Caribbean.

2. Lindley Hoffman Paul Chapin Jr. (1888–1938) was a member of the wealthy and influential Chapin family. His sisters Katherine, a poet, and Cornelia, a sculptor, were also beneficiaries of Mason's patronage. See andré m. carrington, "Salon Cultures and Spaces of Culture Edification," in *A Companion to the Harlem Renaissance*, ed. Cherene Sherrard-Johnson (Hoboken, NJ: John Wiley & Sons, 2015), 251–66.

The Lost Keys of Glory

1. A similar version of the folktale that Hurston expands upon here appears in *Mules and Men* (1935; New York: HarperPerennial, 1990), 31–32, 34. Hurston's typescript treats *Man* and *Woman* as proper nouns, so her capitalization is retained throughout.

2. Clare Boothe Luce (1903–1987) was a white playwright, editor at *Vanity Fair*, journalist for *Life Magazine*, US ambassador to Italy and Brazil, and elected to Congress in 1942. Luce married into wealthy society, divorced her first husband, and later married Henry R. Luce, the publisher of *Time* and *Life Magazine*.

3. Dorothy Celene Thompson (1893–1961) was a white American journalist and radio broadcaster. Thompson is known for interviewing Hitler for *Cosmopolitan Magazine*, which she then parlayed into a book, and cautioning the US against the rise of Nazism. She married three times.

4. As in the passage below, Hurston's reference to the Holy Grail alludes to Arthurian legends in which the elusory Grail, often a cup or chalice, has magical powers.

5. Sir Percival (with the name spelled variously) is a figure from medieval Arthurian romances and a member of King Arthur's round table. Sir Galahad (also spelled in multiple ways) is also a figure from Arthurian legend. Both figures search for the Holy Grail. Here, Hurston probably alludes to Alfred, Lord Tennyson's poem "Morte d'Arthur," although the image of one robed in "white samite," a silk fabric often woven with metallic thread, is clearly much older. For a discussion of the image, see Linda Gowans, "'Clothed in White Samite, Mystic and Wonderful': A Famous Arthurian Image in Tennyson and His Predecessors," *Arthuriana* 26, no. 2 (Summer 2016): 7–24.

The South Was Had

1. Dwight D. "Ike" Eisenhower (1890–1969) served as President of the United States from 1953–61. Prior to that he was a five-star general in the US Army and responsible for planning the Allied assault on Normandy in World War II.

2. John Temple Graves, II (1892–1967) was a white Birmingham, Alabama, newspaper editor and syndicated columnist.

3. Robert A. Taft (1889–1953) was a US Senator from Ohio who lost the Republican nomination for president in 1952 to Eisenhower.

4. Theodore G. Bilbo was a white politician from Mississippi who twice served as governor of the state before serving in the US Senate from 1935–47.

5. Everett McKinley Dirkson (1896–1969) was a white politician who served multiple terms in the House of Representatives and then won a seat in the Senate in 1950.

Take for Instance Spessard Holland

1. Gilbert and Sullivan's *Pirates of Penzance* (1879) is a popular operetta. Hurston once worked as a lady's maid for a performer in a Gilbert and Sullivan troop, so she likely knew the operetta well.

2. An earlier version of this passage includes this text: "a whopping majority of 86,000, which tells who followed the script and who ad libbed." Hurston revised this passage as it appears above. Spessard L. Holland (1892–1971) was a white American attorney and politician who served as governor of Florida from 1941–1945 and US senator from 1946–71. Claude D. Pepper (1900–1989) was also a white attorney, a Democrat, who served Florida in both the US House of Representatives and the US Senate. Although considered a liberal, he opposed making lynching a federal crime.

3. Hurston's manuscript of this essay mistakenly identifies Holland's alma mater as "Clemson" in Atlanta, Georgia. Her error has been corrected. Holland graduated from Emory, which is in Atlanta.

4. Richard Russell Jr. (1897–1971) served as a US senator for Georgia from 1932–1971 after serving as governor of the state. He opposed civil rights legislation and led a like-minded coalition in Washington to prevent such laws from being passed.

5. Florida Agricultural and Mechanical College, known today as Florida A&M University, is a historically Black institution in Tallahassee founded in 1887.

6. John Robert Edward Lee Sr. (1864–1944) was president of Florida Agricultural and Mechanical College, now Florida Agricultural and Mechanical University or FAMU, in Tallahassee, Florida, from 1924–1944.

7. The Freedmen's Bureau (1865–1872) was created just prior to the end of the Civil War to assist with reconstructing the South. It was charged with feeding, housing, and resettling formerly enslaved people and poor whites after the war. Initially the goals for the agency included resettling Black Americans on confiscated or abandoned land. While the federal agency built schools and hospitals and kept people from starving, most land was returned to its original owners.

8. Holland married Mary Agnes Groover in 1919, and they remained together until his death.

9. Robert A. Taft (1889–1953) married Martha Wheaton Bowers (1889–1958) in 1914.

The "Pet Negro" System

1. Japheth and Ham were two of Noah's sons, according to Genesis 5:32 in the Bible. In Genesis 9:25, Noah curses Ham's descendants, stating that they will be servants to the descendants of his other sons. The so-called curse of Ham has since been used as justification for the enslavement of Africans and African Americans.

2. In an effort to end slavery in the United States, John Brown (1800–1859), a devoted white abolitionist, led an ill-fated raid on Harpers Ferry, Virginia, in 1859. Brown was captured and executed for treason.

3. Theodore G. Bilbo (1877–1947) was a white Democratic governor and senator from Mississippi. J. Thomas Heflin (1869–1951) was a white Democratic secretary of state, congressman, and senator from Alabama. Benjamin R. Tillman (1847–1918), also a white Democrat, was governor of South Carolina and served in the Senate.

4. James E. Shepard (1875–1947) was the African American founder and president of the North Carolina State College for Negroes from 1910 until his death. The school's name was changed to North Carolina Central University in 1969.

5. Eartha M. M. White (1876–1974) was a prominent African American singer, teacher, entrepreneur, and humanitarian in Jacksonville, Florida. She spent her business profits on community endeavors, including a home for the elderly, a home for unwed mothers, and an orphanage.

6. Carl Van Vechten (1880–1964) wrote for *Vanity Fair* and befriended (and some said betrayed) a number of Harlem Renaissance writers. He is best remembered for his photographs of the literati and his controversial novel *Nigger Heaven* (1926). Henry Allen Moe (1894–1975) was a World War I veteran, attorney, and administrator. He served as the founding administrator for the Guggenheim Foundation, president of the American Philosophical Society, and the first director of the National Endowment for the Humanities. He was Hurston's principle contact at the Guggenheim Foundation, which awarded her two fellowships for conducting anthropological research in Haiti and Jamaica, which she wrote about in *Tell My Horse* (1938). Both of these men are white. Walter Francis White (1893–1955) was an African American activist who served first as an investigator for and later led the NAACP. He was at the helm when the Legal Defense Fund was founded to fight public segregation, including *Brown v. Board of Education*, which eventually led to the end of segregated schools. Hugo Lafayette Black (1886–1971), a white man, resigned his Ku Klux Klan membership in 1925. He would go on to represent Alabama in the US Senate and serve on the US Supreme Court from 1937–1971. William Edward Burghardt Du Bois (1868–1963) was an African American writer, sociologist, civil rights activist, Pan-Africanist, founding member of the NAACP, and founding editor of the influential *Crisis* magazine from 1910–1934. Joel Spingarn (1875–1939) was a Jewish professor of comparative literature at Columbia University, a soldier in World War I, the second president of the NAACP, and founder of the publishing firm Harcourt, Brace and Company.

Negroes Without Self-Pity

1. Garfield Devoe Rogers Sr. (1885–1951) was a Tampa entrepreneur and philanthropist. His name also appears in advertising as C. D. Rogers. He was a cofounder of the Central Life Insurance Company, alongside C. Blythe Andrews and Mary McLeod Bethune.

Central Life would grow to be one of the largest Black-owned insurance companies in Florida.

2. James Leonard Lewis (1905–1954) served as president of the Afro-American Insurance Company, which his grandfather Abraham Lincoln Lewis founded in 1901. The site of the company's first office in Jacksonville, Florida, is memorialized by a historical marker that identifies it as Florida's first African American insurance company.

3. John Robert Edward Lee Sr. (1864–1944) was president of Florida Agricultural and Mechanical College in Tallahassee, Florida, from 1924–1944.

The Rise of the Begging Joints

1. Fagin is a character in Charles Dickens's serially published novel *Oliver Twist* (1837–1839) who teaches children to be pickpockets.

2. George Schuyler (1895–1977) was an African American fiction writer, journalist, and commentator who is often described as conservative. His satirical novel *Black No More* (1931) skewers American constructions of race. He was a columnist for the influential *Pittsburgh Courier* and published in H. L. Mencken's *American Mercury*, *The Messenger*, *The Crisis*, *The Nation*, and the *Washington Post*.

Crazy for This Democracy

1. President Roosevelt used the phrase "arsenal of democracy" in his Fireside Chat on December 29, 1940, to describe the role he imagined for the United States in fighting the Axis Powers prior to the bombing of Pearl Harbor in 1941. Rather than entering the war directly, he imagined that the US would provide military equipment.

2. Will Rogers (1879–1935) was an American performer and humorist. Known for his quips and ironic humor, he often opened his shows with the line, "All I know is what I read in the papers."

3. President Roosevelt and British prime minister Winston Churchill issued the Atlantic Charter on August 14, 1941. The agreement articulated eight goals for the two nations in World War II. Among them was the right to political self-determination. In the years that followed, the Atlantic Charter served as a means to advance decolonization and human rights. For more on this topic, see Mark Reeves, "'Free and Equal Parts in Your Commonwealth': The Atlantic Charter and Anticolonial Delegations to London, 1941–3," *Twentieth Century British History* 29, no. 2 (June 2018): 259–83.

4. President Roosevelt's State of the Union address on January 6, 1941, outlined "four essential human freedoms": "freedom of speech and expression," "freedom of religion," "freedom from want," and "freedom from fear."

5. Toussaint Louverture (1743–1803) was a powerful figure in the fight for independence in the French colony of Saint-Dominque (now Haiti). He is often described as the leader of the first successful slave revolt in the Americas.

6. Hurston is most likely referring to Sukarno (1901–1970), a key figure in the Indonesian independence movement and considered the first president of Indonesia.

7. Robert Ingersoll (1833–1899), sometimes known as the Great Agnostic, was an American lecturer and an advocate of free thought. He opposed the prevailing doctrine of social Darwinism, which held that some races or nations were inherently inferior to others.

8. FEPC refers to the Fair Employment Practices Committee, which was created in 1941 via executive order. It was created to prevent racial discrimination against African Americans in defense or government jobs. Congress cut funding for the program after World War II, and it was formally dissolved in 1946.

I Saw Negro Votes Peddled

1. Reconstruction was the period from 1863–1877 after the Civil War during which the former Confederacy was rejoined with the Union, slavery was abolished, and attempts were made to establish civil rights for formerly enslaved people.

2. Simply put, the voter casts a ballot in a single race for a specified candidate, who is, in Hurston's example, the second name on the ballot. The voter fails to cast a ballot in any other contest.

3. Harriet Beecher Stowe's influential novel *Uncle Tom's Cabin* first appeared in 1852.

4. Hurston was unwittingly repeating the spurious charges of historians of the Dunning School at Columbia, who saw Reconstruction as a dark period in the history of American democracy, rather than the interracial revolution that it was. Hurston clearly had not read W.E.B. Du Bois's seminal rebuttal to the Dunning School, *Black Reconstruction: An Essay Toward a History of the Part Which Black Folk Played in the Attempt to Reconstruct Democracy in America, 1860–1880* (1935).

5. Rutherford B. Hayes (1822–1893) was the nineteenth president of the United States and oversaw the end of Reconstruction.

6. The clauses Hurston refers to were laws enacted after Reconstruction in Southern states to prevent African Americans from exercising the right to vote. While the Fifteenth Amendment meant that states could not limit the right to vote based on race, the clauses provided other ways to prevent Blacks from voting. The grandfather clauses (overturned by the US Supreme Court in 1909) allowed Blacks and whites to vote only if their grandfathers had been able to. Similarly, property clauses required voters to own property, while literacy clauses required that voters pass a literacy test.

7. See Song of Solomon, chapter 2:12 in the Bible. Though recorded as "turtle" in some translations, the verse refers to the voice of the turtledove.

Mourner's Bench

1. Robert Hemenway cites the title as "Mourners Bench, Communist Line: Why the Negro Won't Buy Communism." *American Legion Magazine* 50, no. 6 (June 1951) lists the essay in the table of contents as it appears above. Notably, Hurston's name and the essay's title appear on the cover of the magazine. Full text available at: https://archive.legion.org /handle/20.500.12203/3965. Henry Winston (1911–1986) was a Mississippi-born organizer, high-ranking Communist Party official, and eventual chairman of the Communist Party USA from 1966 until his death.

2. William Edward Burghardt Du Bois (1868–1963) was an African American sociologist, historian, and activist, a founding member of the NAACP, and a founding editor of the organization's journal, *The Crisis*, from 1910–1934. In 1951, he cofounded the American Peace Crusade, an organization that aimed to end the Korean War. The group's leaders feared a nuclear holocaust would result from the war.

3. Howard Fast (1914–2003) was a Jewish American novelist, member of the Communist Party, and writer for the *Daily Worker*.

4. In 1950 Lt. Leon A. Gilbert (1920–1999), an African American officer, was convicted by an all-white military tribunal of "misbehavior before the enemy" for refusing to rejoin his unit at the front during an incursion of North Korean People's Army soldiers south of the 38th parallel. Although he initially received the death sentence, instead he served five years of a twenty-year prison sentence before being released.

5. Wayland Rudd (1900–1952) was an African American actor who moved to Russia in 1932. While he returned to the US with his wife and daughter, he later moved back to Russia where he lived and worked until the time of his death. William Paterson (1891–1980) was a Black attorney and member of the Communist Party USA. He defended Nicola Sacco and Bartolomo Vanzetti in 1927 and worked on the Scottsboro case. Frank and John Goode, both Black, were Paul Robeson's brothers-in-law (dates unknown). They moved to Russia in 1934 to work on the film *Black and White*, which never materialized. John would return to the US in 1937, while Frank remained until his death. Paul Robeson (1898–1976) was a well-known African American singer, performer, and political activist with Leftist sympathies, although he was not apparently an official member of the Communist Party. He was admitted to the bar in New York, but pursued life as a performer instead.

6. Langston Hughes (1901–1967) was a beloved American writer known for his poetry, fiction, and plays. He and Hurston shared the same literary patron, Charlotte Osgood Mason. At one time Hurston and Hughes were friends and collaborators, but their dispute over a play effectively ended their friendship. Interestingly, Louise Thompson (1901–1999) was the typist who worked on the play with Hughes and Hurston. See *Mule Bone: A Comedy of Negro Life* (New York: HarperPerennial, 2008).

7. Robert A. Taft (1889–1953), son of former president William Howard Taft, was a conservative Republican politician who served as senator from Ohio from 1939–1953; his bids for the presidency were ultimately unsuccessful.

8. The Abraham Lincoln Brigade was a group of Americans recruited by the Communist Party to serve in the fight against Franco in the Spanish Civil War.

9. The Scottsboro case involved nine Black teenagers who were accused of raping two white women in Alabama in 1931. The inflammatory nature of the charges and the location in the Jim Crow South meant a fair trial was virtually impossible.

10. Hurston alludes to the Korean War where Black men were killed in action. Hurston questions the intelligence of expecting people suffering the loss of their loved ones at the hands of Koreans to side with North Korea instead of America.

11. Hurston likely refers to Caroline Schermerhorn Astor (1830–1908). Married to William Backhouse Astor Jr., she is often remembered as a socialite. In the years after her marriage, she emerged as the most powerful figure in New York society circles, setting standards for manners, social customs, and admittance to the New York Four Hundred.

12. Booker T. Washington (1856–1915) was an African American educator, orator, and founder of the famed Tuskegee Institute.

13. The Medicis were a powerful, wealthy family during the Italian Renaissance known for their extensive patronage of the arts.

14. Genghis Khan (1162–1227) was a powerful leader of Mongolian tribes who eventually claimed territory from Poland to the Caspian Sea.

15. Max Schmeling (1905–2005) was a German boxer and heavyweight world champion. He is best known in the United States for his matches against the African American boxer Joe Louis in 1936 and 1938, which Nazi Party leaders attempted to leverage as propaganda.

16. Du Bois applied for membership in the Communist Party in 1961, the same year he moved to Ghana.

17. Baal was a deity in the ancient Canaanite religion. In 1 Kings 18:20–40 Elijah exposes his followers as worshippers of a false god when Baal fails to answer their prayers.

A Negro Voter Sizes Up Taft

1. Robert A. Taft (1889–1953), son of former president William Howard Taft, was a conservative Republican politician who served as senator from Ohio from 1939–1953; his bids for the presidency were ultimately unsuccessful. By "Elephant Boy," Hurston suggests that Taft was fighting to be the Republican candidate in 1952's election.

2. Charles P. Taft II (1897–1983) was Robert A. Taft's younger brother and served as mayor of Cincinnati from 1955–1957.

3. Martha Wheaton Bowers Taft (1889–1958), wife of Robert A. Taft, was an active leader in the Girl Scouts and helped found the League of Women Voters in addition to her work for her husband's campaigns.

4. Abraham Lincoln (1809–1865) was president of the US from 1861–1865 and signed the Emancipation Proclamation that abolished slavery.

5. Daniel Webster (1782–1852) was a politician who served in Congress and as secretary of state during his career, which has been remembered as a remarkable one; along with Robert A. Taft and three others, Webster was named one of the United States' most outstanding senators in 1959 by a Senate committee.

6. Theodore G. Bilbo (1877–1947) served as a US senator from Mississippi from 1935–1947.

7. Alphonso Taft (1810–1891) served as attorney general and ambassador to Russia and Austria-Hungary in addition to his work as secretary of war during the Civil War. Ulysses S. Grant (1822–1885) was president from 1869–1877 after serving as commanding general of the Union army during the Civil War.

8. William Howard Taft (1857–1930) served as US president from 1909–1913 and is currently the only person to have been both president and chief justice of the Supreme Court.

9. Helen Louise Taft (1861–1943) came from a family full of senators and other government officials and was an active First Lady.

10. Lloyd Wheaton Bowers (1859–1910) was a lawyer before his brief tenure as solicitor general from 1909–1910.

11. Thomas Jefferson (1743–1826) was a founding father of the US and served as president from 1801–1809. Martha Washington (1731–1802) was the first First Lady of the United States.

12. Hurston refers to several dynasties often grouped together as the "Boston Brahmins," historically influential upper-class families. The Adams family includes John Adams (1735–1826), second president of the US, his son John Quincy Adams (1767–1848), sixth president, and several other notable members through the country's history. The

Cabot family includes mainly merchants, though Henry Cabot Lodge (1850–1924) and his grandson Henry Cabot Lodge Jr. (1902–1985) were both notable US senators from Massachusetts. The most notable members of the Lodge family are the aforementioned senators also included in the Cabot line.

13. The publications known collectively as the Federalist or the Federalist Papers are eighty-five essays written by John Jay, James Madison, and Alexander Hamilton between 1787 and 1788 that appeared in New York newspapers as the state considered whether to ratify the US Constitution.

14. "Jacksonian democracy" is a term for the trend of policies begun by Andrew Jackson (1767–1845) during his term as president from 1829–1837; the period is characterized by laissez-faire capitalism, emphasis on expanding voting rights to all white men, and manifest destiny across the West.

15. The spoils system, also a characteristic of Jacksonian democracy, refers to the practice of appointing allies to positions of power after they help win an election; Andrew Jackson himself practiced the spoils system liberally after his first presidential win in 1828.

16. Alexander Hamilton (1755 or 1757–1804) was a founding father and the first US secretary of the treasury from 1789–1795.

17. President Franklin Delano Roosevelt (1882–1945) implemented a series of actions known as the New Deal in 1933 to address the crises of the Great Depression. The New Deal included the Works Progress Administration and the Federal Writers' Project, which Hurston worked for in the late 1930s.

18. The term Pinko in Hurston's time was a pejorative term for someone with Communist leanings. With hard-core Marxists being commonly called "reds," a Pinko is someone less so.

19. The Kremlin was the seat of government for the Union of Soviet Socialist Republics from 1922–1991.

20. Henry Agard Wallace (1888–1965) served as the US secretary of commerce, the US secretary of agriculture, and as Roosevelt's vice president. In 1948 the Progressive Party nominated him for president. Frances Perkins (1880–1965) was US secretary of labor from 1933–1945. The first woman appointed to a cabinet post, Perkins led the Department of Labor as it established the first minimum wage, established Social Security, and limited the number of hours that children could work. Anna Eleanor Roosevelt (1884–1962) is remembered for her role in advocating for social causes and for women's rights. Married to Franklin Delano Roosevelt, she served as First Lady from 1933–1945. Vito Marcantonio (1902–1954) represented the state of New York in the US House of Representatives from 1939–1951. He ran unsuccessfully for mayor of New York as the nominee of the American Labor Party in 1949.

21. Aaron Burr (1756–1836) and Thomas Jefferson (1743–1826) were tied in the electoral college vote to become president in 1800. As practices of the time dictated, the winner would become president, while the other would serve as vice president.

22. Hurston is in error here. Alexander Hamilton (ca. 1755–1804) may have persuaded Delaware's lone representative, James A. Bayard, to change his vote, which helped Jefferson win to become president. When Burr ran for governor of New York in 1804, Hamilton again opposed Burr. Burr challenged Hamilton to a duel in which he shot Hamilton, who later died of his wound.

23. A straw man is a logical fallacy. It depends on distorting the opposition's argument only to knock down something that was not part of the opponent's position.

In this case, Hurston suggests that Democrats have been setting up and knocking down straw men in order to attract Black voters.

24. The Smathers-Pepper campaign was the Democratic primary race in Florida in 1950. Incumbent Claude D. Pepper (1900–1989) was challenged by George A. Smathers (1913–2007) in what became an ugly race. Smathers ultimately won the primary and went on to win election to the Senate. The nature of the campaign battered both men's political reputations.

25. Marian Anderson (1897–1993) was a world-renowned African American contralto. When Howard University sought to rent Constitution Hall for a performance by Anderson, the Daughters of the American Revolution (DAR), which owned the hall, denied the request because of her race. First Lady Eleanor Roosevelt, a member of the DAR, resigned from the organization. Anderson would go on to perform on the steps of the Lincoln Memorial before a desegregated crowd.

26. Harold LeClair Ickes (1874–1952) was an American politician appointed by Franklin D. Roosevelt as US secretary of the interior from 1933–1946. Addis Ababa is the capital of Ethiopia.

27. Hurston alludes to the great flood described in Genesis 7 of the Bible. The cloud Hurston describes, however, has been whipped up out of egg whites. That is, it is manufactured.

28. The phrase "chewed tobacco and spit white lime" appears in Hurston's essay "Dance Songs and Tales from the Bahamas" but in an entirely different context. Here the phrase implies anger. In "Dance Songs" it functions as a quip that opens and closes specific folktales. See Hurston, "Dance Songs and Tales from the Bahamas," *Journal of American Folklore* 43, no. 169 (July–September 1930): 294–312, www.jstor.org/stable/534942.

29. Hurston's use of the phrase "Little White Father" probably refers to Nicholas II, Emperor of All the Russias, who was in power in the early years of World War I. Nicholas and his family were executed in 1918.

30. Booker T. Washington (1856–1915) was a Black educator and founder of the famed Tuskegee Institute (now University).

31. Theodore Roosevelt (1858–1919) was president of the US from 1901–1909.

Court Order Can't Make Races Mix

1. Hurston of course refers to the landmark 1954 ruling in *Brown v. Board of Education*, which ended the legality of "separate but equal" public schools.

2. DeWitt Everett Williams (ca. 1899–1975) served as Florida's state agent for Negro schools from 1927–1963.

Which Way the NAACP?

1. Hurston refers to the desegregation of Central High School in Little Rock, Arkansas, in 1957. The governor of Arkansas sent the Arkansas National Guard to bar entry to the nine African American students, often remembered as the "Little Rock Nine." President Dwight D. Eisenhower eventually sent federal troops to escort the students onto the campus.

2. William Edward Burghardt Du Bois (1868–1963) was an African American writer,

sociologist, civil rights activist, Pan-Africanist, founding member of the NAACP, and founding editor of the influential *Crisis* magazine from 1910–1934.

3. Booker T. Washington (1856–1915) was an African American educator, orator, and, as Hurston says here, founder of the famed Tuskegee Institute. While both Washington and Du Bois were considered important leaders and advocates for African Americans, Washington argued for social separation and industrial education that would be seen as less of a threat by whites, while Du Bois advocated for immediate equality and access to a more traditional liberal arts education.

4. Du Bois's autobiographical book is actually titled *Dusk of Dawn: An Essay Toward an Autobiography of a Race Concept* (1940). For a contemporary edition, see *The Oxford W. E. B. Du Bois*, ed. Henry Louis Gates Jr., 19 vols. (New York: Oxford, 2007).

5. Virgil D. Hawkins (1906–1988) was denied admission to the University of Florida law school in 1949, marking the start of a decades-long legal battle to practice law in Florida.

6. Autherine Lucy (b. 1929) was the first African American student to attend the University of Alabama, where she was subjected to degrading attacks by white mobs. She was later expelled after filing a lawsuit against the university for failing to protect her against these attacks. This expulsion was rescinded in 1988, when she was allowed to resume classes.

7. President Harry S. Truman (president 1945–1953) ordered the desegregation of America's military in 1948, but it was not until Dwight D. Eisenhower succeeded him as president (1953–1961) that the order was fully implemented and the last all-Black units were desegregated.

8. Paul Robeson (1898–1976) was a well-known African American singer, performer, and political activist with Leftist sympathies, although he was not apparently an official member of the Communist Party. Langston Hughes (1901–1967) was known for his poetry, fiction, and plays. Hurston knew both men, and at one time she was close to Hughes, with whom she disputed the authorship and ownership of the play *Mule Bone*. See *Mule Bone: A Comedy of Negro Life* (New York: HarperPerennial, 2008). For a study of Hurston and Hughes's relationship, see Yuval Taylor, *Zora and Langston: A Story of Friendship and Betrayal* (New York: W. W. Norton, 2019).

9. Charles Clinton Spaulding (1874–1952) was a highly successful African American businessman. He served as the president of the North Carolina Mutual Life Insurance Company for thirty years, though he was not the founder as Hurston suggests. Under Spaulding's leadership, the company grew to one of the largest Black-owned businesses in the United States. However, his wife, Charlotte Garner Spaulding, did not serve in the Department of Health, Education, and Welfare. Hurston seems to have mistaken her for Jane Morrow Spaulding (1900–1965), a social worker who became the first African American woman to serve as an assistant secretary in a president's cabinet.

10. Oveta Culp Hobby (1905–1995) served as the director of the Women's Army Corps and was the first secretary of the Department of Health, Education, and Welfare.

11. Spaulding opposed funding the hospital because it would not employ African American physicians. A more complete record of Spaulding's career and her dismissal can be found in Jessie Carney Smith, ed., *Notable Black American Women, Book II* (New York: Gale Research, 1995), 609–12.

12. Drew Pearson (1897–1969) was an American journalist best known for his syndicated newspaper column "The Washington Merry-Go-Round."

13. While Spaulding was a Republican at the time of her appointment to the Department of Health, Education, and Welfare, by the time Hurston wrote this article, Spaulding had begun campaigning for Democratic causes.

14. Josephine Baker (1906–1975) was an American entertainer who became famous in her adopted home of Paris. While Baker is remembered by many for her performances in the banana skirt Hurston mentions, she was also active in the French Underground during World War II and received the Croix de Guerre after the war ended.

15. Will Rogers (1879–1935) was an American actor and humorist. Rogers often used the line "All I know is what I read in the papers" in his public appearances, particularly at the beginning of his live performances.

16. Walter Winchell (1897–1972) was an American news and radio journalist, popular in between the 1930s and 1950s.

17. Richard Wright's *Black Power: A Record of Reactions in a Land of Pathos* (1954) is available today in *Black Power: Three Books from Exile* (New York: HarperPerennial, 2008), 1–428.

18. Kwame Nkrumah (1909–1972) was a Ghanaian nationalist leader who served as the country's first prime minister and president after gaining independence from the United Kingdom in 1957.

Zora's Revealing Story of Ruby's 1st Day in Court!

1. In the epigraph, Hurston quotes "Old Folks at Home" (1851), better known as "Suwannee River," by American songwriter Stephen Collins Foster (1826–1864). In the paragraph below, Hurston quotes Foster's 1853 "Old Kentucky Home."

2. The surname LeRoy also appears variously as Leroy. The editors have opted for the capitalization found on Adams's grave marker as indicated by the photograph at www.findagrave.com/memorial/51960369/clifford-leroy-adams. Adams (1908–1952) is buried in the Live Oak Cemetery, Live Oak, Florida.

Ruby McCollum Fights for Life

1. See Acts 16:24–27 in the Bible, which recount Paul and Silas being freed from prison by an earthquake.

Bare Plot Against Ruby

1. Malcolm Johnson (1904–1976) was an investigative reporter who won the 1949 Pulitzer Prize for local reporting for a series of twenty-four stories, "Crime on the Waterfront," published in *The New York Sun*. Bernard Parker remains unidentified.

Ruby's Story: Doctor's Threats, Tussle over Gun Led to Slaying!

1. John "Jack" Barrymore (1882–1942) was a well-known white stage and film star.

2. Spessard L. Holland (1892–1971) was a white American attorney and politician who served as governor of Florida from 1941–1945 and US senator from 1946–1971.

Ruby's Troubles Mount: Named in $100,000 Lawsuit!

1. Ruby McCollum actually had four children, but only three of those were Sam McCollum's.

2. Hurston indicates in part eight of "The Life Story of Mrs. Ruby J. McCollum" that Suwannee County had for the first time included African Americans on a jury, but it is unclear whether she meant they served on Ruby McCollum's jury.

The Life Story of Mrs. Ruby J. McCollum!

1. This is not a proverb as Hurston suggests. Rather, it is a translation of the Qur'an 17:13, which states that each person controls his or her own destiny.

2. Hurston refers to Napoleon Bonaparte (1769–1821), the celebrated military leader and emperor of France from 1804–1814.

3. Victoria Nyansa is a variant spelling of Victoria Nyanza, a large lake crossing the national boundaries of Tanzania, Uganda, and Kenya.

4. Reverend Leonard Francis Morse (1891–1961) was one of three founding members of Phi Beta Sigma Fraternity in 1914 at Howard University.

5. Open range ranching existed in Florida until after World War II when it was discontinued in order to protect new species of cattle from disease harbored and better resisted by older species of "Florida cow." For a discussion of this transition, see W. Theodore Mealor Jr. and Merle C. Prunty, "Open-Range Ranching in Southern Florida," *Annals of the Association of American Geographers* 66, no. 3 (September 1976): 360–76.

6. We believe this to be Leonard Francis Morse, b. 1891

7. Hurston's reference to the pear tree echoes the images she used in *Their Eyes Were Watching God*. For a discussion of these similarities see Roberta S. Maguire, "From Fiction to Fact: Zora Neale Hurston and the Ruby McCollum Trial," *Literary Journalism Studies* 7, no. 1 (Spring 2015): 17–34.

8. An editorial or typesetting error apparently mistakenly cut a passage from the text. The original reads as it does here.

9. The poetic line that Hurston quotes comes from "The Rosary" by Robert Cameron Rogers (1862–1912), which can be found in Edmund Clarence Stedman, ed., *An American Anthology 1787–1900* (Boston: Houghton Mifflin, 1900), 691.

10. Rogers, "The Rosary," 691.

11. Rogers, "The Rosary," 691.

12. Despite the promise of additional stories about Ruby McCollum's life by Hurston, this is the last one the editors have been able to identify.

My Impressions of the Trial

1. This essay was enclosed in Hurston's letter to William Bradford Huie dated May 14, 1954.

2. Hyssop is an herb, precisely which plant is still debated, used in Exodus 12:21–23 of the Bible. Moses advises the Israelites on Passover to dip hyssop into the blood of a sacrificed lamb and mark their doorways with it in order that the Lord might pass them by. It protected the home from the death of the firstborn, a curse upon Pharaoh, who was refusing to emancipate the Israelites.

3. The parable of Dives (a rich man) and Lazarus (a poor man) appears in Luke 16:19–25 in the Bible. While the rich man prospers on earth, he is tortured in hell after his death. Lazarus suffers on earth but in heaven is at Abraham's side. The rags-to-riches Cinderella folktale exists in many cultural traditions. See Jacob Grimm and Wilhelm Grimm, *Grimm's Complete Fairy Tales*, ed. Ken Mondschein, trans. Margaret Hunt (San Diego: Canterbury Classics, 2011), 81–86. Hurston may be alluding to tales in which a peasant presents the king's daughter with an unsolvable riddle and thereby wins the right to marry her. For the Russian version, see "The Princess Who Wanted to Solve Riddles," *The Complete Russian Folktale: v. 6: Russian Tales of Love and Life*, ed. Jack V. Haney (Armonk, NY: M.E. Sharpe, 2006), 4–6. Dick Whittington is a medieval historical figure who served as Lord Mayor of London. In spite of the fact that Whittington himself came from a wealthy family, a tale bearing his name titled "Dick Whittington and His Cat" follows a rags-to-riches plotline, which seems to be the point of Hurston's allusion. A version appears in Marcia Brown, *Dick Whittington and His Cat* (New York: Atheneum, 1988).

4. Hurston alludes here to Isaiah 11:6 in the Bible. The passage is part of Isaiah's prophesy for a world at peace.

5. Millard Fillmore Caldwell (1897–1984), who was white, served as governor of Florida from 1945–1949.

6. Marcus Garvey (1887–1940) was a Jamaican-born activist, publisher, and orator with a Black Nationalist message who led the Universal Negro Improvement Association in the United States. Hurston here alludes to "the lily-maid" of Astolat, a figure from Arthurian legend who dies when Arthur fails to return her love. The story appears in Sir Thomas Mallory's *Le Morte d'Arthur* (1485). See P. J. C. Field, ed., *Sir Thomas Mallory: Le Morte Darthur: The Definitive Original Text Edition* (Suffolk: Boydell & Brewer, 2017).

7. Following this essay, Hurston outlined notes and additional impressions that scholars might find helpful. See that typescript housed in Documents Relating to the Trial of Ruby McCollum for the Murder of Dr. LeRoy Adams, Live Oak, Florida, 1954, Special and Area Studies Collections, George A. Smathers Libraries, University of Florida, Gainesville, Florida.

Index

abolitionism, 243, 252
Abomey, Benin, 39, 40
Abraham, 108, 185, 195
Adams, Doctor C. LeRoy, 18–22, 315, 321–327, 329, 334, 336–342, 344–353, 355, 366, 369, 374–377, 379–391, 394, 395, 399–401
Adams, E. C. L., 112
Adams, Florrie, 351–353
Adams, Judge Hal W., 21, 316–317, 320, 327, 329–331, 333–336, 338, 340, 350, 352, 354, 385, 390–392, 396, 400, 401
Adams, Loretta, 341, 343, 345, 346, 350, 351, 353, 377, 379, 389
Adams family, 287
African Legion, 177
African Methodist Episcopal Church, 97
African nose, 195
African religions, 4
African witch smelling, 96, 97
Afro-American Life Insurance Company, 243
Afro-American newspaper, 165
Alabama, 1, 38, 42, 45, 50–51, 305
Alabama River, 45
Allen, Edward, 92
"Alma Mater," 151
Along This Way (J. W. Johnson), 127
Amazons, 39–41
American Communist Party, 270, 274–283, 288–289, 297, 307–311
American Indians, 111, 145

American Museum of Natural History, 145
American Peace Crusade, 270
Andersen, Hans, 110
Anderson, Marian, 292–293
Anglo-Saxon nose, 184–185
angularity, 51
Armstrong, Louis, 137
Arnold, Della, 348
art, 50–52, 123, 125–128
"Art and Such" (Hurston), 8–10, 123–128
Arthur, King of England, 31, 219
Astor, Caroline Schermerhorn, 279
asymmetry, 52–53
Atlanta University, 246, 251
Atlantic Charter, 254
Autobiography of an Ex-Colored Man (J. W. Johnson), 127

Back Street (Hurst), 118
Bad Lazarus, 54
Bailey, Ellen, 100
Baker, Josephine, 309–310
ballads, 76
Baptist church, 79, 82, 90, 92
"Bare Plot Against Ruby" (Hurston), 329–331
Barnard College, 188
Barrymore, John, 126
Bates, Franke, 94
Bates, Jefferson, 94

Bayou Corne, Alabama, 38
Beaufort, South Carolina, 91–92, 95, 99
Begging Joints, 245, 247–250, 252
Belasco, David, 111
Belo, Jane, 81
Benedict, Ruth, 114
Bennett College, 246
Bethune-Cookman University, 227
big bands, 134–136
"Big Boy Leaves Home" (Wright), 129
Bilbo, Theodore G., 222, 234,
    285–286, 294
Billings, Josh, 111
"Bits of Our Harlem" (Hurston), 25–27
Black, A. K., 348–350, 397–398, 400
Black, Hugo Lafayette, 240
Black Arts Movement, 17
Black Bottom, 60, 63
Black Bourgeoisie (Frazier), 6
Black English, 3–6, 47–50
"Black Lives Matter" movement, 21
Black Power: A Record of Reactions in a
    Land of Pathos (Wright), 311
Black preachers, 57
Black Star Steamship and Navigation
    Line, 175, 178
blackface, 3, 63, 110, 112
Blesh, Rudi, 134–137
blues, 2, 4, 6, 9, 52, 60, 63, 136
Book of American Negro Poetry, The, 7
Bowers, Lloyd Wheaton, 287
Bradford, Roark, 112–114
Brady, Saint Elmo, 167
breathing, 78–79
Brer Rabbit, 4, 30, 33, 112, 192
Bronson, Earl, 200
Brown, Alonzo H., 170
Brown, Gabriel, 136
Brown, George Winston, 155, 171
Brown, John, 221, 234
Brown, Lawrence, 77
Brown, Sterling A., 22, 132
Brown v. Board of Education (1954), 13–15,
    18, 296–298, 305
Burleigh, Harry T., 77
Burr, Aaron, 290
Butter Beans and Susie, 64

Cabot family, 287
Caldwell, Millard Fillmore, 397
Calvin, John, 89
Canada, 256–257
Cannon, Frank T., 21, 336–342, 344–
    350, 352, 378, 397–400
Canterbury Tales (Chaucer), 192
career women, 212–218
carpet-baggers, 259, 264, 265
cattle, 196–203
Central High School, Little Rock,
    Arkansas, 1, 300–301, 305
Central Life Insurance Company, 242,
    352, 353, 370, 380
chanting, 79, 82
Chapin, Paul, 204
"Characteristics of Negro Expression"
    (Hurston), 3, 47–65
charleston, 60
Chaucer, Geoffrey, 14, 192
cheque words, 47
"Chick with One Hen, The" (Hurston),
    3, 131–133
Chickasabogue River (Three Mile
    Creek), 42
Childers, Lulu Vere, 153
China, 272, 277, 280
Chinese people, 147
Christianity, 4–5, 45–46
Cincinnati, Ohio, 284
civil rights movement, 1, 286
Civil War, 31–32, 221, 243, 252, 301
Cleopatra, 184
Clotilda (slave ship), 38, 42–44
Cohen, Octavus, 112, 113
Cohn, David, 4
Coleman, Frank, 167
collectivism, 275
colleges, 118, 124–125, 226–227,
    246–253, 302
Columbia University, 118, 226, 246
Commandment Keeper church, 89
communism, 16, 130, 270–283,
    287, 297
concert singers, 76, 77
Congaree Sketches (Adams), 112
Connelly, Marc, 113

Constitution Hall, Washington, D.C, 292–293

Constitution of the United States, 128, 222, 223, 267, 278, 289, 298

conversion visions, 66–71

"Conversions and Visions" (Hurston), 66–71

Cook, George W., 155

Coolidge, Calvin, 113, 156

"Court Order Can't Make Races Mix" (Hurston), 296–299

Cox, R. H., 347

"Crazy for This Democracy" (Hurston), 13, 20–21, 254–258

Crews, P. Guy, 316–320, 329, 330, 347, 397

Crisis, The (magazine), 307, 308

Crocketville, South Carolina, 99

Cudahy Company, 199

culture heroes, 4, 54, 114

"Culture Heroes" (Hurston), 4

Curry, Samuel Silas, 165

Curry, Thelma, 336, 345, 400

Dabney, John, 45

Dahomey, West Africa, 39–44, 46

Dailey, Carrie, 348

Daily Worker newspaper, 271, 278

dancing, 51–53, 60, 63, 109, 135

Danielson, Jacques, 118–120

Daughters of the American Revolution (DAR), 292

Davis, Jefferson, 320

Davis, Sarah, 358, 361

De Bourg, John Sidney, 179

Declaration of Independence, 137

democracy, 254–256, 258

Democratic Party, 221–223

desegregation, 1, 13–18, 221, 228, 296, 298–301, 305, 307, 310–311

Dett, Nathaniel, 77

Dewey, Thomas E., 221, 223

dialect, 3, 7, 9, 64–65, 78, 128

Dirksen, Everett, 223

"Dixie to Broadway" (show), 61

double descriptive, 49

Douglass, Frederick, 124

Dred Scott Decision (1857), 45

Du Bois, William Edward Burghardt, 2, 175, 178, 240, 270, 282, 302–304, 307

Dugger, B. A., 84

Dunbar, Mary Jane, 96

Durkee, James Stanley, 152–157, 161–163, 165–167, 170–171

Dusk of Dawn: An Essay Toward an Autobiography of a Race Concept (Du Bois), 303–304

Dust Tracks on a Road (Hurston), 11

Dyett, Thomas B., 154

Dying Bed Maker, The, 76

Eatonville, Florida, 186–187

education, 13–18, 226–227, 234–235, 246–252, 286, 296–298, 300–305, 307

Edwards, O. O., 317–319, 331, 338, 341, 347, 348, 397, 400

Eisenhower, Dwight D., 220–223, 307, 309

Elijah, 115

Ellison, Ralph, 22

Emancipation Proclamation, 38, 45, 248

Emory University, 225

"Emperor Effaces Himself, The" (Hurston), 13, 173–181

Ethiopian nose, 185

eye-pops, 109

Fair Employment Practices Committee (FEPC), 258, 261, 268, 286

"Fannie Hurst" (Hurston), 117–122

Fanon, Frantz, 2

Fast, Howard, 271

Federalist Papers, 287

feminism, 10–12

Fernandina, Florida, 126, 127

Ferris, Sir William H., 177

Fessenden Academy, 358–361, 370

Fifteenth Amendment to the Constitution, 128

fighting, 58–59
filibuster, 286
Fisk Jubilee Singers, 63, 64
Fisk University, 77, 147, 226, 246,
    248, 251
Florida, 15, 16, 18–21, 125–128, 186–187,
    191, 193, 196–203, 225–227, 240,
    242, 244, 246, 250, 296–297, 304,
    361, 364–366, 368–390, 397
    primary election of 1950, 259–263,
    267–268, 291
Florida A. and M. University, 246
Florida Agricultural and Mechanical
    College (FAMC), 226, 227, 244
Florida Statewide Negro Defense
    Committee, 242
Florida Writers Project, 123n
folklore, 3, 53–55, 76, 114, 123, 131, 138,
    140, 147, 192, 306, 394
Ford, Henry, 53–55
Ford, Isadora, 99
Foster, Captain Bill, 42–44
Foster, Stephen G., 315
Four Freedoms, 254
Fourteenth Amendment to the
    Constitution, 128
Franklin, Benjamin, 282
Frazier, E. Franklin, 6
free Blacks, 124–125, 264, 303
Freedmen's Bureau, 227
French Antilles, 141
French dances, 135
French language, 192
French revolution, 319
"Full of Mud, Sweat and Blood"
    (Hurston), 4

Gaines, E. L., 177–179
Galahad, Sir, 219
Garvey, Marcus, 13, 173–181, 400
Gates, J. M., 12
Georgia, 140, 223
Gershwin, George, 63
Gettysburg, Pennsylvania, 221
Ghana, 310, 311
Gilbert, Leon A., 271

Gilbert, W. S., 224
glee clubs, 64, 76–77
God Shakes Creation, 4
God's Trombones (J. W. Johnson), 7, 127
Goins, James Garfield, 171
Goins, Suzanne Y., 171
"Gointer Study War No Mo," 151
Goode brothers, 271
Goodwin, Clara, 101
gospel music, 3
Grandfather Clause, 265
Grant, Ulysses S., 287
Graves, John Temple II, 222, 223
Gray, Deputy Sheriff, 347
Grecian nose, 185, 195
Green, Paul, 112
Green Pastures, The (Connelly), 113, 133
Gulf of Guinea, 43

Hagar, 108, 111, 281
Haiti, 255
Hamilton, Alexander, 288
Hamites, 233
Hampton University, 64, 77, 246, 305
Handy, W. C., 136
Hardwick, Harry C., 171
Hardwick, Pezavia Eugene, 171
Harlem, New York, 25–27, 276
Harlem Renaissance, 1, 7
Harrison, Richard B., 113
Harvard University, 124, 225, 302, 303
Hawkins, Virgil D., 304
Hayes, Roland, 57
Hayes, Rutherford B., 265
Hedden, Worth Tuttle, 147
Heflin, J. Thomas, 234
Helen of Troy, 185, 195
Henderson, Odis, 348–350
Henry, Patrick, 320
He's a Mind Regulator, 76
Heyward, DuBose, 112
Higginbotham, Evelyn Brooks, 7
High John de Conquer, 28–37, 140
"High John de Conquer" (Hurston),
    28–37
Hobby, Oveta Culp, 308

Holland, Mary, 229
Holland, Spessard L., 225–227, 229, 349–350
Holmes, Dwight Oliver Wendell, 157
Holy Church of the Living God, 84, 85
Holy Grail, 218, 219
"Honey let yo' drawers hang low," 63
hoodoo dance, 109
"How It Feels to Be Colored Me" (Hurston), 13, 186–190
Howard, Oliver Otis, 159
Howard, Willie James, 19
Howard University, 13, 132, 151–172, 246, 248, 251, 305
"Hue and Cry About Howard University, The" (Hurston), 13, 151–172
Huffman, Ernest, 69
Hughes, Langston, 52, 57, 271, 308
human rights, 262
Humoresque: A Laugh on Life with a Tear Behind It (Hurst), 118, 119
Hurst, Fannie, 117–122, 126
Hurston, Lucy, 14
hymns, 79, 80, 136

"I Saw Negro Votes Peddled" (Hurston), 259–269
Ibsen, Henrik, 193
Ickes, Harold, 292
idiom, 3–5, 8, 9, 128
imitation, 56–57
immigration, 228
Indianapolis Star, 290
Indochina, 255, 256
Ingersoll, Robert, 257
interracial marriage, 240
intimidation, 58
Isaac, 108, 110, 185, 195
Isaiah, 115
Ishmael, 108, 110

Jack, as culture hero, 54
Jackson, Andrew, 288
Jackson, Clemon, 356, 358, 359, 361
Jackson, George Pullen, 136

Jackson, Gertrude, 356–359, 361, 367, 382
Jackson, Hattie, 359
Jackson, Matthew L., 353
Jackson, William, 356–359, 367
Jacksonville, Florida, 187, 191, 193, 240
Jacob, 185, 195
Jamaica Police, 176
James, Jesse, 327
Jamison, R. D., 100
Japan, 273
Japheth, sons of, 233
jazz, 2, 4, 9, 60, 134–137, 188–189
"Jazz Regarded as Social Achievement" (Hurston), 8, 134–137
Jefferson, Jessie, 71
Jefferson, Thomas, 287, 290
Jemison, Dolley, 92
Jenkins, P. S., 99
Jeremiah, 115
Jim Crow, 2, 4, 204–205, 256–258, 293
John Henry, 30, 54, 113–114
Johnson, Bishop R. A. R., 81, 83–89
Johnson, Charles S., 147, 251
Johnson, Hall, 77
Johnson, J. Rosamond, 127
Johnson, James Weldon, 5, 7–8, 127, 175, 240, 307
Johnson, Malcolm, 330
Johnson, Pat, 197, 200
Johnson, Rosamond, 77
Johnson, Sally, 101
Jolson, Al, 113
Jonah, 115
Jonah's Gourd Vine (Hurston), 9, 127–128
Jones, Casey, 113
Jones, Julia, 81, 94–97
Jones, Julian, 94
Jonson, Ben, 192
Jooks, 59–64, 136
Jordan, Frederick Douglass, 155, 171
Journal of American Folk-Lore, 114
Joyce, Peggy Hopkins, 189
Justice, Butts, 88
"Justice and Fair Play Aim of Judge Adams as Ruby Goes on Trial" (Hurston), 334–335

"K-K-K-Katy," 151
Kansas, 221
Kissimmee Prairie, Florida, 196, 200
Korean War, 271, 277
Ku Klux Klan, 126, 174, 265

"Last Slave Ship, The" (Hurston), 38–46
Latin language, 193
"Lawrence of the River" (Hurston), 14, 196–203
Lead Belly, 137
Lee, John Robert Edward Sr., 226, 244
Lee, Robert E., 390
Leonard, Robert Josselyn, 167–169
Lester, B. F., 199, 200
Leveau, Marie, 139, 141
Lewis, Charlotte Beatrice, 153
Lewis, Cudgo (Kossula-O-Loo-Ay), 38–42, 44–46
Lewis, James Leonard, 243, 244
Lewis, Rockford, 140
Lewis, Sinclair, 148
liberals, 289–290
"Life Story of Mrs. Ruby J. McCollum, The!" (Hurston), 354–386
Lincoln, Abraham, 38, 285, 320
Lincoln Brigade, 277
Lincoln University, 246
literacy, 124, 265
Literacy Clause, 265
Little Rock, Arkansas, 1, 300–301, 305, 310
Live Oak, Florida, 18–20, 315–353, 368–401, 387–401
Lloyd George, David, 175
Lochard, Metz T. P., 170
Locke, Alain Leroy, 3, 131–133, 170
Lockwood, Irene, 101
Lodge family, 287
Lomax, Alan, 137
"Long, Long Trail a Winding," 151
Longfellow, Henry Wadsworth, 25–26
Looby, Z. Alexander, 155, 171
Lord's Prayer, 80
"Lost Keys of Glory, The" (Hurston), 10–12, 206–210

Louisiana, 134, 135, 138–142, 223
love-making, 58–59
Luce, Clare Booth, 213–214
Lucy, Autherine, 304–305
Lummox (Hurst), 118
Luther, Martin, 89
Lykes brothers, 197, 199, 200
lynching, 19, 20, 126, 193, 265, 286

Macedonia Baptist church, 90
Mack, Julian William, 179–180
Magic Island (Seabrook), 139, 141
Main Street (Lewis), 148
Mama Jane, 79
mammying-up, 197–198
"Manish Women" (Gates), 12
Marbury, Elisabeth, 120, 121
Marcantonio, Vito, 290
Mark Antony, 182
Martin, Florida, 361, 362, 364, 366, 369
Mason, Charlotte Osgood, 204
McCollough, William H., 316, 317, 320, 325
McCollum, Buck, 366
McCollum, Ruby Jackson, 18–22, 315–319, 321–332, 334–401 (see also Ruby McCollum Series (Hurston))
McCollum, Sam, 19, 20, 325, 326, 340, 345–347, 351, 353, 362, 364–377, 379–383, 388, 389, 391, 395
McCollum, Sam Jr., 367, 370, 372
McCollum, Sonya, 337, 338, 343, 345, 346, 372–374
"McCollum-Adams Trial Highlights" (Hurston), 336
McGhee, Norman L. Sr., 154, 158
McGriff, Releford, 352, 378
"Me Too" movement, 21
Meaher brothers, 42–45
Mencken, H. L., 126
metaphor, 47, 49
Methodist church, 82, 97
Middle Ages, 281
middle class, 6–7, 10, 17
Middle Passage, 29
Miles, John, 153

Miles Mill, South Carolina, 95
military, 1, 221, 307
Miller, Flournoy E., 110
Miller, Kelly, 155–157, 169
Miller, Lyles, 110
Miller, May, 151, 153
mimicry, 56–57
minstrelsy, 3, 110, 112
miscegenation, 228, 240, 393
Mississippi, 222
Mississippi River, 60
Mississippi Sound, 44
Mobile, Alabama, 42, 50–51
Mobile Bay, 38, 44
Mobile Register, 42, 51
Moe, Henry Allen, 240
Mongolian nose, 195
Montgomery Bus Boycott, 1
Moore, Henry, 81, 104
Morehouse College, 226, 246
Morgan State College, 246, 251, 305
Morse, Leonard Francis, 359, 360
"Mother Catherine" (Hurston), 5
motherhood, 214
"Mourner's Bench: Why the Negro Won't
    Buy Communism" (Hurston),
    270–283
mulattos, 301, 303
Mules and Men (Hurston), 128, 140
Murphy, Carl J., 165
Musgrave, Vero, 336
music, 4, 7, 9, 34–37, 52, 53, 56, 57, 59–
    60, 63–64, 123, 127, 128, 134–137,
    188–189
"My Impressions of the Trial" (Hurston),
    387–401
"My Most Humiliating Jim Crow
    Experience" (Hurston), 13–14,
    204–205

Napoleon Bonaparte, 174, 255, 355
Nashville, Tennessee, 60
National Association for the
    Advancement of Colored People
    (NAACP), 16, 277, 300–311
Neal, Claude, 193

Negro History Week, 271
"Negro National Anthem," 127
"Negro Voter Sizes Up Taft, A"
    (Hurston), 284–295
Negro World (newspaper), 175
"Negroes Without Self-Pity" (Hurston),
    242–244
neologisms, 3, 5
neurasthenics, 214
New Chapel, Florida, 361
New Deal, 288, 289, 293, 298
New Negro, The: An Interpretation
    (Locke), 131, 133
New Orleans, Louisiana, 134, 135,
    138–142
New Zion church, 90
Niagara Falls, 120–121
Niagara Movement, 302, 307
Nigger Heaven (Van Vechten), 147
Nkrumah, Kwami, 310, 311
Normans, 192
North Carolina, 239, 240, 308
North Carolina Mutual Life Insurance
    Company, 308
North Carolina State College for
    Negroes, 239
Northwestern University, 226
noses, 184–185, 194–195
"Noses" (Hurston), 13, 184–185
"Now Take Noses" (Hurston), 13,
    194–195

Ocala, Florida, 356, 362, 365, 369
Ohio, 284–286
Ol' Man Adam an' His Chillun
    (Bradford), 113
"Old Folks at Home," 315
Old Testament, 3, 6, 51, 80, 113
oleomargarine, 107, 109
Opportunity magazine, 131
originality, 56
Other Room, The (Hedden), 147

"Paramour Rights" (Hurston), 18
Parker, Bernard, 330

Parks, Edward L., 155
Patterson, William, 271
Paul, 328
Paul Bunyan, Paul, 114
Peace Information Center, The, 270
Pearson, Drew, 309
Pennington, Ann, 63
Pepper, Claude, 225, 227, 260, 268, 291
Percival, Sir, 219
Perkins, Frances, 290
"'Pet Negro' System, The" (Hurston), 13, 233–241
pet system, 234–241
Peter the Apostle, 54
Peterkin, Julia, 112, 145
Pickens, William, 175
Pilgrim Baptist Church, 92
Pirates of Penzance, The (Gilbert and Sullivan), 224
Pittsburgh Courier, 18, 277–278, 390
Plateau, Alabama (African Town), 38, 45
poetry, 3, 4, 5
polka, 135
poll taxes, 286
prayers, 3, 4, 6, 51, 78–80, 82, 83
preachers, 2, 5, 7, 67, 68, 72, 78–79, 81–83, 89–93
presidential elections
    of 1800, 290
    of 1952, 220–223, 275, 284, 294–295
Price, Doctor, 316, 320, 325
Priestly, Alfred Cromwell, 171
Princeton University, 246
privacy, concept of, 58–59
Property Clause, 265
Prospero, Ada, 128
publishing, 144, 148

quadrille, 135

"Race Cannot Become Great Until It Recognizes Its Talent" (Hurston), 3, 14, 191–193
Race Champions, 124–126, 128, 302

race relations, 2, 10, 13–14, 115, 233–241, 245–253, 291–293 (see also McCollum, Ruby)
Rand School, 154
Range, Melrose, 94
Reconstruction, 123, 125, 187, 222, 227, 243–245, 259, 264, 265, 276, 297, 305
Reid, O. Richard, 126
religion, 4–6, 51, 65–80, 82–104
Republican Party, 221–223, 284–286, 295, 309
reversion to type, doctrine of, 147
"Review of Voodoo in New Orleans by Robert Tallant" (Hurston), 138–142
revival meetings, 66
rhythm, 52–53, 72–73, 82, 135
"Rise of the Begging Joints, The" (Hurston), 13, 245–253
"Ritualistic Expression from the Lips of the Communicants of the Seventh Day Church of God" (Hurston), 5, 81–104
Roberts, Carrie, 94
Robeson, Paul, 271, 282, 308
Robinson, Bill (Bo-Jangles), 52, 64
Robinson, George, 99, 101
Robinson, Izora, 81, 82, 93, 95, 97–102
Robinson, Liza, 94
Robinson, Willie, 94, 97
Rockefeller, John D., 53–55
Rogers, Garfield Devoe Sr., 242–244
Rogers, Will, 254, 309
"Roll Jordan Roll," 151
Rollins College, 227
Roman nose, 184, 195
Roosevelt, Eleanor, 290, 292–293
Roosevelt, Franklin D., 254, 255
Roosevelt, Theodore, 127, 295
Rosie the Riveter, 12
Royal African Guards, 176
"Ruby Bares Her Love Life" (Hurston), 337–343
"Ruby McCollum Fights for Life" (Hurston), 327–328
Ruby McCollum Series (Hurston)
    "Bare Plot Against Ruby," 329–331

"Justice and Fair Play Aim of Judge Adams as Ruby Goes on Trial," 334–335

"Life Story of Mrs. Ruby J. McCollum, The!," 354–386

"McCollum-Adams Trial Highlights," 336

"Ruby Bares Her Love Life," 337–343

"Ruby McCollum Fights for Life," 327–328

"Ruby Sane!," 324–326

"Ruby's Story: Doctor's Threats, Tussle over Gun Led to Slaying!," 344–350

"Ruby's Troubles Mount: Named in $100,000 Lawsuit," 351–353

"Trial Highlights," 332–333

"Victim of Fate!," 321–323

"Zora's Revealing Story of Ruby's 1st Day in Court!," 315–320

"Ruby Sane!" (Hurston), 324–326

"Ruby's Story: Doctor's Threats, Tussle over Gun Led to Slaying!," 344–350

"Ruby's Troubles Mount: Named in $100,000 Lawsuit" (Hurston), 351–353

Rudd, Wayland, 271

Runnin' Wild (musical), 110

Russell, Richard Jr., 225

St. Johns River, 196, 199

St. Peters, Florida, 364

Santo Domingo, 139–141

Sarah, 108

Saunders, Doctor, 340

Savage, Augusta, 126–127

scalawags, 259, 264–265

Scarlet Sister Mary (Peterkin), 145

Schmeling, Max, 282

Schuyler, George, 250

Scott, Emmett J., 155–157, 165, 167, 171

Scottsboro Case, 277

Seabrook, William, 139

Second Zion Baptist Church, New Orleans, 79

segregation, 220, 221, 225, 227, 229, 237, 256, 265, 301

Selma, Alabama, 45

Senate Appropriations Committee, 154

sermons, 3, 4, 7, 51, 79, 82, 83

Seventh Day Church of God, 81–104

Sex (Mae West), 63

Shakespeare, William, 2, 14, 56, 141, 191, 192

Shepard, James E., 239

Shining Trumpets: A History of Jazz (Blesh), 134–137

shouting, 72–75

"Shouting" (Hurston), 72–75

Shuffle Along (musical), 110

Silas, 328

Silas, Lawrence, 196–203

Silas, Sarah, 200

simile, 47, 49

Simmons, William J., 173

singing, 7, 37, 60, 63–64, 76–79, 151–153, 170

single-shotting, 259, 267–269

slavery, 2, 4, 18, 28–36, 38–45, 76, 107, 123, 128, 139, 141–144, 152, 187–188, 221, 228, 243, 244, 248, 263, 265, 301, 390, 392

Slowe, Lucy Diggs, 157

Smathers, George, 260, 268, 291

Smith, Emory B., 167

Smith, Georgia, 94

Smith, Helen, 361

Smokey Joe, 54

Smoot, Reed, 153–154, 156, 170

Snake Hips, 52, 64

Socrates, 194

Sothern, Edward Hugh, 152

Souls of Black Folk, The (Du Bois), 2

South Africa, 257

South Carolina, 91–92, 95, 99, 140

"South Was Had, The" (Hurston), 220–223

Southern University, 246

Soviet Union, 271–275, 282, 297

Spanish America, 135

Spaulding, Charles Clinton, 308

Spaulding, Charlotte Garner, 308–310

Spingarn, Joel, 240

spirituals, 3, 4, 6, 7, 8, 57, 63–64,
    76–79, 125, 127, 132–133, 136, 152,
    153, 170
"Spirituals and Neo-Spirituals"
    (Hurston), 5–7, 76–80
Spoils System, 288
Springer, Susanna, 69
Stacker Lee, 54
Stalin, Joseph, 272
Stefansson, Vilhjalmur, 121
Stephenson, Reuben, 94
Stony, W. M., 97
"Stories of Conflict" (Hurston),
    129–130
Stork Club, New York, 309
Stribling, T. S., 112
Sullivan, Arthur, 224
Supreme Court of the United States,
    15, 178, 220, 223, 228, 287,
    296–298, 305
survival of the fittest, 58
Sutton, Aunt Shady Anne, 31–32
Suwanee County Courthouse, Live Oak,
    Florida, 315–353, 387–401
Swift and Company, 199
swing, 134, 136

Taft, Alphonso, 287
Taft, Charles P. II, 284
Taft, Helen Louise, 287
Taft, Martha Bowers, 229, 285, 287
Taft, Robert A., 220, 222–223, 229, 275,
    284–291, 293–295
Taft, William Howard, 287
Taft-Hartley Bill, 286
"Take for Instance Spessard Holland"
    (Hurston), 224–229
Takkoi, West Africa, 38–40, 45–46
Talladega College, 246
Tallant, Robert, 138–142
Taney, Roger B., 45
Tell My Horse (Hurston), 128, 141
"Ten Commandments of Charm, The"
    (Hurston), 10, 182–183
Texas, 220–223, 308
theatre, 61–62, 64

Their Eyes Were Watching God (Hurston),
    3, 8, 9, 128, 131, 132
Thompson, Brooks, 127
Thompson, Dorothy, 218
Thompson, Louise, 271
Thornton, Orlando C., 170
Three Precious Keys, 206–212
Thurman, Wallace, 111
Tibbs, Roy W., 153
Tillman, Benjamin R., 234
Tombigbee River, 45
Touisssant Louverture,
    François-Dominique, 255
trade unions, 275, 286
"Trial Highlights" (Hurston), 332–333
Truman, Harry, 221
Tudors, 192
Tunnell, William V., 155–156
Tuskegee Institute, 64, 77, 156, 180, 246,
    252, 302, 303
Twelve Mile Island, 44

Uncle Remus, 108
Uncle Tom's Cabin (Stowe), 263–264
Uncle Tom's Children (Wright),
    129–130
Universal Negro Improvement
    Association (UNIA), 175, 177
universities, 118, 124–125, 226–227,
    246–253, 302
University of Alabama, 305
University of Berlin, 302, 303
University of Chicago, 226
University of Florida, 304

Valentino, Rudolph, 56
Van Vechten, Carl, 147, 240
vaudeville, 3, 110
verbal nouns, 50
Vertical City, The (Hurst), 118
Virginia State University, 246
virtue belt factor, 247, 248
Voodoo (Hoodoo), 138–142, 147, 192
Voodoo in New Orleans (Tallant), 138–142
voting, 259–269, 284–295, 304

Walker, Alice, 9
Wallace, Henry, 290
waltz, 135
Ward, Tommie Lee, 358, 361
Washington, Booker T., 178, 252, 281,
     302–303, 310
Washington, Carrie, 81, 103–104
Washington, Carrie Belle, 89
Washington, D.C, 292–293
Washington, Elise, 89
Washington, Eloise, 89
Washington, George, 290
Washington, Hattie, 94
Washington, Hattie Mae, 81, 103
Washington, Hugh, 81, 89, 102–103
Washington, Martha, 287
Washington, May Belle, 89
Washington, Rev. George, 81, 89–93, 95
Waters, Ethel, 136
Webster, Daniel, 285
Wesley, Charles H., 151–153
Wesley, John, 89
West, Mae, 63
"What a Friend We Have in Jesus," 103
"What White Publishers Won't Print"
     (Hurston), 143–148
"Which Way the NAACP?" (Hurston),
     14–16, 300–311
White, Eartha M. M., 240
White, Josh, 136
White, Walter, 240, 309
White High School, Washington, D.C,
     292–293
white mare, doctrine of, 276, 297–299

white supremacy, 2, 237
Whiteman, Paul, 56
Whittier, Florida, 199
Whydah, Dahomey, 42, 43, 45
Wienstein, 151
Wilkinson, Frederick D., 157, 165, 167
will to adorn, 48–51
William, the Conqueror, 191
Williams, DeWitt Everett, 298
Williams, Roger, 89
Wilson, Woodrow, 58
Winchell, Walter, 309–310
Winston, Henry, 270, 277, 282
Woodward, Dudley Weldon, 167
Work, John, 77
Workman, Doctor, 316, 320, 325
Works Progress Administration
     (WPA), 140
World War I, 282
World War II, 221, 307, 368
Wright, Richard, 3, 5, 8, 22, 129–130,
     310, 311
Writers Project of the WPA, 140

Yale University, 124
Yates, Roland J., 353
"You Don't Know Us Negroes"
     (Hurston), 1, 3, 107–116

"Zora's Revealing Story of Ruby's
     1st Day in Court!" (Hurston),
     315–320